E.A. Lloyd

SYNTHESE LIBRARY

STUDIES IN EPISTEMOLOGY,

LOGIC, METHODOLOGY, AND PHILOSOPHY OF SCIENCE

Managing Editor:

JAAKKO HINTIKKA, *Florida State University, Tallahassee*

Editors:

DONALD DAVIDSON, *University of California, Berkeley*
GABRIËL NUCHELMANS, *University of Leyden*
WESLEY C. SALMON, *University of Pittsburgh*

VOLUME 161

DISCOVERING REALITY

Feminist Perspectives on Epistemology, Metaphysics,
Methodology, and Philosophy of Science

Edited by

SANDRA HARDING

Dept. of Philosophy, University of Delaware

and

MERRILL B. HINTIKKA

Dept. of Philosophy, Florida State University

D. REIDEL PUBLISHING COMPANY

DORDRECHT : HOLLAND / BOSTON : U.S.A.

LONDON : ENGLAND

Library of Congress Cataloging in Publication Data

Main entry under title:

Discovering reality.

(Synthese library ; v. 161)
Bibliography: p.
Includes index.
1.   Feminism—Addresses, essays, lectures.   2.   Philosophy—
History—Addresses, essays, lectures.   3.   Science—Philosophy—
Addresses, essays, lectures.   4.   Social Sciences—Philosophy—Addresses,
essays, lectures.   I.   Harding, Sandra G. 1935–   .   II.   Hintikka,
Merrill B., 1939–   .
HQ1154.D538   1983   305.4'2   82–16507
ISBN 90-277-1496-7
ISBN 90-277-1538-6 (pbk)

Published by D. Reidel Publishing Company,
P.O. Box 17, 3300 AA Dordrecht, Holland.

Sold and distributed in the U.S.A. and Canada
by Kluwer Boston Inc.,
190 Old Derby Street, Hingham, MA 02043, U.S.A.

In all other countries, sold and distributed
by Kluwer Academic Publishers Group,
P.O. Box 322, 3300 AH Dordrecht, Holland.

D. Reidel Publishing Company is a member of the Kluwer Group.

Printed in The Netherlands

TABLE OF CONTENTS

We dedicate this volume to the memory of Jane English, whose brilliance, inspiration, courage and warmth are sorely missed by her colleagues in philosophy and by her friends.

SANDRA HARDING AND MERRILL B. HINTIKKA

INTRODUCTION

During the last decade, feminist research has attempted to add understandings of women and their social activities to what we all thought we knew about nature and social life. However, from the very beginning of this project, it has appeared to be in tension with some of the most fundamental insights of the Second Women's Movement. Only recently has the nature of this tension become clear. Within the theories, concepts, methods and goals of inquiry we inherited from the dominant discourses we have generated an impressive collection of "facts" about women and their lives, cross-culturally and historically — and we can produce many, many more. But these do not, and cannot, add up to more than a partial and distorted understanding of the patterns of women's lives. We cannot understand women and their lives by adding facts about them to bodies of knowledge which take men, their lives, and their beliefs as the human norm. Furthermore, it is now evident that if women's lives cannot be understood within the inherited inquiry frameworks, than neither can men's lives. The attempts to add understandings of women to our knowledge of nature and social life have led to the realization that there is precious little reliable knowledge to which to add them. A more fundamental project now confronts us. We must root out sexist distortions and perversions in epistemology, metaphysics, methodology and the philosophy of science — in the "hard core" of abstract reasoning thought most immune to infiltration by social values.

When we called for papers for this collection, we formulated our project in the following way:

In the last decade feminist thinkers have provided brilliant critiques of the political and social beliefs and practices of patriarchal cultures. But less attention has been given to the underlying theories of knowledge and to the metaphysics which mirror and support patriarchal belief and practice. Are there — can there be — distinctive feminist perspectives on epistemology, metaphysics, methodology and philosophy of science?

While it would do a disservice to the richness and variety of the papers we selected to suggest that a single theme emerged from responses to our question, the reader will find that in different ways, all of the contributors examine the nature and implications of the one discovery of the decade

ix

Sandra Harding and Merrill B. Hintikka (eds.), Discovering Reality, ix-xix.
Copyright © 1983 by D. Reidel Publishing Company.

which enables us to understand and to move beyond the tension between
the discursive inheritances of Western thought and the feminist perspective.
What counts as knowledge must be grounded on experience. Human experience differs according to the kinds of activities and social relations in which
humans engage. Women's experience systematically differs from the male
experience upon which knowledge claims have been grounded. Thus the
experience on which the prevailing claims to social and natural knowledge
are founded is, first of all, only partial human experience only partially
understood: namely, masculine experience as understood by men. However,
when this experience is presumed to be gender-free — when the male experience is taken to be the human experience — the resulting theories, concepts,
methodologies, inquiry goals and knowledge-claims distort human social
life and human thought. Several contributors to this collection argue that
masculine perspectives are not only distorting because they are partial; they
are inherently distorting because they must invert some of the real regularities
of social life and their underlying causal tendencies. These essays identify
some of the philosophical consequences of relying on this partial and distorting experience as the foundation for all knowledge claims, and they begin
to describe the kind of experience upon which could be grounded maximally
scientific human understanding.

Consequently, contributors to this volume pursue two complementary
projects. On the one hand, they contribute to the feminist "deconstructive
project." They identify how distinctively masculine perspectives on masculine
experience have shaped the most fundamental and most formal aspects of
systematic thought in philosophy and in the social and natural sciences —
the aspects of thought supposedly most gender-neutral. They show how
men's understanding of masculine experience shape Aristotle's biology
and metaphysics, the very definition of "the problems of philosophy" in
Plato, Descartes, Hobbes and Rousseau, the "adversary method" which is
the paradigm of philosophic reasoning, contemporary philosophical psychology, individuation principles in philosophical ontology, functionalism
in sociological and biological theory, evolutionary theory, the methodology
of political science, marxist political economy, and conceptions of "objective
inquiry" in the social and natural sciences. On the other hand, many of the
contributors also begin the feminist "reconstructive project". They identify
distinctive aspects of women's experience which can provide resources for
the construction of more representatively human understanding. Some
of the essayists focus extensively on this reconstructive project, showing us
what is required in social practice and in scientific inquiry to make women's

experience into a foundation for a more adequate and truly human epistemology, metaphysics, methodology and philosophy of science. The following paragraphs offer a brief glimpse of the arguments in these essays.

Most of the "Great Philosophers" have produced passages, or even whole treatises, where they explicitly explain the "natural" superiority of men and inferiority of women. As Lynda Lange points out in 'Woman is Not a Rational Animal: On Aristotle's Biology of Reproduction,'

It has been the practice of twentieth-century scholars and educators, in the face of the greater equality of women, simply to disregard these works, which they view as minor or peripheral, or perhaps crankish, like Berkeley's late essays on the value of tar water. (1–2)

In contrast, Lange argues that Aristotle's theory of sex difference is consistently interwoven into the fabric of his philosophy, and that it cannot be cut away without affecting the rest of his thought. She shows how Aristotle's biological theory leads him to the unavoidable implication that the female's "most perfect functioning involves creating the conditions for the male element to prevail." (p. 11) It is the "male element" which provides the fully human element, and so we find Aristotle holding that women are nothing else but "matter set in motion by and for the soul of the unified male, for the ends of the (male) species." (p. 12) Thus in Aristotle's ethics and politics, it is an act of positive virtue for a man to attempt to dominate a woman and compel her to be obedient to him, and a man who treated a woman as an equal would be acting in a shameful manner. Challenging Aristotle's sexism requires that we re-evaluate the soundness of the rest of his thought.

In 'Aristotle and the Politicization of the Soul,' Elizabeth V. Spelman also argues that it distorts Aristotle's thought to separate his sexism from the rest of his philosophy. She points out that Aristotle justifies his political theory on the grounds that it is consistent with his theory of the soul. As the rational part of the soul should have certain kinds of authority over the irrational part of the soul, the rational elements in society should have authority over the irrational elements. However, Spelman's close reading of the text shows that his argument is circular:

both understanding what it means to talk about relations of authority between parts of the soul, and establishing that one part has authority over another, depend on understanding what it means to talk about relations of authority between classes of persons (including those between men and women), and on establishing or assuming that certain classes do have authority over others – in particular, that men have authority over women. (p. 17)

SANDRA HARDING AND MERRILL B. HINTIKKA

Hence Aristotle's political "conclusions" drawn from his metaphysics are in fact the same claims upon which his metaphysics was grounded in the first place. Central parts of Aristotle's philosophy are permeated with a masculine perspective on masculine social experience.

In 'The Unit of Political Analysis: Our Aristotelian Hangover,' Judith Hicks Stiehm demonstrates how Aristotle's assumptions about natural order still inform and distort some of the most respected work in political science today. Although political scientists often believe they are thinking in terms of individuals, groups, or classes, they often in fact think about the polity in terms of families. And they do so principally when they think about women. "The result is a mixed analysis and confused thinking." (p. 31) Her essay demonstrates how this confusion appears in the 1972 book which won an American Political Science Association award as the best new book in its field, and which was cited especially for its excellent methodology. Philosophers will be especially interested in her observation that John Rawls' *A Theory of Justice* is structured by the same confusion. Rawls discusses "the family", which ostensibly contains individuals, as the environment of pre-adults (children), and he also discusses the problem of justice between generations. However, Stiehm points out that in both discussions

his focus seems actually to be on male heads of households and those who will become such heads. Wives, mothers and daughters are of slight interest. Even in thinking about generations his vision is restricted. He considers the next generation (of sons) while *ignoring* the previous generation (with its numerous widowed mothers). (p. 40)

The next two papers initiate a critical focus pursued by many of the other papers in this collection. They argue that we need to reexamine the assumption that the physical sciences have in general succeeded in eliminating political "values" from their descriptions and explanations of nature's regularities and their underlying causal determinants, and that the best scientific inquiry is therefore "value-free." Ruth Hubbard, in 'Have Only Men Evolved?', points out that "Science is made by people who live at a specific time in a specific place and whose thought patterns reflect the truths that are accepted by the wider society." (p. 45) Consequently, we should expect scientific thought to have wide areas of congruence with the social and political ideology of the societies which produce science. Examining evolutionary theory and sociobiology, Hubbard goes on to show how while science defines reality, its definitions are created against a background of what is socially accepted as real. What has been socially accepted as real in societies producing science, as in Aristotle's society, is an androcentric perspective which interprets as women's biological heritage the sexual and social stereotypes feminists reject. From the perspective of this essay we can begin to see how the knowledge claims of the physical sciences, no less than those of the social sciences, are founded on distinctive and often perverse masculine understandings of only masculine social experience.

In 'Evolution and Patriarchal Myths of Scarcity and Competition,' Michael Gross and Mary Beth Averill show how two related themes in the patriarchal image of nature — scarcity and competition — came into evolutionary theory and how they appear in contemporary evolutionary and ecological thought. They note that we should not be surprised to find the theme of competition appearing in accounts of nature at the same time that modernized societies extolled the virtues of competition between economic individuals to insure that only the "fittest" would survive to direct capitalist accumulation projects. But Gross and Averill point out that the theme of scarcity also reflects the interests of capitalists, for it was perceived as virtually the sole brake on unbridled exploitation of natural resources. The authors show how these two themes remain central in the writings of leading biologists and ecologists today, providing an incomplete and distorted understanding of nature. They suggest that a more reliable understanding of nature would be gained by thinking in terms of concepts central to feminist thought which also accurately reflect the natural order — concepts such as plenitude and cooperation.

Ann Palmeri directly challenges the claim that social science should be "value-free." In 'Charlotte Perkins Gilman: Forerunner of a Feminist Social Science' she examines the conception of social science advanced by this early Twentieth Century socialist feminist theorist. Where Gilman's critics have dismissed her work on the grounds that it is directed by a moral commitment to equality between the sexes, Palmeri argues that it is not moral argument that should cast doubt upon lines of scientific reasoning, but only evidence of improper moral argument. She shows us how Gilman's moral-political ideal of adrogyny made interesting and plausible her claim that patriarchal societies' abnormal exaggeration of sex-difference was damaging to the human species. Palmeri proposes that

through a look at Gilman, we can begin to comprehend more fully the claim of present feminists that adopting a view of women as fully human and as actors in history leads to a more well-founded social science. While a commitment to the equality of women and men by no means commits any social scientist to a particular theory, the acceptance of this moral assumption does preclude the consideration and acceptance of certain sorts of theories. (p. 97)

The next two papers challenge the charge that feminist research and analyses lead us to any new philosophical, methodological or other theoretical insights. As psychological research has repeatedly revealed (see, for example, the Broverman study cited in the Hintikkas' paper), this kind of judgment is to be expected of any characteristics or accomplishments as soon as they are identified as women's. In 'The Trivialization of the Notion of Equality,' Louise Marcil-Lacoste argues that feminist research does not merely make women "equal" to men by adding information about women to male categories of thought, or by simply repeating in a mirror-reflection the concepts of male thought. Drawing on a broad selection of feminist and philosophical writings from Europe as well as North America, she argues that "in introducing historicity, materiality, and values as fundamental epistemological categories, feminist writings represent a forceful challenge to criticial thought seen as formal or meta-discourse." (p. 127) In the Kuhnian sense, feminists introduce a new paradigm for human understanding. From Marcil-Lacoste's perspective, the sex-specificity of the new feminist categories is but the prelude to the creation of logics and languages which come closer to human modes of thinking.

Merrill Hintikka and Jaakko Hintikka argue that feminist research and analyses suggest several ways in which the principles of individuation and identification proposed in the history of philosophical ontology and the philosophy of language are sex-biased. Drawing examples from Quine, Kripke, Lewis, Aristotle and Leibniz, they suggest that "Western philosophical thought has been overemphasizing such ontological models as postulate a given fixed supply of discrete individuals, individuated by their intrinsic or essential (non-relational) properties." (p. 146) Feminist research suggests that these are distinctively masculine modes of identifying and individuating parts of the world, for "women are generally more sensitive to, and likely to assign more importance to, relational characteristics (e.g., interdependencies) than males, and less likely to think in terms of independent discrete units." (p. 146) This kind of philosophical "case study" indicates just one of the ways in which feminist concerns raise theoretical problems for the most abstract (and presumably, therefore, most value-free) aspects of Western thought.

Both philosophers and scientists argue that it is the fact that they systematically seek severe criticism of their claims which makes these kinds of knowledge-claims so much more reliable than our common-sense, every day beliefs. The next two papers point out some problems with the kind of criticism philosophers and scientists value so highly. In 'A Paradigm of Philosophy: The Adversary Method,' Janice Moulton analyzes the model of

philosophic methodology which accepts a positive view of aggressive behavior
and then uses it as the paradigm of philosophic reasoning. She suggests that
the primacy of this model seems based on the perception that aggression is
"natural" for males and good for them to exhibit, and "unnatural" for
females and bad for them to exhibit. Philosophic methodology thus restricts
itself to, and legitimates, attributes thought natural only for males. Moulton
points out that even feminist theorists often only criticize the assumption
that aggression is more natural to one sex than the other; they often fail to
examine the conditions under which aggressive behavior is unproductive
and self-defeating regardless of who exhibits it. In philosophy, aggression is
uniquely legitimated in the "adversary method," which requires that all
beliefs and claims be evaluated only by subjecting them to the strongest,
most extreme opposition. Moulton shows how the restriction of critical
thought to the adversary paradigm misinterprets the history of philosophy,
unjustifiably permits programmatic claims to remain viable, restricts the
legitimate range of philosophic argument, and leads to bad reasoning. She
argues that discarding the adversary method as *the* single legitimate paradigm
of philosophic method would encourage the development of alternative and
more productive methods of evaluating thought.

Kathryn Pyne Addelson argues that the scope of scientific criticism is too
narrow. In 'The Man of Professional Wisdom,' she points out that a central
project of mainstream philosophy of science in this century has been to
separate scientific method from metaphysics — from prior assumptions
about the nature of the world and of ourselves which scientists cannot
help bringing to inquiry and which, it is claimed, always distort our understanding. Because scientific method insists on severe criticism of itself,
as well as of the knowledge-claims it produces, it is believed that science can
and does succeed in providing understandings of nature and of ourselves
which are free of these prior metaphysical assumptions. But Addelson points
out that because the scope of criticism is restricted only to the intellectual
procedures and products of scientific method, science's faith in the efficacy
of its critical method is unjustified. Scientists necessarily bring distorting
metaphysical commitments from their class and sex-dominating roles in
social life, and these are reinforced in the social arrangements within science.
Since the social arrangements of the scientific enterprise as well as of the
larger society are considered outside the legitimate domain of scientific
criticism, these metaphysical commitments systematically escape critical
scientific examination, and thus pass through the scienific process to distort
our understandings of nature and ourselves. Her example is functionalist

assumptions in biology and in social theory, and she points out how functionalism's fit with distinctively masculine, professional, and western social experience distorts the experience of nature and social life which women, non-professionals, and non-westerners have. Addelson argues that science's "cognitive authority" to define reality for us is unjustified and illegitimate as long as science fails to include as a legitimate and necessary focus of scientific criticism the class and sex hierarchies which exist both within the scientific enterprise and within the larger society.

The next two papers explore the origins, functions and some of the consequences of the sexual metaphors which permeate our thinking about science and about knowledge in general. In 'Gender and Science', Evelyn Fox Keller argues that both contemporary social research and a re-examination of the history of science reveal that our conceptions of scientific objectivity are deeply distorted by their linkage to cultural stereotypes of masculinity as well as to the real requirements of achieving masculinity. First she reviews the recent discoveries of widespread cultural images containing the curious juxtaposition of scientists as super-masculine and simultaneously as "less sexual" than other men. Within the history of science, this same juxtaposition is reflected in Francis Bacon's metaphors of Nature as the true "bride" of the scientist who should want to control and dominate "her". Turning to post-Freudian psychoanalytic theory, Keller explains how the masculine gendering process requires the development of rigid boundaries between the masculine "self" and all "others." The masculine model of "self" must maintain separation from and control over "others" in order to retain its own identity – just the dynamic of "objectifying" nature and social others required of distinctively scientific methodologies. Keller points out that discovery of these social roots for this distorting conception of scientific methodology has implications for our understanding of, and for our programs for restructuring, both science and social life. We must understand and overcome the

circular process of mutual reinforcement . . . in which what is called scientific receives extra validation from the cultural preference for what is called masculine, and conversely, what is called feminine – be it a branch of knowledge, a way of thinking, or woman herself – becomes further devalued by its exclusion from the special social and intellectual value placed on science. (p. 202)

Contemporary feminist thought has often been suspicious that the pervasive reliance on a visual metaphor for knowledge marks Western philosophy as patriarchal. In 'The Mind's Eye,' Evelyn Fox Keller and Christine R.

Grontkowski examine the role Plato, Newton, and Descartes assign to the
visual in their theories of knowledge. They point to the assumption of

two different, even paradoxical, functions of the visual . . . – a connective and a dissociative. Vision connects us to truth as it distances us from the corporeal. As we trace
the use of the visual metaphor through history, we find that these functions, originally
intertwined, become quite distinct – splitting, finally, into functions of two different
eyes, the body's eye and the mind's eye. This split is paralleled by the division of the
functions of science into the objectifiability and knowability of nature. (p. 209)

They conclude their analysis by raising some questions about both our
reliance in Western thought on the hierarchy of the senses, and about feminist
suspicions of this reliance.

In the first part of 'Individualism and the Objects of Psychology' Naomi
Scheman points out how such twentieth century philosophers as Hilary
Putnam, David K. Lewis, Donald Davidson, W. V. O. Quine, and Ludwig
Wittgenstein devote a great deal of discussion to questions about what sorts
of "things" emotions, beliefs, intentions, virtues and vices are. However,
these philosophers never examine the assumption in all of these discussions
that psychological states can be assigned and theorized about on an individualistic basis. There has been no examination of the assumption that the self
is sharply distinguishable from its social surroundings. She shows how this
unexamined assumption structures the ideology of individualism not only
as it appears in philosophical psychology, but also as it appears in classical
and contemporary liberal social theory, and even in liberal feminist thought.
The assumption of a sharply distinguishable self is not a "natural fact" she
argues. The ways in which the assumption

permeates our social institutions, our lives, and our senses of ourselves are not unalterable. It is deeply useful in the maintenance of capitalist and patriarchal society and
deeply embedded in our notions of liberation, freedom and equality. (p. 226)

In the last half of her paper she shows how this assumption has its social
roots in the psychosexual development of males mothered by women in a
patriarchal society, in the development in males of ego and of ego-boundaries.
She argues that this assumption "is fundamentally undercut by an examination of female experience, if that experience is seen in its own terms and not
as truncated male experience." (p. 226)

In 'Political Philosophy and the Patriarchal Unconscious: A Psychoanalytic
Perspective on Epistemology and Metaphysics,' Jane Flax examines the
thought of Plato, Descartes, Hobbes and Rousseau to show how distinctively
masculine experience has contributed to the definition of *the* "problems

of philosophy." She argues that "the denial and repression of early infantile experience has had a deep and largely unexplored impact on philosophy. This repressed material shapes by its very absence in consciousness the way we look at and reflect upon the world." (p. 245) Flax is not making claims about the particular psychological makeup of individual philosophers, nor is she suggesting that philosophy can or should be treated as the mere rationalization of unconscious impulses and conflicts. Instead, she argues that a feminist theory of knowledge must begin by examining how and why the dilemmas projected onto the world as the "human condition" in fact have their social roots in a virtually universal kind of masculine experience. The apparently irresolveable dualisms of subject-object, mind-body, inner-outer, reason-sense reflect real but repressed dilemmas originating in masculine infantile experience of the sexual division of labor. The epistemologies of all bodies of knowledge which claim to be emancipatory, including psychoanalysis and Marxism, must be analyzed from this perspective, for the female dimensions of experience tend to be lost in philosophies developed under patriarchy. However, she argues that women's experience "is not in itself an adequate ground for theory. As the other pole of the dualities it must be incorporated and transcended" (p. 270), and this requires political transformation of both individuals and of social life more generally.

Nancy Hartsock argues that feminists must create the epistemological standpoint from which an adequate feminist social theory can emerge. In "The Feminist Standpoint: Developing the Ground for a Specifically Feminist Historical Materialism,' she shows how feminists can use for their own purposes the epistemological tools provided by Marx's explanation of the enlarged vision available to the proletariat as a consequence of its position in the structural division of labor by class. With these tools, we can grasp how women's structural position in the institutionalized sexual division of labor allows the creation of a standpoint from which feminists can systematically understand patriarchal institutions and ideologies as perverse inversions of more human social relations. The material conditions for the creation of the feminist standpoint begin in the sexual division of labor in childrearing, for this results in systematic differences in male and female experience. "These different (psychic) experiences both structure and are reinforced by the differing patterns of male and female activity required by the sexual division of labor, and are thereby replicated as epistemology and ontology." (p. 296) However, in order to develop the epistemology and ontology required for a feminist standpoint, we must grasp as did Marx "that there are some perspectives on society from which, however well-intentioned one may be, the real relations of humans with each other and with the natural world are not visible." (p. 285) Sorting this contention into five distinct epistemological and political claims, she shows how women's life activity as structured by the sexual division of labor provides unique possibilities for understanding nature and social life and for grasping and opposing all forms of domination.

In 'Why Has the Sex/Gender System Become Visible Only Now?', Sandra Harding argues that feminist inquiry during the last decade has produced a new object for scientific scrutiny: the sex/gender system. While it is widely recognized that the feminist perspective calls for revisions in our scientific understandings, in morals and in politics, Harding argues that it also calls for a revolution in epistemology. The need for a new theory of knowledge can be seen if we ask an intuitively reasonable question: why has the sex/gender system emerged into visibility only now? That is, what "causes" its "discovery" at this moment in history rather than in 1776, 1848, or 1919? This question is one which must be asked if we are to understand new ideas as more than "intellectual" achievements − if we are to understand scientific advances as in part the product of changes in historical social relations. While this question appears intuitively reasonable, Harding shows why it is unintelligible − it cannot even be formulated − from the perspective of the three dominant existing epistemologies. Thus asking the question reveals the self-imposed limitations of empiricist, functionalist/relativist, and Marxist epistemologies. She suggests that it is because feminist researchers in the social sciences have relied too heavily on these inherited epistemological discourses that the title question remains relatively unexamined. Until it is answered we will not able to understand the feminist standpoint as one which history has made possible.

The Women's Movement has created the impetus and direction for vast changes in the beliefs and social activities of modern societies. We hope these essays contribute to our understanding of how women's experience can be made into a valuable resource in the creation of more complete and less distorted belief and in the design of emancipatory social life.

LYNDA LANGE

WOMAN IS NOT A RATIONAL ANIMAL:
ON ARISTOTLE'S BIOLOGY OF REPRODUCTION*

> Aristotle . . . pretends that women are but monsters.
> Who would not believe it, upon the authority of so
> renowned a personage? To say, it is an impertinence;
> would be, to choak his supposition too openly.
> If a woman, (how learned soever she might be),
> had wrote as much of men, she would have lost all
> her credit; and men would have imagined it sufficient,
> to have refuted such a foppery; by answering, that
> it must be a woman, or a fool, that had said so.
> From *De l'égalité des deux sexes* (Paris, 1673)
> François Poulain de la Barre (1647–1723), anony-
> mously translated into English as *The Woman as Good
> as the Man* (London, 1677).

The conservatism of Aristotle has long been a subject of discussion among philosophers. His belief in the superiority of the male sex, however, while it has not entirely escaped their notice,[1] has not thus far been carefully examined. In a path-breaking article,[2] Christine Garside-Allen brought to our attention the possibility that the work of Aristotle is in fact the study of the male human, rather than the human species, and pointed to the possibility that this may be true of most influential philosophers. This task of hers was an explicit necessity because in most cases philosophers' ideas about sex difference are not now widely known. In what are known as the "main" theories of various political or moral philosophers, distinctions of sex are not often mentioned, or they are alluded to briefly in a way that makes them appear inessential to the theory. I want to suggest that the reason for this is not that their views of sexual differences are incidental to the theory but that in almost every case they are considered to be a question which is prior to general ethical or political issues. This may be the case regardless of whether or not these views are actually discussed in any detail. For most of these thinkers, however, there will be found a treatise of some sort on the subject. It has been the practice of twentieth-century scholars and educators in the face of the greater equality of women, simply to disregard these works, which they view as minor or peripheral, or perhaps crankish, like Berkeley's

1

Sandra Harding and Merrill B. Hintikka (eds.), Discovering Reality, 1–15.
Copyright © 1983 by D. Reidel Publishing Company.

late essays on the value of tar water. Unfortunately, this policy is not soundly
based on the actual role of these views in most political and social philosophy.
I believe that in the case of Aristotle, these views are more than an "analogy
between the biological and ethical relations of man and woman", as Garside-
Allen suggested.[3] According to her, Aristotle draws this analogy but does
not explain *why* a woman is, as he claims, a privation of man.[4] I want to
argue that, however unacceptable his characterization of women may be,
within the framework of his own thought he actually does explain it to
perfection, in terms of the four types of cause. In fact, judging by the number
of references to the question, Aristotle considered the existence and nature
of women to be one of the features of life that most compellingly called
for an explanation. To Aristotle, it was obvious, as we shall see, that women
are inferior, and did not actualize the unique human potential of self-
governance by reason. In terms of final causes, there was for him a question
as to why they existed at all as separate individuals, rather than there being
one type of human capable of reproducing itself in a hermaphroditic fashion.

Aristotle's biological writings as a whole, and not only what he writes
about women, have also been treated by many as dispensable for the study
of his philosophy, a fact which tends to aid and abet the practice of ignoring
his sexism. It has been traditional to approach Aristotle through the logical
and metaphysical writings, yet there is evidence that Aristotle himself considered the biological works of great importance. According to J. H. Randall,
"his most characteristic distinctions and emphases grow naturally out of the
intellectual demands of the subject matter of living processes."[5] This is a
controversial claim, but regardless of whether or not it is true, the fact
remains that the important Aristotelian distinctions between "form" and
"matter", "mover" and "moved", "actuality" and "potentiality", are all used
by Aristotle to distinguish male and female. His theory of sex difference is
at the very least interwoven in a consistent manner into the fabric of his
philosophy, and it is not at all clear that it can simply be cut away without
any reflection on the status of the rest of the philosophy.

In this paper, I shall first present Aristotle's theories of generation and
sex distinction, and then proceed to a philosophical examination of their
basis and their implications. The outline of the more empiricist skeleton of
"the biology" helps to clarify the discussion of the issues, although it must
be borne in mind that this is a gross modernization of the concept of biology.
The unified Aristotelian view of science, however, ought to emerge in the
subsequent section on Aristotle's methods and assumptions.

THE BIOLOGY

Aristotle's initial definition of "male" and "female" in *De Generatione Animalium* is the following: "For by a male animal we mean that which generates in another, and by a female animal that which generates in itself". (716 a 13) To Aristotle this indicates not only a difference of anatomical parts but a difference in their "ability or faculty". (716 a 18) "The distinction of sex", he writes, "is first principle". (716 b 10) As such, it has many other differences consequent upon it.

The Aristotelian theory of generation must have been developed in a milieu of considerable biological speculation, judging by the amount of discussion of rival theories. The question of whether or not both male and female produce semen, where semen is loosely conceived of as "whatever-it-is" that initiates the movement of growth of a new living individual, appears to have been a major controversy. The central question of generation for Aristotle is the explanation of the transfer or creation of soul to give life to the material of flesh and blood, for, as he puts it, a hand is not a hand in a true sense if it has no soul. (726 b 25)

Another question was whether or not "semen" (in the sense in which Aristotle was using the term) comes from the whole of the body of the parent, or only from some part. Aristotle poses both of these questions and states that if "semen" does not come from the whole of the body, then "it is reasonable to suppose that it does not come from both parents either". (721 b 8) It is apparent that this does not follow rigorously, a fact which the translator attempts to explain by saying that Aristotle wishes to reject the Hippocratic view which combined two distinct theories, and appears to assume that oversetting one of them affords a presumption against the other. However, I think this is a somewhat naive underestimate of Aristotle's dialectical discussion, for reasons which will appear below. After numerous arguments against the view that semen comes from the whole body of the parent(s)[6] Aristotle reiterates, "if it does not come from all the male it is not unreasonable to suppose that it does not come from the female", (724 a 8) to which the translator observes in a note "I do not follow this argument"! Indeed, it seems the reverse position is just as plausible: that if the "semen" *does* come from the whole of the body, it would need to do so from only one parent to create a new individual. Conversely, if it comes from a specialized part, both parents might make a contribution. The latter view would be consistent with modern biology, according to which half the genetic endowment is from the male in the sperm and half from the female in the ovum. The problem of explaining why offspring are sometimes female and sometimes male remains with all of these views, and is dealt with by Aristotle separately. What *does* follow from the view that the "semen" does not come from all the body, is that it is *possible*, from a strictly logical point of view, that it does not come from both parents. If it is a specialized function of some part to begin with, there is no *logical* reason why both parents would have to have the same function, or share aspects of the function between them. Aristotle chooses to proceed on the hypothesis that it does not come from both parents, and encounters nothing to contradict this view. He then addresses himself to the question of what "semen" is.

In Aristotelian terms, "semen" must be either the material from which the offspring is made, or the form which acts upon something else, or both at once. (724 b 5) Since semen "contributes to natural growth" it must be a secretion of useful nutriment, rather than a waste product or other excretion. Aristotle reasons that semen is a secretion of the blood, because all other parts of the body are from the blood when "concocted and somehow divided up". (726 b 5) It is the last and finest concoction of the blood because it is produced best by healthy individuals in the prime of life, and absent or infertile in the old and sick. In the case of children he reasons that they use up all their nutriment for growth before this stage of concoction is reached. Further proof that semen is highly refined and concentrated nutriment, says Aristotle, is the fact that the emission of even a small quantity of semen is exhausting, like the loss of nutriment as a result of bleeding! (725 b 6) The view that "semen" is a concoction of the blood was also used to explain resemblance, since the "semen" is a sort of quintessence of the blood which comes from all parts of the parent's body. Thus the "semen" "is already the hand or face or whole animal" potentially, either "in virtue of its own mass" or because it has a certain power. (726 b 13)

Aristotle argues that the catamenia (menses) in females are analogous to "semen" in males. His argument here is based on several principles of biology that he terms "necessary", which all involve the concept of vital, or soul, heat. According to Aristotle: (1) "the weaker animal should have a secretion greater in quantity and less concocted"; (2) "that which Nature endows with a smaller portion of heat is weaker"; and, (3) woman has less vital heat than man. (726 b 30) It follows from these principles that the catamenia are this secretion, and the fact that their presence is associated with generation in women further supports the conclusion. It is clear that if the catamenia are a less thorough concoction of the blood than "semen", on account of lesser heat, the female cannot have "semen" also. (727 a 27) From this

Aristotle concludes that "the female contributes the material for generation and this is in the substance of the catamenia". (727 b 31) The fact that a woman may conceive without the sensation of pleasure in intercourse is for Aristotle further proof that she has no "semen". (727 b 8) The suppressed premise, of course, is that the emission of "semen" is accompanied by pleasure, which is one of the more superficial examples of male bias in Aristotle's biology. The ovum of the female would in fact be "semen" in the sense in which Aristotle is using the term, but of course its release is not usually accompanied by any sensation at all.

Since the female contributes the material of the new individual, it cannot be the case that she also has the power to infuse soul into it, for then she could reproduce herself without a male. However, there are males, and according to Aristotle, "Nature does nothing in vain". Aristotle therefore concludes by a process of elimination that the male must impart the motion and the form, and the male semen is the means of doing it, analogous to a craftsman's tool. The form of the child exists potentially in the male soul, and the semen, as the tool, possesses "motion in actuality". (730 b 21) What causes the semen to be productive is vital heat. It is interesting to note that the observations of Aristotle could not possibly have suggested any role for the male except the mere addition of material in the form of semen, so that he has assigned the male a reproductive function for which, by definition, there could be no observation-base, namely the transmission of the form of the soul. Had he set out consciously to formulate a theory of male superiority that would be difficult to disprove, given the intellectual milieu of his age, he could scarcely have done better than he did.

As to the existence of separate sexes, there are two distinct questions. The first is: why are there males and females of a species, rather than individuals with the faculty of self-generation? Second, what causes a particular individual to be male or female?

The first question concerns final causes, and the determination of why it is better that there be two separate sexes. Aristotle's explanation for this is succinct:

. . . as the first efficient or moving cause, to which belong the definition and the form, is better and more divine in its nature than the material on which it works, it is better that the superior principle should be separated from the inferior. Therefore, wherever it is possible, and so far as it is possible, the male is separated from the female. (732 a 6)

The separation of the sexes is not required for generation itself, since as

Aristotle knew, there are hermaphroditic animals. These, however, are all lower forms of life, a fact that is not without significance.

The question of why a particular individual is female or male concerns efficient causes. According to Aristotle, the material secretion of the female contains, potentially, all the parts of the animal, including the parts of both male and female that differentiate them from each other. The sexual parts of both are included, since the male, by Aristotle's analogy with the craftsman, contributes no material from his body to the offspring. (737 a 25) The material of the offspring prior to conception is therefore potentially either male or female. The essential distinction between male and female is the capacity or incapacity to concoct semen containing vital heat. According to Aristotle, the male, who can do this, is hotter than the female, and therefore the cause of maleness must be greater vital heat. The amount of vital heat a copulating male has varies with his nature, his age, his state of health, and the weather! Semen is said to be less well concocted if there is greater moisture, hence if a moist south wind is blowing at the time of copulation, the offspring are more likely to be female. (766 b 34)

The female has some effect on the sex of the embryo. The catamenia as well may be better or worse concocted, and, not surprisingly, "If the generative secretion in the catamenia is properly concocted, the movement imparted by the male will make the embryo in the likeness of itself". (767 b 17)

According to Aristotle, the determination of sex occurs at the very beginning of embryonic development, at which time nature gives both the faculties and the organs of the sexes at the same time.

There is a further refinement of the theory of generation in connection with the rational soul. This is necessary because the theory of generation concerns all animals, and only humans have a share in reason. The rational soul, according to Aristotle, may exist without body, "for no bodily activity has any connexion with the activity of reason". (736 b 29) The nutritive and sensitive souls, however, are not entities, but "the actuality of something that possesses a potentiality of being besouled". (*De Anima*, 414 a 28) In other words, they are the living organization of matter. Since they do not exist without matter, they cannot logically be transferred without the transference of matter, which the craftsman-father does not do. Aristotle resolves the difficulty by concluding that these souls are present potentially in the female. "It remains then", he writes, "for the reason alone so to enter and alone to be divine." (736 b 29) The rational soul, of course, is the type of soul that is distinctive of human personhood.

THE *ARCHAI* OF GENERATION

Aristotle's notion of science includes not only a collection of "the facts", but also "the reasons why" the facts are as they are observed to be. These reasons why consist in each case of formal and/or final causes. The full explanation usually also includes some principles, which may be either metaphysical or generalizations of a more or less empirical nature. Each subject matter has its own "starting points", among which are the distinctive *archai*, or "principles", which guide the investigation, and according to J. H. Randall, "always function very much as what we should today call 'hypotheses'".[7] The *archai* themselves are arrived at by a process which is a form of induction. However, this process, as it appears in Aristotle, is undeniably weighed down by familiar epistemological problems in the philosophy of science that are associated with the analysis of induction. It raises at least as many questions as it answers. The process appears to be a form of reflection on the observation of phenomena that is meant to result in "understanding" it. According to Randall "those *archai* themselves are established and validated, not by reasoning and demonstration, but by *noûs*: by "seeing" that it is so, that this is the way in which the facts can be understood".[8] These *archai* lead to the determination of causes which ought to form the premises from which the "observed facts" can be demonstrated as a conclusion. According to Aristotle, the *archai* of the sciences are determined by the rational order of the world.

Considered as a description of the *practice* of science this view is one that many would consider today to be basically correct, as far as it goes, with respect to hypothesis formation and the relation of general hypotheses to causal laws and observation. This is largely because hypothesis formation is still considered essentially unexplained. The "causes", in modern terms, are of course always "efficient", i.e., physicalist. The observation of a contradiction between the *archai* and "the facts" counted for Aristotle as disconfirmation of the *archai*, and the logic of this too is essentially the same as that of the modern practice. It is just that Aristotle did not devote most of his energy to looking for disconfirmation, which we now consider it the obligation of a scientist to do.

In view of the stature and influence of Aristotle as a philosopher of science, and scientist, it is of some interest to feminist scholarship to determine how, in the use of his method, Aristotle arrived at the views he did, especially since, as a practitioner of it, Aristotle is not held to be among the worst.[9] It is not enough to say that Aristotle was limited by the lack of a microscope, or, what Aristotle himself would have considered much more relevant, a thermometer. With all the hardware of modern science, the differences in higher "nature", if any, between female and male, are still virtually unknown, and who can say what it is we currently lack that limits our vision? Nor is it sufficient, I think, to brush the theory aside as obviously inferior, unless we are willing to be equally cavalier about other aspects of the history of science and philosophy. Aristotle's theory of generation went largely unchallenged for about as long as his logic.[10] And the intuitions which appear to have given rise to, and been reinforced by, the theory, especially in its ethical and political implications for women, are by no means dead yet. It is useful, therefore, to attempt to identify precisely at least some of the points in Aristotle's pursuit of biology where specific bias is introduced.

The initial task for Aristotle was to identify the subject matter as to kind. This identification is what Hintikka calls in Aristotle a "generic premise", which is a definition of immediate terms which "consists in an indemonstrable assumption of what they are". (*Posterior Analytics*, II, 94 a 9)[11] Of kinds of things, there are, for example, the living and the non-living. Among the living, there are plants and animals, in a graded series as to amount of vital heat, the lowest having the least. Aristotle writes in *Historia Animalium*, "So throughout the entire animal scale there is a graduated differentiation in amount of vitality". (588 b 21) Within species there is female and male. Aristotle's initial identification of the female as being the kind of thing "which generates in itself" seems unobjectionable. Even this, however, is immediately followed by the observation that this must be why the poets call the earth Mother, and the sun and heaven (i.e., the gods) Father! This poetic analogy is unimportant, however, compared with other characterizations of the female that appear later on in the discussion. These have serious implications, but they seem to be for Aristotle equally a matter of simple identification of the type, rather than the outcome of rational discourse.

It appears that Aristotle does something with the study of humans that in general terms survives as a controversy in our own day. He takes a scientific method that works well for the study of the natural world and applies it to the study of human nature (and, by implication, social relations), expecting it to work equally well there. Yet, like his modern colleagues, he ends with questionable and controversial results. Why does this occur in Aristotle's biology? The "generic premises", according to Hintikka, are supposed to be an account of how a term should be defined in view of an exhaustive knowledge of the relevant facts.[12] Although this approach works well with fish (Aristotle's classification of fish still stands), it may be observed that with reference to people this approach is inevitably biased in favour of what is the case at the time the analysis is made. Furthermore, since Aristotle's view of soul (i.e., life) was teleological, he saw the nature of living things in terms of function or purpose. For example, if the eye were an animal its soul would be sight. Thus the type of soul of such social groups as women or slaves, according to Aristotle, fitted them, not surprisingly, for the functtion which they happened to be fulfilling. Aristotle's method does not draw attention to the differences between a biological function and a social function. For example, if it is a matter of fact that gestation occurs in a (socially) inferior being, this method makes it natural to assume that gestation is itself an inferior function to impregnation.

As we have seen, at the end of his critical examination of existing opinions as to whether or not the female contributes any "semen" to generation, Aristotle introduces a set of assertions which enable him to resolve the question. Among these are the general statement that what "Nature endows with a smaller portion of heat is weaker", and "it has already been stated that such is the character of the female". (726 b 32) This prior statement concerning women is not in the De Generatione Animalium itself, and appears to be a reference to De Partibus Animalium. In the discussion of blood, Aristotle writes that "Noblest of all are those whose blood is hot", and that in these respects the male is superior to the female. (648 a 10–15)

While the amount of heat in the blood appears to be considered a matter of fact by Aristotle, its greater warmth is an indication of greater faculties of soul, and it is these gradations of faculties, rather than the amount of heat per se, that perform the actual job of explanation. Given this scheme, the superiority of maleness is naturally said to have the form of an increment of faculties. The female is therefore quite literally a privation of the male, or as Aristotle puts it:

The woman is as it were an impotent male, for it is through a certain incapacity that the female is female, being incapable of concocting the nutriment in its last stage into semen. (728 a 18)

In the De Generatione Animalium the assertions concerning the weakness and lesser heat of women are listed along with another "fact" of observation. He writes that the weaker animal will have a reproductive secretion greater in quantity and less concocted, and being less concocted, it will be more like blood. This the female has. Since Aristotle ranked animals according to amount of vital heat, the female was identified at the start as a notch below the male in the graduated differentiation of animals. The alleged inferiority of the female, therefore, is for Aristotle one of the "starting points" of science, based on a feature of experience which it is the task of natural philosophy to make intelligible, and not the outcome of rational discourse.

Hintikka uses a phrase which neatly captures the nature of a principle such as the one that women have less vital heat than men. It is a "discussion-stopper". Thus a brief questioning of an Aristotelian about the status of women might run as follows:

"Every woman is (and ought to be) a non-citizen."

"Why?"

"Because every woman is inferior to every male citizen."

"Why?"

"Because every woman is devoid of the highest form of human reason."

"Why?"

"Because every woman has inadequate vital heat for the exercise of the highest form of human reason."

"Why?"

Discussion-stopper: "Because having less vital heat than the male is what a woman *is*."

or "That is how "woman" ought to be defined."

According to Hintikka, "the finitude of the elements in question is according to Aristotle a consequence of the knowability of essences."[13] Since the world is rationally ordered the nature of things must be rationally determinable. As we have seen, however, an approach which has done yeoman service in the history of natural science is full of pitfalls when it comes to the study of human nature and social relations.

There is another, equally consequential, source of bias. According to Aristotle, the forms of soul are distributed in a sort of pyramid structure, the lowest (nutritive soul), being common to all living things. Each form of soul has its function, the functions of the nutritive being the use of food and reproduction. For living things, after eating, "the most natural act is the production of another like itself". (415 a 28) For this there is a final cause, for the sake of which all animals do whatever their nature makes possible. They seek to "partake in the eternal and the divine". Since as perishable living things they cannot be eternal as individuals, they seek to reproduce another like themselves, so as to be members of an eternal species. (415 b 1–8) Their final cause is to perpetuate their species because "soul is better than body, and the living, having soul, is better than the lifeless which has none, and being is better than not being, living than not living."

(731 b 29) These *archai* of generation are meant to explain, at the most basic level, why there is "that which generates" and "that from which it generates", regardless of how these principles might be manifested. These final causes have a direct application to the efficient causes of maleness and femaleness. Since the reproduction of the same type of individual is the goal of reproduction, Aristotle operates in *De Generatione Animalium* on the hypothesis that *exact* reproduction of the parent in the offspring is the most "natural". The existence of such a norm which is "always a perfected activity", is said to be characteristic of Aristotle's practice of science.[14]

Aristotle refers loosely to mother and father as parents, but the real parent is clearly the male, since it is he who contributes the human soul. Also, as we have seen, the most perfect functioning of the reproductive process is said to produce a child "like the father and not like the mother" in non-sexual characteristics as well as sexual. Any departure from the likeness of the father is a deficiency in terms of this norm. (767 b 23) A further characterization of the nature of the female emerges when Aristotle writes:

Even he who does not resemble his parents is already in a certain sense a monstrosity; for in these cases Nature has in a way departed from the type. The first departure indeed is that the offspring should become female instead of male; (. . .) And the monstrosity, though not necessary in regard of a final cause and an end yet is necessary accidentally. (767 b 5ff)

We have the paradox that although the most natural act for an animal is said to be the reproduction of another like itself, the most perfect functioning of the female is said to be to produce a male like her partner. What does this mean? While the production of males by the male is a sign of superiority, *vide* "Hercules, who among all his two and seventy children is said to have begotten but one girl", (585 b 23), the production of females by the female is not, being caused by "improper" concoction of the catamenia. On the other hand, the production of *males* by a female is no woman's glory either, because of Aristotle's notion of conflict between the male and female elements in the determination of sex, whereby her femaleness is "prevailed over" if she produces a male. The unavoidable implication is that her most perfect functioning involves creating the conditions for the male element to prevail.[15]

It is apparent that the logic of the reproductive norm only holds for the male. The individuals who seek to partake in the eternal and the divine by reproducing themselves are by implication only the males. *Her* partaking is quite different. She is instrumental to species eternity, and potentially rather than actually human. The actuality of her human potential is the incubation of a male child.

Aristotle regarded the "monstrosity" of the female as an accidental necessity *of the species*, the norm of which species is obviously male. This position is very different from that wherein maleness and femaleness are considered as accidents of individuals. The latter view makes it possible to assign full human ends to all individuals, whereas Aristotle's view does not. Final causes he found operant only in relation to the whole species, male and female, or in relation to the male alone. The final cause of the female individual, *qua* deficient human, was quite literally outside herself, and was that of being instrumental to the reproduction of male humans. Given the function of the female, and the sense in which Aristotle thought of the organs of the body as "instruments" of the unified person, the female has been defined as virtually an organ of the male body. The organs of the body are matter set in motion by and for the soul of the unified person, for the ends of the species. These organs, or parts, are "material causes" of animals, and it may be noted that the female is no more than a material cause of the animal. Reading 'male' for 'person', what else are women but "matter set in motion by and for the soul of the unified male, for the ends of the (male) species"?

CONCLUSION

According to Randall, Aristotle did not look to knowledge – not even to what a modern would call scientific knowledge – to *do* anything other than give understanding. Randall writes "we can say that for Aristotle the highest power a man can exercise over the world is to understand it – to do, because he sees why it must be done, what others do because they cannot help themselves."

Aristotle's political philosophy, in which he includes what are called the ethics, is meant to tell us how to be good, rather than how to be free, although it is in the nature of his concept of virtue that its mainspring must be within the individual, and not imposed externally. It requires autonomy because its basis is the having of certain dispositions or habits. Insofar as freedom was a concern of Aristotle's, I think he can be interpreted as believing that human freedom of action is only authentic if based upon a solid understanding of the ends and the limitations that arise out of the nature of humankind and the rest of the living and non-living world.

Since virtue for a woman must also be founded on the understanding of the causes (of all sorts) that explain her nature, the biology of Aristotle has an obvious relation to his ethical and political views concerning women. This relation may be regarded as foundational or it may be that the "starting points" of the biology and the politics have in common the "fact" of the inferiority of women. The woman herself, of course, could not have knowledge of this, but only true opinion, based on the acceptance of male authority. Given this scheme, it would be an act of positive virtue for a man to attempt to dominate a woman and compel her to be obedient to him. Conversely, a man who treated a woman as an equal would be acting in a shameful manner.

The best support for the relevance of the biology to the ethics and politics comes from Aristotle himself. In the *Eudemian Ethics* he writes:

... we should not think even in political philosophy that the sort of consideration which not only makes the nature of the thing evident but also its cause is superfluous, for such consideration is in every enquiry the truly philosophic method. (Book I, Ch. 6)

According to Randall, "What is often felt as the "conservatism" of Aristotelian spirit and temper is the conservatism inherent in trust in experience, in facts long encountered".[15] Randall finds Aristotle willing to build on the observations and opinions of others, and suggests that this is because Aristotle viewed science as a gradual accretion of knowledge to which successive thinkers could add. I think this may be misleading, since Aristotle does not merely build on previously acquired "knowledge". In the biology, he invariably opens the discussion of each issue by refuting the theories of other thinkers, before proceeding to argue for his own theory. It *is* true, however, that Aristotle seldom challenges an observation of "facts" because he did not regard their determination as involving any difficulty, other than the practical. Science was not for him the discovery of "facts". According to Marjorie Grene,

Natural science, as (Aristotle) understands it, remains within the framework of everyday perception and makes more precise, within that framework, our formulation and understanding of the essential natures of quite ordinarily accessible entities.[16]

It is quite consistent with this approach to science that Aristotle simply accepted the "observation" of his time and place that women are inferior, although he did not accept the explanations for it that had thus far been developed.

For many people today the inferiority of women is a "fact long encountered". However, since the advent of the concept of theory-laden observational data and "facts", even those who vehemently disapprove of the concept have been forced to be more self-conscious about whether or not they dress their "facts" in the costumes of a particular theory. Some people believe there are unadorned facts (known as "brute" or "raw" facts), and some believe there is really no such thing as a fact at all: there is only the "conceptual framework", a sort of theoretical clotheshorse. Aristotle's facts, it seems clear, come dressed in the full regalia of Greek philosophy and social practices. Thus he explains all, but challenges nothing, and all heaven and earth is marshalled in interlocking hierarchies patterned after the structure of Greek society.

My purpose is not to malign Aristotle as an individual. What is called, with deserved contempt, the phallocentric world view (i.e., the conflation of "male" and "human"), is, however, still very much alive, and there is a presumption that knowing the history of an idea is always useful in attaining full command of it, or, effectively opposing it.[17]

Calgary Institute for the Humanities
University of Calgary

NOTES

* All references to Aristotle are to *The Works of Aristotle*, ed. by J. A. Smith and W. D. Ross (Oxford University Press: Oxford, 1908–1952).

[1] In his preface to the 1910 translation of *De Generatione Animalium*, Arthur Platt remarks on the "curious depreciation of the female sex."

[2] Garside-Allen, Christine, 'Can a Woman be Good in the Same Way as a Man?', *Dialogue* 10 (1971), 534–544.

[3] Garside-Allen, p. 536.

[4] Garside-Allen, p. 537.

[5] Randall, J. H., *Aristotle* (New York, 1960), pp. 220, 242.

[6] Aristotle makes, for example, the ingenious argument that this does not in fact explain ressemblance, which it was intended to do, for "if really flesh and bones are composed of fire and the like elements, the semen would come rather from the elements than anything else, for how can it come from their composition? Yet without this composition there would be no resemblance. If again something creates this composition later, it would be *this* that would be the cause of the ressemblance, not the coming of the semen from every part of the body." (722 a 34)

[7] Randall, p. 41.

[8] Randall, p. 46.

[9] Darwin wrote "Linnaeus and Cuvier have been my two gods, though in very different ways, but they were mere schoolboys to old Aristotle" (*Life and Letters*, vol. iii, p. 252.)

[10] Even Randall, writing in 1960, remarks on Aristotle's errors in connexion with spontaneous generation, and refers to his "generally correct" theory!

[11] Hintikka, Jaakko, 'On the Ingredients of an Aristotelian Science", *Noûs* 6, 55–69 MR 72.

[12] Hintikka, p. 59.

[13] Hintikka, p. 58.

[14] "The object of investigation is always an end, or function of the subject matter: what that kind of thing does, how it operates. And the problems of that science are how everything in the subject matter, all the facts there displayed, are related to and involved in that function. The inquiry thus seeks to analyze the factors involved in a certain function. Hence it is important at the outset to establish norms: in the *De Anima*, knowing; in the *Ethics*, the "prudent" or intelligent man. The norm is always a perfected activity, . . . " Randall, p. 52.

[15] Should it be doubted that such a concept reflects any significant tendancy, consider the admiration of Nietzsche for Greek civilization, where "Woman had no other mission than to produce beautiful, strong, bodies, in which the father's character lived on as unbrokenly as possible". 'Human All Too Human', in *The Complete Works of Nietzsche*, trans. O. Levy (New York, 1964), p. 238, #259.

[15] Randall, p. 52.

[16] Grene, Marjorie, 'Aristotle and Modern Biology', *Journal of the History of Ideas* 33, 395–424.

[17] It has been objected to me that this entire paper is a mistake, on the ground that terms like "active" and "passive", "masculine" and "feminine", and even "inferior" and "superior", are mere metaphysical terms, and have nothing to do with actual inferiority and superiority. In this connection, it may be noticed that the translator, Arthur Platt, remarks that throughout *De Generatione Animalium* "the male" and "the female" are in *the neuter*, and their force cannot be conveyed precisely in English. These tortured ideas are the basis of the notion that Aristotle is not really a sexist, and that he meant only that "the female" (in the neuter, of course) was not a bearer of the highest form of rationality, and not necessarily that actual females (in the female) were such. Whatever this may mean, it's supposed to be a good thing for women. But if this were the case, why does Aristotle go ahead and conclude that women are in fact inferior, and ought to have an inferior position in society? Perhaps it is only "the female" or "the female principle" he meant to exclude from political life. If so, actual women have nothing to worry about. Leaving their "principles" at home, they may presumably involve themselves in politics as men do!

ELIZABETH V. SPELMAN

ARISTOTLE AND THE POLITICIZATION OF THE SOUL

In Book I of the *Politics* Aristotle argues that men are by nature the rulers of women. The conclusion of the argument, which has to do with relationships *between* people — in particular, political relationships between men and women — is said to be based on what is known about relationships *within* people: in particular, relationships between the rational and irrational elements of the human soul. That is, this part of Aristotle's political theory is said to rest on his metaphysics or theory of the soul. I hope to show that not the least of the reasons for examining Aristotle's argument is that doing so sheds light on the question of whether metaphysical positions are politically innocent. To ask this question is a defining if not necessarily a distinguishing characteristic of a feminist perspective in philosophy.

Aristotle's argument is outlined briefly in Part I. In Part II I begin examination of the argument by describing Aristotle's theory of the soul, noting especially the kind of authority which, according to Aristotle, the rational part of the soul has over the irrational part. In Part III I observe that when he tries to make use of his view about the authority of the rational part of the soul over the irrational part, to defend his view about the authority of men over women, Aristotle ends up contradicting his view about the authority of the rational part. In Part IV I argue that Aristotle's attempt to justify the authority of men over women by reference to the authority of the rational part over the irrational part is in any event circular: a close reading of the texts shows that both understanding what it means to talk about relations of authority between parts of the soul, and establishing that one part has authority over another, depends on understanding what it means to talk about relations of authority between classes of persons (including those between men and women), and on establishing or assuming that certain classes do have authority over others — in particular, that men have authority over women. Aristotle makes clear to us what the relation between the rational and irrational parts of the soul is, by reference to the very same political relationships he hopes to justify by reference to the soul. Part V concludes with some comments on the nature of Aristotle's argument and the nature of my response to him.

17

Sandra Harding and Merrill B. Hintikka (eds.), Discovering Reality, 17–30.
Copyright © 1983 *by D. Reidel Publshing Company.*

Aristotle's argument about the natural authority of men over women is very close to his arguments about the natural authority of masters over slaves, fathers over children, "intellectuals" over laborers, and is offered simultaneously with those arguments in the *Politics*. Though my examination focuses mainly on Aristotle's view of women, the scope of Aristotle's argument is a reminder that oppressive attitudes towards women have close connections to oppressive attitudes towards other groups or classes, that the oppression of women is related in theory as well as in practice to the oppression of other groups.

I

One of the requirements of a state is that some rule and some be ruled:

there must be a union of those who cannot exist without each other; . . . of natural ruler and subject, that both may be preserved. (P, 1252a25–32)

And this means in particular, Aristotle says, that men are to rule women, masters are to rule slaves, fathers are to rule children. But why? The mere principle that some are to rule and some are to be ruled doesn't itself tell us who is to rule whom. Aristotle is untroubled by the idea that humans are to rule animals, because he believes that animals' lack of reason establishes their inferiority to humans and disqualifies them from eligibility to rule. But all humans *qua* humans have reason and "share in the rational principle" (P 1259b27). So to what grounds must one move in order to establish the inferior and subordinate status of women vis-à-vis men, slaves vis-à-vis masters, children vis-à-vis fathers?

Well, says Aristotle, fortunately "the very constitution of the soul has shown us the way" (P 1260a5). The soul has two main parts or elements, the rational and the irrational, and it is "natural and expedient" for the rational to rule over the irrational (P 1254b4ff.).[1] Just so, men are to rule women, for in women the deliberative capacity of the rational element is without authority – it is easily overruled by the irrational element. In similar fashion, masters are to rule slaves, for while slaves, in virtue of the rational element in their souls, can hear and obey orders, they really don't have the capacity to deliberate. Indeed all that distinguishes slaves from non-human beasts of burden is that they, unlike beasts, have just enough reason to understand the results of the masters' deliberations; otherwise their capacities are identical to those of the beasts (P 1254b19ff.). Fathers are to rule children, because although children have the capacity to deliberate that is associated with the rational element of the soul, this capacity is immature (P 1260a6–15).

It is, then, by reference to the relationships between the rational parts of the soul that Aristotle tries to justify his view that certain classes of beings are naturally subordinate to others. Just as one part of the soul stands in a certain relationship to another, so one class of beings stands in a certain relationship to another class. But this is a bare outline of the argument. In order to understand Aristotle's argument thoroughly, we have to understand in more detail how he describes the workings of the parts of the soul and their relationship to one another (Part II). We also have to understand just how he moves from a description of the parts of the soul to a description of the parts of the state (Part III).

II

We have to turn to the *Nicomachean Ethics* as well as to parts of the *Politics* to fill in the details of Aristotle's description of the relationship between the rational and irrational parts of the soul. A central feature of his depiction of that relationship is that it is a relationship of authority. The rational part is supposed to rule the irrational part. This is an authority intended for it and vested in it by nature (P 1254b7–8), though Aristotle both explicitly and implicitly allows that the rational part is not always fully empowered to exercize that authority: as we've seen, Aristotle says that in the case of women, slaves and children, the rational part does *not* rule the irrational part (as we shall soon see, we have to ask whether it is even *supposed* to, in the case of women, slaves and children). Even in adult male masters, sometimes the irrational part is not ruled by the rational part; if that weren't so, Aristotle presumably would not have thought it necessary, as he does in the *Ethics* and the *Politics*, to give instructions about the importance of the rational part remaining in control and command. Hence when Aristotle talks about the rule of the rational part of the soul over the irrational part, he cannot be said to be merely pointing out that what happens in one part of the soul determines what happens in another part in some mechanical fashion. In fact, if this is what Aristotle meant by the rule of the rational over the irrational part, then he would have to say that the irrational part sometimes rules over the irrational part; but he explicitly resists this when he suggests that sometimes the irrational part "appears" to rule over the rational part even when it really doesn't (P 1254b). So the rule or authority he ascribes to the rational part must have to do with entitlement: the rational part has the right to, or ought to, or is intended by nature to, rule the irrational part, even if that isn't always what happens.

The first thing, then, to note about Aristotle's description of the relation between the rational and irrational parts of the soul is that it is a relationship of ruler and subject. But, again, this relationship is not described in merely mechanical terms, as if something in one part of us moves or clicks or fires, and as a result something in another part of us moves or clicks or fires. As soon as he begins to characterize the rule of the soul over the body, or of the rational part of the soul over the irrational part, Aristotle turns to the language of persons and politics: such rule is despotical, or constitutional, or royal (P 1254b3–5); when the parts of the soul are properly aligned, the rational part dictates or commands, and the irrational part obeys – in the way in which a child obeys its father (NE 1102b30–1103a4). Indeed the relationship between the parts of the soul is treated by Aristotle in most of the *Nicomachean Ethics* and the *Politics* as if it were a relationship between political entities, not as if it were between impersonal, quasi-organic parts. He not only describes the parts of the soul in language applicable to persons or agents; he sometimes speaks of the parts of the soul by analogy to *particular* persons or kinds of persons:

as the child should live according to the direction of his tutor, so the appetitive element should live according to rational principle (NE 1119b14)

the nature of appetite is illustrated by what the poets call Aphrodite, 'guile-weaving daughter of Cyprus' (NE 1149b17)

The relationship of master and servant or that of husband and wife show us

the ratios in which the part of the soul that has a rational principle stands to the irrational part (NE 1138b5–12)

In fact Aristotle goes so far as to suggest the *identification* of the person with a part of the soul:

the things men have done on a rational principle are thought most properly their own acts and voluntary acts. That this is the man himself, then, or is so more than anything else, is plain . . . (NE 1168b35ff.; cf. 1166a17ff.)

Indeed, sometimes he depicts reason as another person to whom the person owes obedience:

reason in each of its possessors chooses what is best for itself and the good man obeys his reason (NE 1169a17–18)

In sum, Aristotle's depiction of the relationship between the parts of the soul is a highly personalized (or anthropomorphized) and politicized one. One reason this is so noticeable is that Aristotle here fails to respect his own admonitions in the *De Anima* [2] about not personalizing the parts or functions of the soul. There, in a manner familiar to students of 20th-century Anglo-American philosophy, Aristotle suggests we remember that predicates which apply to persons in virtue of having souls nevertheless apply to the *persons* and not to their souls: the person is angry, not the soul or a part of the soul; the person thinks, not the soul or a part of it (*De Anima*, 408b14, 408b27). But his own warning is not heeded here, for Aristotle treats the parts of the soul as if they were persons or agents themselves, and in particular as if they were persons standing in political relations to one another: relations apparently best described in terms such as constitutional or royal ruler; obedient or resistant subject; master and slave.

Having looked at Aristotle's description of the soul, we must now look at the way in which he makes use of this description to justify his view that women are naturally subordinate to men.

III

As we have seen, in the *Politics* Aristotle turns to the constitution of the soul in order to justify his view that certain classes of beings are by nature to rule over other classes. He wants us to see that just as the irrational part of the soul is subordinate by nature to the rational part, so women are subordinate by nature to men.

We must remember here that Aristotle's claim about the natural subordination of the irrational part to the rational part, about the authority of the rational part over the irrational part, can only be understood as a claim about entitlement, for sometimes, even in free men, the irrational part in fact is not controlled by the rational part. It is unnatural for the irrational part not to listen to and obey the rational part, for it is intended by nature to do so (P 1254b4ff.). With this in mind, let's spell out Aristotle's argument. Aristotle's claim about the soul is that

(a) In the soul, the rational part by nature rules or has authority over (but does not always control) the irrational part.

Now Aristotle wants to use (a) to argue that understanding the nature of the authority of the rational part over the irrational part shows us the nature of the authority of men over women. So he wants to argue, on the basis of (a), that

(b) In the state, men by nature rule or have authority over women

But why does Aristotle associate men with the rational part and women with the irrational part of the soul? Because, he holds,

(c) In men's souls, the rational part by nature rules or has authority over the irrational part

while

(d) In women's souls, the rational part by nature does not rule or have authority over the irrational part

Or is it that

(e) In women's souls, the irrational part by nature rules or has authority over the rational part?

While (c) seems straightforward, both (d) and (e) seem to be uninvited guests, given Aristotle's insistence and dependence on (a). Does Aristotle hold that in the case of women, the lack of authority in the rational part is tantamount to the assumption of authority in the irrational part (as in (e))? Aristotle certainly must hold at least (d), for in the case of women, he says, the deliberative capacity is without authority and the irrational part actually controls or overwhelms the rational part. So even if Aristotle is not committed here to the view (e) that in women the irrational part is supposed to rule (parallel to his claim that in free men the rational part is supposed to rule), he must be saying (d) that in the case of women the rational part is not supposed to rule, it is not supposed to have authority. For his claim is not just that women *happen* to be subordinate to men; they are intended by nature to be subordinate to men. (Just as those who are to rule are rulers by nature, so those who are to be ruled are subjects by nature.) And nature could not succeed in this intention with respect to those who are subordinate unless nature at least intended the rational part to be without authority in women. On Aristotle's own reckoning, women are subordinate to men by nature; that by virtue of which women are naturally subordinate to men must be, of course, something intended by nature; so the lack of authority in women's rational element must be intended by nature. In short, women are by nature unnatural.

In light of this, the eager remark by some Aristotelian scholars,[3] that Aristotle's view about the relation between men and women is not merely a comment on or reflection on the *status quo*, takes on special significance. For in one sense that is quite right: in saying that women are subordinate to men, Aristotle was not merely making an observation on the world around him. For insofar as that observation is correct, nature, according to Aristotle, has gotten her way (recall that Aristotle doesn't think this always happens, e.g., P 1254a–b). The rational element in women doesn't just happen to be without authority; for if that were the case, there would be no way to distinguish between the natural condition of a woman and the unnatural condition of a free man who is, for example, overindulgent: in the latter case, the rational part of the man's soul doesn't happen to exercize its authority, though it is intended to have authority; while in the former case, the rational part of the woman's soul is without authority but is not intended to have authority; if it were intended to have authority it wouldn't be in a woman!

But now Aristotle's argument is in deep trouble. For he *begins* his argument by saying that nature intends the rational part of the soul to have authority over the irrational part. But in order to get from there to the claim that nature intends men to rule over women, he has *also* to say that nature intends the rational part of the soul not to have authority over the irrational part, in the case of women. The merely contingent lack of authority of the rational part of someone's soul would not establish the claim that that person is naturally subordinate to someone else. Those who are naturally subordinate must be, on Aristotle's own reckoning, those in whom we can say that nature intended their rational part not to have authority.

Aristotle's problem is not merely that in order to generate the conclusion he wishes to reach, he has to deny one of his central premises. That is, the problem is not merely that in order to reach the conclusion that women are by nature subordinate to men, he has to deny that the rational part of the soul by nature rules over the irrational part. Let us look at the argument once again. He says that just as the rational part of the soul is to rule the irrational part, so men are to rule women. As we've seen, this requires the association of rationality with men and irrationality with women. But on what grounds does he make these associations? Well, as we noted, he makes them on the grounds that nature intended rationality to rule in men, and intended rationality not to rule in women. But this is merely begging the question, unless he can explain how he knows that nature intended this. His only possible reply to this is that in men rationality prevails and in women it does not. But even if true this wouldn't justify the view that this is what nature intends – for as we've seen, Aristotle himself points out that the mere fact of the dominance of one part of the soul over the other, or of one group of people over another, is never proof that this is the ways things ought to be according to nature.

So Aristotle's argument for the natural subordination of women to men is, to put it charitably, wobbly: he holds an inconsistent view about the natural relationship between the rational and irrational parts of the soul, and he begs the question when he claims that the rational element by nature rules in men but does not in women.

IV

What I've been analyzing above is Aristotle's attempt to come to a position in political theory on the basis of a metaphysical position. As noted in Part II, that metaphysical position is itself a politicized one insofar as it is deeply etched in the language of political theory. The *dramatis personae* of the soul, and the drama itself, are modelled on the human persÒns and the human drama found in the political realm. If this is so, then we have to ask what business Aristotle has referring to the kind of relation that exists between metaphysical entities to clarify the kind of relation existing between political entities, if the relationship between the metaphysical entities itself is modelled on the relationship between political entities. In short, is Aristotle's argument from metaphysics to politics circular?

As we saw in Part II, Aristotle refers to relationships between kinds of persons to describe by analogy the relationships that hold between parts of the soul. And on more than one occasion, Aristotle explains why we need analogies to describe intra-psychic relationships. For example, in explaining the kind of relationship there is between the irrational element and the rational element, he says:

while in the body we see that which moves astray, in the soul we do not. No doubt, however, we must none the less suppose that in the soul too there is something contrary to the rational principle, resisting and opposing it. . . . Now even this seems to have a share in a rational principle, as we said; at any rate in the continent man it obeys the rational principle – and presumably in the temperate and brave man it is still more obedient; for in him it speaks, on all matters, with the same voice as the rational principle. (NE 1102b27–30)

Aristotle says it is hard to tell what kind of soul a man has – whether that of a freeman or slave – because it doesn't always happen that a man with a freeman's soul also has a freeman's body and we cannot view the soul directly.

And doubtless if men differed from one another in the mere forms of their bodies as much as the statues of the Gods do from men, all would acknowledge that the inferior class should be slaves of the superior. And if this is true of the body, how much more just that a similar distinction should exist in the soul? but the beauty of the body is seen, whereas the beauty of the soul is not seen. (P 1254b32–1255a2)

As he also says, "to gain light on things imperceptible we must use the evidence of sensible things" (NE 1104a13)[4] So Aristotle seems to take it as a given that in order to describe the soul, in order to make sense of relationships among parts of the soul, he has to rely on reference to and analogy to visible things. The particular visible things he relies on, as we've seen, are human beings standing in relationships of power and authority.

To point this out is not to ignore the fact that in general Aristotle's descriptions of the many parts of the world are in hierarchical terms. For example, Aristotle refers to ruler/subject relations as existing not only in the soul and in the state:[5]

in all things which form a composite whole and which are made up of parts, whether continuous or discrete, a distinction between ruling and subject element comes to light. Such a duality exists in living creatures, but not in them only; it originates in the constitution of the universe; even in things which have no life there is a ruling principle, as in a musical mode. (P 1254a29–34)

However, not all ruling and subject elements need be conceived of as being like persons in political relationships. For example, although Aristotle spends a good bit of time in the De Anima discussing the hierarchy of functions in the soul, one is hard put to find evidence of the highly personalized and politicized language that appears both in the NE and the Politics. Indeed, as mentioned above (p. 21), he explicitly discourages his readers from thinking of the parts or functions of the soul as if they were themselves persons.[6]. In a somewhat similar fashion in the Metaphysics he advises us not to take seriously the myths according to which the gods are thought of in anthropomorphic terms (Meta. 1074b6ff.). But when he describes the ruling and subject elements in the soul, he immediately recurs to the language of persons and politics. He speaks of the kind of rule of the rational part over the irrational part in terms that have to do with the rule of a person or persons over other persons − "constitutional" or "royal" rule (P 1254b5; cf. NE 1138b5ff.). This fact is very significant. For presumably Aristotle is "bringing to light" the distinction between the ruling and subject elements among humans by pointing to the distinction between ruling and subject elements in the soul. But the kind of distinction between ruling and subject elements in the soul itself "comes to light" only through relationships of authority among human beings. As described in Part II, Aristotle brings to light what he takes to the appropriate relationship — the relationship intended by nature — between parts of the soul, by analogy to relationships of authority among humans:

as the child should live according to the direction of his tutor, so the appetitive element should live according to rational principle (NE 1119b14)

Metaphorically and in virtue of a certain resemblance there is a justice, not indeed between a man and himself, but between certain parts of him; yet not every kind of justice but that of master and servant or that of husband and wife. For these are the ratios in which the part of the soul that has a rational principle stands to the irrational part. (NE 1138b5–12)

We have to note that Aristotle does *not* try to justify his view about the natural rule of men over women by reference to a general principle about ruling and subject elements, for he quite explicitly refers us in particular to the constitution of the soul. There we find ruling and subject elements, but they are highly personalized entities whose relationships are described in terms of political relationships among human beings. In light of this, we must conclude that Aristotle's argument for the natural rule of men over women is circular. He argues for the position that men by nature rule women. How do we know that they do? We know this because the rational element of the soul by nature rules the irrational element. And how to do we know this? This is where we come full circle: Because men rule women (and also because masters rule slaves, because tutors rule children). In fact the rule of men over women provides us with a means of understanding the kind of relationship among parts of the soul; and, coupled with the assumption that men represent the rational element and women represent the irrational element, it provides us with a means of establishing that in the soul the rational element rules the irrational.

Aristotle took a short-cut in his journey from his metaphysics, from his philosophical psychology, to his political theory: he built the particular relationships of authority he wished to justify on the basis of the metaphysics, into the metaphysics itself. For first of all, the terms used to describe the kind of authority the rational part has over the irrational part are unapologetically borrowed from the terminology of political relationships; yet presumably we are supposed to be relying on an understanding of the kind of authority the rational part has over the irrational part to understand the kind of authority men in the *polis* have over women. And secondly, the location of authority in the soul is supposed to tell us about the location of authority in the *polis*. Yet we are told that the clue that such authority is located in the rational part of the soul over against the irrational part, is that authority is located in men over against women, masters over against slaves, etc.

None of this is to say that if Aristotle hadn't thought men were naturally the rulers of women, he never would have suggested that the rational part of the soul rules the irrational part. Had he believed — ah, the power of counterfactuals to stretch the imagination — had he believed and wished to justify the belief that women were by nature the rulers of men, he might have used such a relationship between them as a model for the relationship between parts of the soul, and have ended up with the same problems of circularity.

I've said that Aristotle's political conclusion is built into his metaphysics. The political conclusion is sexist. Does that mean the metaphysics is sexist? Is the view that the rational part of the soul by nature rules over the irrational part a sexist view? No. What is sexist is not the assertion of the authority of the rational part over the irrational part, but his association of rationality with men and irrationality with women.[7] The metaphysics *is* politicized, however, and this primes it for use to defend political positions that are sexist.

V

There are several reasons why I've thought it important to focus on Aristotle's argument for the natural subordination of women to men. It is not just that the argument doesn't work.

First of all, in taking a close look at the metaphysical position Aristotle relies on, we begin to see how thoroughly drenched it is with political language and imagery. Even if this were all we could say it would be important to say it, because according to what might be called the theory of philosophical cleanliness, metaphysics and philosophical psychology are supposed to be separate from and cleansed of political considerations: we may be entitled to draw political conclusions from metaphysics or psychology, but partic- ular conclusions are not supposed to shape the metaphysics and psychology itself. So we see Aristotle wanting us to draw conclusions about politics from his psychology — a disingenuous request, and an absurd operation, if he believed that the psychology really was a version of politics to begin with. In a very similar way, I take it, we are nowadays asked to think about what consequences for social and political relations might be drawn, for example, from the studies called sociobiology, and from studies on intelligence: what do studies of human biology, or the animal kingdom, or the human psyche, have to tell us about the appropriateness or inappropriateness of particular social and political relations? ("Dear Dr. Freud, tell us what women's psyches are like so we'll know what to do with their lives!") We are expected to assume, however, that these studies are not themselves influenced by the political or social conclusions which will be said to follow from them. I leave it to students of sociobiology, psychology, and intelligence theory to judge whether and how such studies are biased by political considerations; for example, is it true, as Marshall Sahlins suggests, that "the theory of sociobiology has an intrinsic ideological dimension"; is it true that sociobiology involves "the grounding of human social behaviour in an advanced or scientific notion of organic evolution, which is in its own terms the representation of a cultural form of economic action"?[8] What I hope to have shown here is just how Aristotle's psychology is infused with the language and imagery of politics and how the political conclusions Aristotle wished to draw from his psychology get attached to his psychological premises.

It is important to note in this connection that this paper is not written from the viewpoint of a social historian or an historian of ideas. I have not been attempting here to try to show the connection between Aristotle's philosophical views, on the one hand, and the historical and political context in which he lived, on the other. Rather, I have been trying to show the conceptual and logical connections between Aristotle's political theory and his psychological theory. As mentioned above, the question of their relative dependence on one another has been raised before. Most recently a version of the question has been raised by W. W. Fortenbaugh, in an article called 'Aristotle on Women and Slaves'[9]. I shall conclude this essay with a response to Fortenbaugh.

Fortenbaugh holds that Aristotle's argument about women and slaves is a "political application of . . . philosophical psychology" (p. 135). He considers the possible charge that Aristotle's use of his psychology to defend a political position is merely an *ad hoc* device to defend the status quo. In response to this, Fortenbaugh represents Aristotle as holding *not* that women are subordinate to men, and that's how we know their deliberative capacity is without authority; but *rather* that women's deliberative capacity is without authority and that is how we know they are subordinate to men. According to Fortenbaugh, Aristotle's claim about the lack of authority which characterizes women's deliberative capacity is based not on "inter-personal relationships" but on "an intra-personal relationship" (*ibid.*):

Aristotle "looks within the slave to explain his social position . . . [and] looks within the woman to explain her role and her virtues" (p. 138).

Now since, as we've seen, Aristotle himself said that we can't "look within" to see the soul, we have to ask how, according to Fortenbaugh, we know that in women's soul the deliberative capacity lacks authority. Fortenbaugh's answer would seem to be, "because it is often overruled by her emotions or alogical side" (p. 138). But doesn't that happen in the case of free men also? Yes, the reply must go (see p. 19 above), but that doesn't mean that free men are to occupy subordinate roles, for in them the deliberative capacity is by nature with authority even if it doesn't always exercize it. So it must be that in women, on the other hand, the deliberative capacity is by nature without authority. But how do we know that? Psychology can't tell us that, because it tells us − that is the foundation for Aristotle's political argument, according to Fortenbaugh − that the rational part by nature *does* have such authority. The only reason we've been given for believing not just that women's deliberative capacity lacks authority, but *by nature* lacks this authority, is that if it weren't true women would not be by nature subordinate to men (at least not in terms of the argument of the *Politics*). Hence it is a requirement of Aristotle's politics that in women the deliberative capacity be without authority; it is not a conclusion of his psychology. In fact Fortenbaugh himself says this repeatedly throughout his short article without realizing the damaging import of it for his argument: e.g., he says that Aristotle "demands of women a virtue which reflects their domestic role" (p. 137); "it also seems proper to assign slaves [this applies to women as well] a virtue limited by the demands of their subordinate role" (p. 136). Fortenbaugh is saying here that Aristotle has a certain role in mind for women and slaves, and must thus posit for them and in them a psychological condition befitting their position. This isn't, however, what Fortenbaugh describes his own view as proposing!

Fortenbaugh fails to consider what gets built into Aristotle's psychological theory to make it seem even a plausible basis from which to argue for a political conclusion. He does not here consider the *kind* of authority the rational part is supposed to have over the irrational part. He also seems to think that the question of the role of irrationality in women is an empirical one (p. 139); but if it were treated by Aristotle as an empirical question, then it would have to be conceivable that women's souls could be like what Aristotle calls the souls of free men. But if that is conceivable, women could not be said to be *by nature* subordinate to men.

Aristotle's argument deserves far more serious attention than Fortenbaugh has given it. I hope to have provided some of that attention – not only to show that the argument suffers from difficulties Fortenbaugh doesn't even imagine, but to point out that the movement from a metaphysical position to a political position is not always as innocent as it seems.

Smith College

NOTES

[1] Cf. *Nicomachean Ethics*, Bk. I, Ch. 13.

[2] Even in the *De Anima*, as Hamlyn has pointed out, "Aristotle does not often live up to this remark". D. W. Hamlyn, *Aristotle's De Anima* (Oxford: Clarendon Press, 1968), p. 81.

[3] Most recently, in W. W. Fortenbaugh's 'Aristotle on Slaves and Women', in *Articles on Aristotle: 2. Ethics and Politics*, eds. Jonathan Barnes, Malcolm Schofield, Richard Sorabji (London: Duckworth, 1977), pp. 135–139.

[4] I think the view adumbrated here by Aristotle, about the necessity of referring to publicly observable beings or things or activities to describe things or activities that are not publicly observable, is complemented by fairly recent developments in the philosophy of mind – in particular, by the view roughly associated with Wittgenstein and with Strawson according to which the concept of mind or soul is parasitic upon concepts of publicly observable things. I shall not elaborate on that view here, but wish to point out that it may enable us to see how easily it may happen that we use anthropomorphized language to describe the soul or mind.

[5] Barker thinks that the fact that Aristotle refers to "a general principle of rule and subordination" saves him from the charge that he appears to argue in a circle. Ernest Barker, *The Politics of Aristotle* (Oxford: Clarendon Press, 1952), p. 35, fn. 1. In what follows I explain why I don't think Barker's view can be sustained.

[6] As mentioned in fn. 2 above, Aristotle sometimes uses personalized language in the *De Anima*. But interestingly it is not also politicized language.

[7] That association, as we saw in Section II, is perfectly arbitrary. Moreover, to maintain it, Aristotle has to hold that in one class, the class of men, the rational by nature rules, and in another class, that of women, the rational by nature does not rule. But if there is any class in whom the rational by nature does not rule, then the original premise of Aristotle's argument – that the rational by nature rules the irrational – is contradicted.

[8] Marshall Sahlins, *The Use and Abuse of Biology: An Anthropological Critique of Sociobiology* (Ann Arbor: University of Michigan Press, 1977), pp. xii and xv. See also Donna Haraway, 'Animal Sociology and a Natural Economy of the Body Politic' (two parts), *Signs* 4 (1978), 21–60, and 'The Biological Enterprise: Sex, Mind and Profit from Human Engineering to Sociobiology', *Radical History Review* 20 (1979), 206–237; Paul Thom, 'Stiff Cheese for Women', *Philosophical Forum* 8 (1976), 94–107.

[9] Fortenbaugh, *op. cit.* Further page references in text are to this article.

JUDITH HICKS STIEHM*

THE UNIT OF POLITICAL ANALYSIS:
OUR ARISTOTELIAN HANGOVER

INTRODUCTION

Every scholar knows that assumptions shape conclusions. In particular, students of politics know that the unit chosen for analysis has a crucial effect on what is seen and recommended. At different times and places the analytical unit has been the family, the tribe, the corporation, the individual, the group, social class, the mass (and or the elite), the nation, and even "the globe."

Debate between critics using different analytic units is often unproductive. For example, those who see a world composed of competing individuals and those who see a world of competing ethnic groups have rather different views about the need for affirmative action. Similarly, those who view the globe as an arena for competition between super-powers understand a change of regime in Angola differently from those who focus separately on nation states and the responsiveness of governments to their own populations.

Today in the U.S. most citizens think of political action as individual, and to most the fundamental political act is that of one man's casting one vote for one of two other men. In college courses, however, students are taught about reference and interest groups, political parties and global strategy. This enables them to think more like the political actors who make and carry out policy.

Sometimes it is noted that women and men participate differently in politics and government; explanations for that differential have so far been inadequate [1]. In this paper it will be argued that the failure to explain must partly be attributed to a collective Aristotelian Hangover. By this it is meant that although we often think we are thinking in terms of either individuals, groups, or classes, we in fact often slip over into thinking about the polity in terms of families. We do so principally when we think about women. The result is a mixed analysis and confused thinking. Centuries after, our thoughts too often resemble those of a morning-after.

THE ARISTOTELIAN HANGOVER

Aristotle's theory of the household is described in *The Politics* [2]. There

31

Sandra Harding and Merrill B. Hintikka (eds.), Discovering Reality, 31–43.
Copyright © 1983 by D. Reidel Publishing Company.

Aristotle argues that the *polis* is composed of villages which are composed of households which are the result of two elementary associations — that of the male and female and that of the master and slave. For Aristotle there is no smaller social or political unit than the household. There the adult male rules "naturally" over the slave, who is without the faculty of deliberation, the child who has it in a yet immature form, and the woman whose capacity "remains inconclusive." Because the freeman, slave, wife, and child "complete one another" they have a common interest and the freeman can act appropriately for all of them in the public arena.

While today's state has assumed certain responsibilities *vis-à-vis* children and can intervene if parents become abusive, few argue that children should act politically on their own behalf. In contrast none argue that "natural" slaves must be directed and acted for by "natural" masters. Indeed, as will be shown, U.S. political commentators are quite sensitive to the after-effects of slavery and carefully consider the different political experience of black and white citizens.

With women it is different. Discussion sometimes treats women as individuals and sometimes demonstrates how hard it is to describe them as a group. However, in both analysis and policy-making, women are often treated as members of a household — as either wives or daughters. Further, this is often done unconsciously. Aristotle lives but he is not necessarily acknowledged.

SEEING THE DATA

Data to illustrate the arguments in this essay will be drawn from Sidney Verba and Norman Nie's *Participation in America* [3]. This volume won an American Political Science Association award as the best new book in its field; it was especially commended for its methodology. It is not a straw man.

First let us illustrate the analysts' sensitivity to the slavery hangover (and to contemporary racism) and their insensitivity to the "Aristotelian Hangover," (and contemporary sexism). Verba and Nie develop six measures of participation and score a variety of groups for "over" and "under" representation in political participation. They conclude that "men are somewhat over-represented in the most activist political groups but not to a very great degree," and that is the extent of their discussion of male-female differences [4]. In contrast black-white differences are found to be both important and interesting. A full chapter is devoted to their analysis. But what did the data show? What did the tables look like which produced these conclusions?

Black-white differences on the six measures varied from 4–27%. There was a 15% average difference. Female-male differences ranged from 11 to 28% with a 19% average difference. Female-male differences were clearly greater than black-white yet the female-male data were essentially disregarded while the black-white data were carefully discussed.

MEASURING CLASS

The principle finding of Verba and Nie was that there is a "close relationship among social status, participation and (governmental) responsiveness" [5]. That is, that even in a country (the U.S.) committed to individualistic competition, social class seems to be the principle determinant of political participation. Since class or SES (socio-economic standing) is the crucial variable, its definition would seem to be of fundamental importance. Yet on investigation one finds that the measurement of SES was *not* treated as a matter of importance but as one of convention, and that convention, in fact, handled the measurement of women's and men's status differently.

Survey data is collected from individuals and one might assume that individuals would be assigned an SES based on their individual characteristics. Reasonable as this may sound, it is not the conventional way of doing this chore. Instead sociologists and political scientists regularly assign women an SES based (at least in part) upon either their father's or husband's characteristics rather than upon their own. In the study in question, for example, women are ranked according to an SES index derived from data on (1) education, (2) family income, and (3) occupation of head of household.[1] To repeat, this is not unusual. The rationale is that SES is not a measure of individual standing but a measure of social access or of offsprings' economic potential. The assumption is that for social activities and economic prediction the family functions as a unit and that the adult male's influence is primary [6]. However, most political action is individual action. Only individuals vote or are selected for office. It would seem appropriate, then, that political scientists consider individuals as individuals. After all, a male lawyer, a female lawyer, and the wife of a male lawyer do not enjoy equal access to political power even if they do enjoy a similar life-style and even if their children do have similar economic opportunities.[2] Therefore, even if one accepts conventional measures of SES as adequate for certain kinds of social analysis, one can still argue that those measures are inappropriate for the prediction of political and/or governmental participation [7].

The conventional way of assigning SES is designed to yield similar male-female status distributions. It is the argument of this paper that if women were assigned SES independent of their male relatives, male and female status distributions would be found to be dissimilar. Specifically, women would almost disappear from the highest SES categories; they would also move out of the lowest categories. Both facts are relevant to thinking about political participation and social policy.

The data collected by Verba and Nie for their *Participation in America* will be used in the demonstrations below. There three different methods will be used to produce and then compare male-female SES distributions. In some cases two distributions will be produced. In one individuals will be assigned a status based on the occupation of the head of house. In the second individuals will be assigned a status based on their own occupation. The SES methods which will be used (and modified) are: (1) that used by Verba and Nie, (2) one based on Duncan's occupation-scoring, and (3) one developed by Valerie Oppenheimer based on U.S. Census Occupation Categories.

SES DISTRIBUTION BY SEX USING THE METHOD OF VERBA
AND NIE

The Verba and Nie index for SES combines data on education, occupational prestige and family income. While the precise method employed remains ambiguous, the index seems to be built on individual education, family income and head of house occupation scores.[3] Z scoring was used to produce six SES categories with approximately 17% of the population in each category. This method yields very similar SES distributions for women and men although there are a few more women at the very bottom and a few less at the very top. When respondents are assigned their own occupational status, there is little change in the men's SES distribution. For women, however, the change is substantial. (See Table I)

Table I indicates that when women are assigned their own occupation a shift away from the mean occurs. This shift is to both the uppper and lower ends of the occupational scale and is due to the fact that while many women are found in the unskilled and semi-skilled (clerical and sales) positions, relatively few are found in the skilled and independently employed categories. The shift does not appear so dramatic when occupation of the respondent is computed into the composite SES index, however. This is because of the centralizing effects of the inclusion of family (not individual) income and of education variables both of which more closely approximate a normal curve.

TABLE I
Occupational Rankings and SES of Women Respondents
in Verba-Nie Sample, Computed on Basis of
Occupational Status of Head of Household and Respondent

	OCCH		OCCR		SES (Using OCCH)		SES (Using OCCR)	
	(N)	(%)	(N)	(%)	(N)	(%)	(N)	(%)
1.	134	11.0	82	14.6	231	17.7	85	15.1
2.	325	26.8	127	22.6	171	12.9	95	16.8
3.	108	8.9	13	2.3	208	15.8	58	10.3
4.	224	18.5	54	9.6	246	18.7	73	12.9
5.	233	19.2	215	38.3	245	18.6	134	23.8
6.	189	15.6	70	12.5	216	16.4	119	21.1
	1213	100.0	564	99.9	1317	100.1	564	100.0

The most dramatic feature of Table I, however, is that the female sample is reduced by more than half to 564 cases. This is because the occupation of most women is assigned to a "O" category. It is not calculated. It is not considered. Housewife, the most-common U.S. occupation, is not quantified at all. This will be discussed further, but the gross reduction in sample size should serve as a vivid reminder that there may be other elements of sex bias built into SES measures. For example, even though the family is treated as an economic unit for consumption and tax purposes, family power studies suggest that allocation and control of resources is related to who generates the income. Thus a woman who produces $10,000 of a family's $25,000 income may play quite a different role both at home and in the community than does the woman who produces none of the family's $25,000 [8]. Present data conventions do not permit sensitivity to this phenomenon.

THE HOUSEWIFE PROBLEM

How should the prestige of the occupation "housewife" be assessed? In her study *Jobs and Gender: Sex and Occupational Prestige* Christine Bose specifically examined the occupational prestige of the "housewife" [9]. By using her findings one can first compute a reasonable prestige value for housewives and then assign all individuals their own, individual, SES.

The Verba and Nie occupational categories were:

Unskilled	1
Semiskilled	2
Independently employed	3
Skilled	4
Clerical and sales	5
Professional and Managerial	6

Where does housewife belong?

Articles available about the work of housewives describe their work in so many ways that reasons could be offered for assigning them to several of the above groups. Washing clothes requires little skill, cooking some, and tailoring a good deal. Managing family records, being an intelligent shopper, arranging for lessons and social life, and buying and selling stock as well as practicing good public and mental health can involve the most sophisticated managerial and professional skills. However, the category which seems most appropriate is "Independently Employed." This is because the major occupations in the category are "proprietors" and "farmers" − occupations which are like "housewife" in crucial ways. First, each involves a limited operation in which the individual asumes important responsibilities and performs a wide variety of tasks at a variety of skill levels. Second, each individual works in relative isolation, setting his or her own goals. Third, each may generate only a limited amount of cash income although managing a substantial amount of money and/or property. Last, each occupies a position in the U.S. political/economic/social myth which is out of propor- tion to his or her effective role. The last is important. Wives and mothers, tillers of soil, and small entrepreneurs all occupy an unusual position. Like ministers, society assigns them high honor and prestige but low income. They figure large in campaign rhetoric but small in campaign calculations; they are a part of descriptive and prescriptive economics but not a significant part of quantified economics. It seems sensible, then, to group these rather special occupations together.

With housewives coded as "3" and all employed individuals assigned a status based on their own education, their own occupation, and their combined family income, the SES distributions for women and men differ. There are fewer women in both the top and bottom categories. (See Table II)

TABLE II
Socio-Economic Status for Men and Women,
Computed on the Respondent's Occupational Status,
with Housewife's Occupation Computed as "3"

	MEN		WOMEN	
	(N)	(%)	(N)	(%)
1.	232	18.8	118	9.0
2.	249	20.2	318	24.1
3.	163	13.4	234	17.8
4.	166	13.5	173	13.1
5.	184	14.9	279	21.2
6.	238	19.2	195	14.8
	1232	100.0	1317	100.0

ANOTHER METHOD: DUNCAN AND BLAU

Another way to approach "the housewife problem" is to consider it as an occupation and to scale it in a way approximating the Duncan-Blau method used to scale men's prestige in male-dominated occupations.

Bose did this and found that the status of the occupation "housewife" varied a great deal with context and sex. Specifically, she found that the occupation was scored at 60 (on a hundred point scale) by women and at 40 by men. When it was suggested that housewife was a job not a role, and that it was a job which could be held either by a male or a female, women lowered their rating to 36 while men raised theirs to 50 [10]. Overall, Bose found the housewife score was 51; (that of "all jobs" was 45 [11].) To give women housewives the full benefit of the prestige society accords them, the Duncan-Blau range has been reduced from 100 to 10 and the SES distribution has been calculated with housewives assigned a "6" on a scale of 10 (see Table III). Note that there are more than three times as many men as women in the top three categories (8, 9 and 10).

SES DISTRIBUTION BY SEX USING THE
OPPENHEIMER/MODIFICATION OF CENSUS CATEGORIES METHOD

The occupation measures used above have difficulties. The major groups used by Verba-Nie are extremely heterogeneous. A "professional," for

TABLE III
Socio-Economic Status Computed by Sex Based on Duncan and
Blau's Occupational Prestige Ratings
Using Respondent's Occupation

	MEN		WOMEN (with housewives)	
	(N)	(%)	(N)	(%)
1.	48	3.9	71	5.4
2.	37	3.0	44	3.3
3.	175	14.2	72	5.5
4.	107	8.7	28	2.1
5.	191	15.5	38	2.9
6.	79	6.4	794	60.3
7.	167	13.6	128	9.7
8.	148	12.0	39	3.0
9.	164	13.3	92	7.0
10.	116	9.4	11	.8
	1232	100.0	1317	100.2

instance, includes physicians, teachers, and dancers, while a manager includes international bankers and foodstore managers. The Duncan scale assigns SES strictly on the basis of the named occupation but it is based on all-male data so that women's jobs and women in men's jobs are not taken into account. Thus, one more method will be used to determine how the SES distribution varies when housewives are included and when everyone is assigned their own occupational ranking.

The method is one developed by Valerie Oppenheimer and it is designed to distinguish occupations within the major groupings on the basis of peak median earnings [12]. What this does is to separate potentially high income occupations (surgeons) from average income occupations (nursery school teachers) within a "kind-of-work" category. This is especially important because "peak median earnings' seem to separate the women from the men. One example will illustrate. When professionals are divided into peak-median income categories, lawyers (who are 97% male) and nurses (who are 98% female) fall into different instead of the same categories.

In applying Oppenheimer's method eighteen categories were collapsed to six so that a high prestige but low income occupation would be grouped with the next lower prestige but high income occupation. For instance,

"professionals" with a peak median income of $5–7,000 were classed with
"managers" with a peak median of $9+ or $7–9,000. The "housewife"
problem was handled in the following way. First, high status work is usually
associated with higher education. A major anomaly occurs in the "service"
category, however, where a high education level is combined with an apparently low occupational prestige. This category, 2, includes occupations
such as police officer, and fireman, and this is where we assigned housewives
on the grounds that their principle role is service, their income potential is
low, but their education is average and sometimes high. The results are shown
in Table IV.

TABLE IV
Socio-Economic Status by Status-by-Sex based on Oppenheimer's Occupational
Divisions Computed on the Basis of Respondent's Occupation

	MEN		WOMEN (with housewives)	
	(N)	(%)	(N)	(%)
1.	395	32.1	236	17.9
2.	319	25.9	777	59.0
3.	209	17.0	147	11.2
4.	95	7.7	48	3.7
5.	155	12.6	63	4.8
6.	59	4.8	46	3.5
	1232	100.1	1317	100.0

Twice as many men are in the top two categories. On the other hand almost
twice as many men are in the lowest category.

IMPLICATIONS OF THE ARISTOTELIAN HANGOVER

By convention most social scientists partly treat women as the associates of
their (male) "head of house." They have no number (language) for the
occupation of "housewife" at all.[4] Thus, while most social scientists consider
class (SES) a crucial variable they *de facto* handle their data on women and
men differently and in a way which obscures differences between them.

Philosophers sometimes make the same error. For example, John Rawls'
tour de force, A Theory of Justice, ostensibly concerns individuals. "The

family" is discussed as the environment of pre-adults (children) and "the problem of justice between generations" is discussed too. But for Rawls the latter is a matter of ascertaining "how much they [fathers] should set aside for their sons," and the former is a problem in providing "fair opportunity" to individuals [13]. Rawls gives no attention to the different roles adult family members play, although he notes that in the family "the principle of maximizing the sum of advantages is rejected" and he also refers to the [presumed different] virtues of a "good son" and a 'good daughter" and of "a wife and husband" [14]. Mostly adult women are invisible. One begins to grasp that Rawls has had a little too much Aristotle, however, when he describes the "parties" of his analysis as "heads of families" and as representatives of family lines, and then proceeds to say that "What men want is meaningful work in free association with others" [15]. Like the social scientists, then, his focus seems actually to be on male heads of households and those who will become such heads. Wives, mothers, and daughters are of slight interest. Even in thinking about generations his vision is restricted. He considers the next generation (of sons) *while ignoring* the previous generation (with its numerous widowed mothers.)

Thinking about men as individuals who direct households and about women as family members has implications for public policy as well as for political thought. Social Security is an old program; its inequitable treatment of women and men might be considered a hangover from a period when men did not marry until they could support a wife, and women quit work when they married. However, new policy decisions are being made concerning unemployment benefits and job training, and these, too, have the effect of hurting women by not treating them as individuals. The language used does not categorize women separately from men but it uses categories like "family income" and the "principal wage earner" with the result that women and men are affected by the law differently.

Comprehensive Employment Training Act (CETA) programs serve as an excellent example. This massive Federal program provides temporary paid training/employment to individuals who meet certain criteria. The criteria are individual *except* for the income requirement. That requirement is based on the individual's *family's* income. The level is so low that, in effect, if one family member is employed, other family members are ineligible for CETA. Again, if a woman or a teenager has an employed husband or father, they are not eligible for job training no matter how long they have been unemployed. The result is that women and teens, who have higher unemployment rates than men, are the groups least likely to be eligible for training and employment.

Then President Jimmy Carter specifically stated that "Our goal is for every single family to have a guaranteed job by the government." Yet most women including married women with children work, and 47% of families with $15,000 or more income have two persons contributing to that income [16]. The surest way out of poverty, then, is not to have one individual obtain an M.D. or Ph.D. or play quarterback for the Los Angeles Rams. It is to have *two* persons working. A family's standard of living is not likely to improve by assisting one individual's social mobility; it is more likely to improve when each member has a chance to participate in productive work. CETA's goal is said to be to make the program available to the "most needy" (families not individuals). However, to make it more palatable to voters (i.e., to enable some individuals from middle and upper class families to participate), provisions are sometimes made to exempt a certain percent of the slots from family income requirements. Even so, the unit considered in determining eligibility is the family not the individual.

A final example concerns a U.S. Supreme Court decision of June 1979. In Califano vs. Westcott the court held that Federal Aid to Families with Dependent Children–Unemployed (AFDC–U) could not limit payment to families with unemployed fathers. The remedy proposed by the state of Massachusetts, which Congress may adopt in new legislation, was not to extend aid to families with unemployed mothers, but to base eligibility on the state of employment of the "principal wage earner" regardless of his or her income and regardless of the desire and availability for work of any other family member.

Arguments for the "principal wage earner" criteria are economic – that it would otherwise cost too much – "why pay for all those doctor's wives." Of course, one might point out that doctor's wives are also citizens who eat, sleep, study, work, pay taxes, vote, and are entitled to be considered as discrete, whole persons by their government. But the argument is only an (intended) diversion. The real point is that it would be expensive to treat women as citizens instead of as appendages. It would be expensive because the need is great. Most women are poor or middle income women (whether measured by their own SES or that of the "head of house.") Doctor's wives are few and fewer are seeking unemployment or job training assistance. It is *most* women, not a few, who suffer from policies which do not treat them as individuals – policies made mostly by men who, like Aristotle, think of themselves as heads of houses, and their task one of economizing not equalizing.

University of Southern California

NOTES

* The computer calculations for this work were done by Ruth Scott Halcrow, Associate Professor of Health Science, California State University, Dominguez Hills.
1 The U.S. Census has not permitted a woman to identify herself as "the head of the house" if there is any adult male in residence. Nevertheless, one in eight U.S. families is headed by a woman. (Los Angeles Times, August 8, 1974, Part 1, p. 24.)
2 Clearly their female and male children do not enjoy the same opportunities.
3 Even though a close relationship between SES and political activity is their central finding, Verba and Nie present an inadequate discussion of the construction of their SES index. The two brief descriptions given in their book appear to be contradictory; also, apparently neither description fits the method actually used. In the appendix (p. 366) it is said that the index uses data on (1) respondent's education, (2) family income, and (3) occupation of head of household. In the text (p. 130) the SES is said to be based on (1) respondent's education, and (2) the occupation and (3) income of the head of the household. Telephone efforts to clarify this matter eventually produced a letter from the data analyst which stated that the index also included a measure based on the number of credit/financial devices used by the family. It indicated that the other data employed were family income, head of house occupation and head of house education. However, it appeared that the question on education used in constructing the index was one which queried the respondent only about his or her own education.
This muddle dramatically makes the point. The distinction between the head of house and other adults resident with him or her was simply not made. Individuals were not examined as individuals. The SES index was not justified. Convention rather than craft controlled the analysis of data.
4 While the limitations of SES scales seem to be forgotten by the users, their authors (creators) freely acknowledge their limitations. Albert Reis (*Occupations and Social Status*, Free Press, New York, 1961) notes that there is no evidence on SES for female occupation, and that topic remains a "conceptual obscurity." He also notes (p. 149) that only 38% of the population is "classifiable." Omitted are wives (29%), workers under nineteen (7%), women heads of house (7%), unemployed men (10%), unemployed single women (6%), and men in the military or in institutions (2%).

REFERENCES

[1] Efforts have been made in Jane Jaquette (ed.), *Women in Politics* (New York: John Wiley and Sons, 1974); Marianne Githens and Jewel L. Prestage (eds.), *A Portrait of Marginality* (New York: David McKay Company, Inc., 1977) Jeanne Kirkpatrick, *The New Presidential Elite* (New York: Russell Sage Foundation and the Twentieth Century Fund, 1976); Marjorie Lansing and Sandra Baxter, *Women and Politics* (Ann Arbor: University of Michigan Press, 1980).
[2] *The Politics of Aristotle*, ed. by Ernest Barker (London: Oxford University Press, 1961), pp. 1–8 and 32–38.
[3] Sidney Verba and Norman Nie, *Participation in America* (New York: Harper and Row, 1972).
[4] *Op. cit.*, pp. 98–101.
[5] *Op. cit.*, p. 339.

[6] Criticism of various SES measures is offered in Joan Acker 'Women and Social Stratification: A Case of Intellectual Sexism,' *American Journal of Sociology* 78 (1973), 936–945, and Walter B. Watson and Ernest A. T. Barth, 'Questionable Assumptions in the Theory of Social Stratification,' *The Pacific Sociological Review* 7 (1964), 10–16.

[7] Marie R. Haug and Marvin B. Sussman discuss the importance of using different masures for different purposes (such as social class and social status) in 'The Indiscriminate State of Social Class Measurement,' *Social Forces* 49 (1961), 549–63. In the same way potential for political impact may require a distinct and separate measure.

[8] Family power studies have been reviewed by Constantina Safilios Rothschild, 'The Study of Family Power Structure: A Review 1960–69,' *Journal of Marriage and the Family* 32 (1970), 539–52. In general women seem to have more power when they are income producers and when their income is a substantial portion of the family's total income. Both circumstances decrease as family income rises. See other issues of the the same journal for further discussion.

[9] Christine M. Bose, *Jobs and Gender: Sex and Occupational Prestige*, (Baltimore Center for Metropolitan Planning and Research, The Johns Hopkins University, August 1973).

[10] *Op. cit.*, pp. 143–48.

[11] *Op. cit.*, p. 94.

[12] Valerie Oppenheimer, 'The Life-Cycle Squeeze: The Interaction of Men's Occupational and Family Life Cycles,' *Demography*, May, 1974, pp. 237–245.

[13] John Rawls, *A Theory of Justice*, (Cambridge: Harvard University Press, 1971), pp. 74 and 284.

[14] *Op. cit.*, pp. 105 and 467–68.

[15] *Op. cit.*, pp. 128–9, 292, and 290.

[16] U.S. Bureau of the Census, *Statistical Abstract of the United States: 1977*, 98th ed. (Washington, D.C.: Government Printing Office, 1977), p. 391; and U.S. Bureau of the Census, *Current Population Reports*, Series p–60, no. 105, 'Money Income in 1975 of Families and Persons in the United States' (Washington D.C.: Government Printing Office, 1977), p. 36.

RUTH HUBBARD

HAVE ONLY MEN EVOLVED?*

> ... with the dawn of scientific investigation it might
> have been hoped that the prejudices resulting from
> lower conditions of human society would disappear,
> and that in their stead would be set forth not only
> facts, but deductions from facts, better suited to the
> dawn of an intellectual age . . . .
> The ability, however, to collect facts, and the
> power to generalize and draw conclusions from them,
> avail little, when brought into direct opposition to
> deeply rooted prejudices.
> Eliza Burt Gamble, *The Evolution of Woman* (1894)

Science is made by people who live at a specific time in a specific place and
whose thought patterns reflect the truths that are accepted by the wider
society. Because scientific explanations have repeatedly run counter to
the beliefs held dear by some powerful segments of the society (organized
religion, for example, has its own explanations of how nature works), scientists are sometimes portrayed as lone heroes swimming against the social
stream. Charles Darwin (1809–82) and his theories of evolution and human
descent are frequently used to illustrate this point. But Darwinism, on the
contrary, has wide areas of congruence with the social and political ideology
of nineteenth-century Britain and with Victorian precepts of morality,
particularly as regards the relationships between the sexes. And the same
Victorian notions still dominate contemporary biological thinking about
sex differences and sex roles.

SCIENCE AND THE SOCIAL CONSTRUCTION OF REALITY

For humans, language plays a major role in generating reality. Without
words to objectify and categorize our sensations and place them in relation
to one another, we cannot evolve a tradition of what is real in the world.
Our past experience is organized through language into our history within
which we have set up new verbal categories that allow us to assimilate present
and future experiences. If every time we had a sensation we gave it a new

45

Sandra Harding and Merrill B. Hintikka (eds.), Discovering Reality, 45–69.
Copyright © 1979 by Schenkman Publishing Company.

name, the names would have no meaning: lacking consistency, they could not arrange our experience into reality. For words to work, they have to be used consistently and in a sufficient variety of situations so that their volume — what they contain and exclude — becomes clear to all their users.

If I ask a young child, "Are you hungry?", she must learn through experience that "yes" can produce a piece of bread, a banana, an egg, or an entire meal; whereas "yes" in answer to "Do you want orange juice?" always produces a tart, orange liquid.

However, all acts of naming happen against a backdrop of what is socially accepted as real. The question is *who* has social sanction to define the larger reality into which one's everyday experiences must fit in order that one be reckoned sane and responsible. In the past, the Church had this right, but it is less looked to today as a generator of new definitions of reality, though it is allowed to stick by its old ones even when they conflict with currently accepted realities (as in the case of miracles). The State also defines some aspects of reality and can generate what George Orwell called Newspeak in order to interpret the world for its own political purposes. But, for the most part, at present science is the most respectable legitimator of new realities.

However, what is often ignored is that science does more than merely define reality; by setting up first the definitions — for example, three-dimensional (Euclidian) space — and then specific relationships within them — for example, parallel lines never meet — it automatically renders suspect the sense experiences that contradict the definitions. If we want to be respectable inhabitants of the Euclidian world, every time we see railroad tracks meet in the distance we must "explain" how what we are seeing is consistent wih the accepted definition of reality. Furthermore, through society's and our personal histories, we acquire an investment in our sense of reality that makes us eager to enlighten our children or uneducated "savages," who insist on believing that railroad tracks meet in the distance and part like curtains as they walk down them. (Here, too, we make an exception for the followers of some accepted religions, for we do not argue with equal vehemence against our fundamentalist neighbors, if they insist on believing literally that the Red Sea parted for the Israelities, or that Jesus walked on the Sea of Galilee.)

Every theory is a self-fulfilling prophecy that orders experience into the framework it provides. Therefore, it should be no surprise that almost any theory, however absurd it may seem to some, has its supporters. The mythology of science holds that scientific theories lead to the truth because they operate by consensus: they can be tested by different scientists, making their own hypotheses and designing independent experiments to test them. Thus, it is said that even if one or another scientist "misinterprets" his or her observations, the need for consensus will weed out fantasies and lead to reality. But things do not work that way. Scientists do not think and work independently. Their "own" hypotheses ordinarily are formulated within a context of theory, so that their interpretations by and large are sub-sets within the prevailing orthodoxy. Agreement therefore is built into the process and need tell us little or nothing about "truth" or "reality." Of course, scientists often disagree, but their quarrels usually are about details that do not contradict fundamental beliefs, whichever way they are resolved.[1] To overturn orthodoxy is no easier in science than in philosophy, religion, economics, or any of the other disciplines through which we try to comprehend the world and the society in which we live.

The very language that translates sense perceptions into scientific reality generates that reality by lumping certain perceptions together and sorting or highlighting others. But what we notice and how we describe it depends to a great extent on our histories, roles, and expectations as individuals and as members of our society. Therefore, as we move from the relatively impersonal observations in astronomy, physics and chemistry into biology and the social sciences, our science is increasingly affected by the ways in which our personal and social experience determine what we are able or willing to perceive as real about ourselves and the organisms around us. This is not to accuse scientists of being deluded or dishonest, but merely to point out that, like other people, they find it difficult to see the social biases that are built into the very fabric of what they deem real. That is why, by and large, only children notice that the emperor is naked. But only the rare child hangs on to that insight; most of them soon learn to see the beauty and elegance of his clothes.

In trying to construct a coherent, self-consistent picture of the world, scientists come up with questions and answers that depend on their perceptions of what has been, is, will be, and can be. There is no such thing as objective, value-free science. An era's science is part of its politics, economics and sociology: it is generated by them and in turn helps to generate them. Our personal and social histories mold what we perceive to be our biology and history as organisms, just as our biology plays its part in our social behavior and perceptions. As scientists, we learn to examine the ways in which our experimental methods can bias our answers, but we are not taught to be equally wary of the biases introduced by our implicit, unstated and often unconscious beliefs about the nature of reality. To become conscious of these is more difficult than anything else we do. But difficult as it may seem, we must try to do it if our picture of the world is to be more than a reflection of various aspects of ourselves and of our social arrangements.[2]

DARWIN'S EVOLUTIONARY THEORY

It is interesting that the idea that Darwin was swimming against the stream of accepted social dogma has prevailed, in spite of the fact that many historians have shown his thinking fitted squarely into the historical and social perspective of his time. Darwin so clearly and admittedly was drawing together strands that had been developing over long periods of time that the questions why he was the one to produce the synthesis and why it happened just then have clamored for answers. Therefore, the social origins of the Darwinian synthesis have been probed by numerous scientists and historians.

A belief that all living forms are related and that there also are deep connections between the living and non-living has existed through much of recorded human history. Through the animism of tribal cultures that endows everyone and everything with a common spirit; through more elaborate expressions of the unity of living forms in some Far Eastern and Native American belief systems; and through Aristotelian notions of connectedness runs the theme of one web of life that includes humans among its many strands. The Judaeo-Christian world view has been exceptional – and I would say flawed – in setting man (and I mean the male of the species) apart from the rest of nature by making him the namer and ruler of all life. The biblical myth of the creation gave rise to the separate and unchanging species which that second Adam, Linnaeus (1707–78), later named and classified. But even Linnaeus – though he began by accepting the belief that all existing species had been created by Jehovah during that one week long ago ("Nulla species nova") – had his doubts about their immutability by the time he had identified more than four thousand of them: some species appeared to be closely related, others seemed clearly transitional. Yet as Eiseley has pointed out, it is important to realize that:

Until the scientific idea of 'species' acquired form and distinctness there could be no dogma of 'special' creation in the modern sense. This form and distinctness it did not possess until the naturalists of the seventeenth century began to substitute exactness of definition for the previous vague characterizations of the objects of nature.[3]

And he continues:

. . . it was Linnaeus with his proclamation that species were absolutely fixed since the beginning who intensified the theological trend. . . . Science, in its desire for classification and order, . . . found itself satisfactorily allied with a Christian dogma whose refinements it had contributed to produce.

Did species exist before they were invented by scientists with their predilection for classification and naming? And did the new science, by concentrating on differences which could be used to tell things apart, devalue the similarities that tie them together? Certainly the Linnaean system succeeded in congealing into a realtively static form what had been a more fluid and graded world that allowed for change and hence for a measure of historicity.

The hundred years that separate Linnaeus from Darwin saw the development of historical geology by Lyell (1797–1875) and an incipient effort to fit the increasing number of fossils that were being uncovered into the earth's newly discovered history. By the time Darwin came along, it was clear to many people that the earth and its creatures had histories. There were fossil series of snails; some fossils were known to be very old, yet looked for all the world like present-day forms; others had no like descendants and had become extinct. Lamarck (1744–1829), who like Linnaeus began by believing in the fixity of species, by 1800 had formulated a theory of evolution that involved a slow historical process, which he assumed to have taken a very, very long time.

Possibly one reason the theory of evolution arose in Western, rather than Eastern, science was that the descriptions of fossil and living forms showing so many close relationships made the orthodox biblical view of the special creation of each and every species untenable; and the question, how living forms merged into one another, pressed for an answer. The Eastern philosophies that accepted connectedness and relatedness as givens did not need to confront this question with the same urgency. In other words, where evidences of evolutionary change did not raise fundamental contradictions and questions, evolutionary theory did not need to be invented to reconcile and answer them. However one, and perhaps the most, important difference between Western evolutionary thinking and Eastern ideas of organismic unity lies in the materialistic and historical elements, which are the earmark of Western evolutionism as formulated by Darwin.

Though most of the elements of Darwinian evolutionary theory existed for at least hundred years before Darwin, he knit them into a consistent theory that was in line with the mainstream thinking of his time. Irvine writes:

The similar fortunes of liberalism and natural selection are significant. Darwin's matter was as English as his method. Terrestrial history tuned out to be strangely like Victorian history writ large. Bertrand Russell and others have remarked that Darwin's theory was mainly 'an extension to the animal and vegetable world of laissez faire economics.' As a matter of fact, the economic conceptions of utility, pressure of population, marginal fertility, barriers in restraint of trade, the division of labor, progress and adjustment by competition, and the spread of technological improvements can all be paralleled in *The Origin of Species*. But so, alas, can some of the doctrines of English political conservatism. In revealing the importance of time and the hereditary past, in emphasizing the persistence of vestigial structures, the minuteness of variations and the slowness of evolution, Darwin was adding Hooker and Burke to Bentham and Adam Smith. The constitution of the universe exhibited many of the virtues of the English constitution.[4]

One of the first to comment on this congruence was Karl Marx (1818–83) who wrote to Friedrich Engels (1820–95) in 1862, three years after the publication of *The Origin of Species*:

It is remarkable how Darwin recognizes among beasts and plants his English society with its division of labour, competition, opening up of new markets, 'inventions,' and the Malthusian 'struggle for existence.' It is Hobbes's 'bellum omnium contra omnes,' [war of all against all] and one is reminded of Hegel's *Phenomenology*, where civil society is described as a 'spiritual animal kingdom,' while in Darwin the animal kingdom figures as civil society.[5]

A similar passage appears in a letter by Engels:

The whole Darwinist teaching of the struggle for existence is simply a transference from society to living nature of Hobbes's doctrine of 'bellum omnium contra omnes' and of the bourgeois-economic doctrine of competition together with Malthus's theory of population. When this conjurer's trick has been performed . . . the same theories are transferred back again from organic nature into history and now it is claimed that their validity as eternal laws of human society has been proved.[5]

The very fact that essentially the same mechanism of evolution through natural selection was postulated independently and at about the same time by two English naturalists, Darwin and Alfred Russel Wallace (1823–1913), shows that the basic ideas were in the air – which is not to deny that it took genius to give them logical and convincing form.

Darwin's theory of *The Origin of Species by Means of Natural Selection*, published in 1859, accepted the fact of evolution and undertook to explain how it could have come about. He had amassed large quantities of data to show that historical change had taken place, both from the fossil record and from his observations as a naturalist on the Beagle. He pondered why some forms had become extinct and others had survived to generate new and different forms. The watchword of evolution seemed to be: be fruitful and modify, one that bore a striking resemblance to the ways of animal and plant breeders. Darwin corresponded with many breeders and himself began to breed pigeons. He was impressed by the way in which breeders, through careful selection, could use even minor variations to elicit major differences, and was searching for the analog in nature to the breeders' techniques of selecting favorable variants. A prepared mind therefore encountered Malthus's *Essay on the Principles of Population* (1798). In his *Autobiography*, Darwin writes:

In October 1838, that is, fifteen months after I had begun my systematic enquiry, I happened to read for amusement Malthus on *Population*, and being well prepared to appreciate the struggle for existence which everywhere goes on from long-continued observation of the habits of animals and plants, it at once struck me that under these circumstances favourable variations would tend to be preserved and unfavourable ones to be destroyed. The result of this would be the formation of new species. Here, then, I had at last got a theory by which to work.[6]

Incidentally, Wallace also acknowledged being led to his theory by reading Malthus. Wrote Wallace:

The most interesting coincidence in the matter, I think, is, that I, *as well as Darwin*, was led to the theory itself through Malthus. . . . It suddenly flashed upon me that all animals are necessarily thus kept down – 'the struggle for existence' – while *variations*, on which I was always thinking, must necessarily often be *beneficial*, and would then cause those varieties to increase while the injurious variations diminished.[7] (Wallace's italics)

Both, therefore, saw in Malthus's struggle for existence the working of a natural law which effected what Herbert Spencer had called the "survival of the fittest."

The three principal ingredients of Darwin's theory of evolution are: endless variation, natural selection from among the variants, and the resulting survival of the fittest. Given the looseness of many of his arguments – he credited himself with being an expert wriggler – it is surprising that his explanation has found such wide acceptance. One reason probably lies in the fact that Darwin's theory was historical and materialistic, characteristics that are esteemed as virtues; another, perhaps in its intrinsic optimism – its notion of progressive development of species, one from another – which fit well into the meritocratic ideology encouraged by the early successes of British mercantilism, industrial capitalism and imperialism.

But not only did Darwin's interpretation of the history of life on earth fit in well with the social doctrines of nineteenth-century liberalism and individualism. It was used in turn to support them by rendering them aspects of natural law. Herbert Spencer is usually credited with having brought Darwinism into social theory. The body of ideas came to be known as social Darwinism and gained wide acceptance in Britain and the United States in the latter part of the nineteenth and on into the twentieth century. For example, John D. Rockefeller proclaimed in a Sunday school address:

The growth of a large business is merely the survival of the fittest. . . . The American Beauty rose can be produced in the splendor and fragrance which bring cheer to its beholder only by sacrificing the early buds which grow up around it. This is not an evil tendency in business. It is merely the working-out of a law of nature and a law of God.[8]

The circle was therefore complete: Darwin consciously borrowed from social theorists such as Malthus and Spencer some of the basic concepts of evolutionary theory. Spencer and others promptly used Darwinism to reinforce these very social theories and in the process bestowed upon them the force of natural law.[9]

SEXUAL SELECTION

It is essential to expand the foregoing analysis of the mutual influences of Darwinism and nineteenth-century social doctrine by looking critically at the Victorian picture Darwin painted of the relations between the sexes, and of the roles that males and females play in the evolution of animals and humans. For although the ethnocentric bias of Darwinism is widely acknowledged, its blatant sexism − or more correctly, androcentrism (male-centeredness) − is rarely mentioned, presumably because it has not been noticed by Darwin scholars, who have mostly been men. Already in the nineteenth century, indeed within Darwin's life time, feminists such as Antoinette Brown Blackwell and Eliza Burt Gamble called attention to the obvious male bias pervading his arguments.[10, 11] But these women did not have Darwin's or Spencer's professional status or scientific experience; nor indeed could they, given their limited opportunities for education, travel and participation in the affairs of the world. Their books were hardly acknowledged or discussed by professionals, and they have been, till now, merely ignored and excluded from the record. However, it is important to expose Darwin's androcentrism, and not only for historical reasons, but because it remains an integral and unquestioned part of contemporary biological theories.

Early in *The Origin of Species*, Darwin defines sexual selection as one mechanism by which evolution operates. The Victorian and androcentric biases are obvious:

This form of selection depends, not on a struggle for existence in relation to other organic beings or to external conditions, but on a struggle of individuals of one sex, generally males, for the possession of the other sex.[12]

And,

Generally, the most vigorous males, those which are best fitted for their places in nature, will leave most progeny. But in many cases, victory depends not so much on general vigor, as on having special weapons confined to the male sex.

The Victorian picture of the active male and the passive female becomes even more explicit later in the same paragraph:

the males of certain hymenopterous insects [bees, wasps, ants] have been frequently seen by that inimitable observer, M. Fabre, fighting for a particular female who sits by, an apparently unconcerned beholder of the struggle, and then retires with the conqueror.

Darwin's anthropomorphizing continues, as it develops that many male birds "perform strange antics before the females, which, standing by as spectators, at last choose the most attractive partner." However, he worries that whereas this might be a reasonable way to explain the behavior of peahens and female birds of paradise whose consorts anyone can admire, "it is doubtful whether [the tuft of hair on the breast of the wild turkeycock] can be ornamental in the eyes of the female bird." Hence Darwin ends this brief discussion by saying that he "would not wish to attribute all sexual differences to this agency."

Some might argue in defense of Darwin that bees (or birds, or what have you) do act that way. But the very language Darwin uses to describe these behaviors disqualifies him as an "objective" observer. His animals are cast into roles from a Victorian script. And whereas no one can claim to have solved the important methodological question of how to disembarrass oneself of one's anthropocentric and cultural biases when observing animal behavior, surely one must begin by trying.

After the publication of *The Origin of Species*, Darwin continued to think about sexual selection, and in 1871, he published *The Descent of Man and Selection in Relation to Sex*, a book in which he describes in much more detail how sexual selection operates in the evolution of animals and humans.

In the aftermath of the outcry *The Descent* raised among fundamentalists, much has been made of the fact that Darwin threatened the special place Man was assigned by the Bible and treated him as though he was just another kind of animal. But he did nothing of the sort. The Darwinian synthesis did not end anthropocentrism or androcentrism in biology. On the contrary, Darwin made them part of biology by presenting as "facts of nature" interpretations of animal behavior that reflect the social and moral outlook of his time.

In a sense, anthropocentrism is implicit in the fact that we humans have named, catalogued, and categorized the world around us, including ourselves. Whether we stress our upright stance, our opposable thumbs, our brain, or our language, to ourselves we are creatures apart and very different from all others. But the scientific view of ourselves is also profoundly androcentric. *The Descent of Man* is quite literally *his* journey. Elaine Morgan rightly says:

It's just as hard for man to break the habit of thinking of himself as central to the species as it was to break the habit of thinking of himself as central to the universe. He sees himself quite unconsciously as the main line of evolution, with a female satellite revolving around him as the moon revolves around the earth. This not only causes him to overlook valuable clues to our ancestry, but sometimes leads him into making statements that are arrant and demonstrable nonsense. . . . Most of the books forget about [females] for most of the time. They drag her on stage rather suddenly for the obligatory chapter on Sex and Reproduction, and then say: 'All right, love, you can go now,' while they get on with the real meaty stuff about the Mighty Hunter with his lovely new weapons and his lovely new straight legs racing across the Pleistocene plains. Any modifications of her morphology are taken to be imitations of the Hunter's evolution, or else designed solely for his delectation.[13]

To expose the Victorian roots of post-Darwinian thinking about human evolution, we must start by looking at Darwin's ideas about sexual selection in *The Descent*, where he begins the chapter entitled 'Principles of Sexual Selection' by setting the stage for the active, pursuing male:

With animals which have their sexes separated, the males necessarily differ from the females in their organs of reproduction; and these are the primary sexual characters. But the sexes differ in what Hunter has called secondary sexual characters, which are not directly connected with the act of reproduction; for instance, the male possesses certain organs of sense or locomotion, of which the female is quite destitute, or has them more highly-developed, in order that he may readily find or reach her; or again the male has special organs of prehension for holding her securely.[14]

Moreover, we soon learn:

in order that the males should seek efficiently, it would be necessary that they should be endowed with strong passions; and the acquirement of such passions would naturally follow from the more eager leaving a larger number of offspring than the less eager.[15]

But Darwin is worried because among some animals, males and females do not appear to be all that different:

a double process of selection has been carried on; that the males have selected the more attractive females, and the latter the more attractive males. . . . But from what we know of the habits of animals, this view is hardly probable, for the male is generally eager to pair with any female.[16]

Make no mistake, wherever you look among animals, eagerly promiscuous males are pursuing females, who peer from behind languidly drooping eyelids to discern the strongest and handsomest. Does it not sound like the wish-fulfillment dream of a proper Victorian gentleman?

This is not the place to discuss Darwin's long treatise in detail. Therefore, let this brief look at animals suffice as background for his section on Sexual Selection in Relation to Man. Again we can start on the first page: "Man is more courageous, pugnacious and energetic than woman, and has more inventive genius."[17] Among "savages," fierce, bold men are constantly battling each other for the possession of women and this has affected the secondary sexual characteristics of both. Darwin grants that there is some disagreement whether there are "inherent differences" between men and women, but suggests that by analogy with lower animals it is "at least probable." In fact, "Woman seems to differ from man in mental disposition, chiefly in her greater tenderness and less selfishness,"[18] for:

Man is the rival of other men; he delights in competition, and this leads to ambition which passes too easily into selfishness. These latter qualities seem to be his natural and unfortunate birthright.

This might make it seem as though women are better than men after all, but not so:

The chief distinction in the intellectual powers of the two sexes is shown by man's attaining to a higher eminence, in whatever he takes up, than can women — whether requiring deep thought, reason, or imagination, or merely the use of the senses and hands. If two lists were made of the most eminent men and women in poetry, painting, sculpture, music (inclusive both of composition and performance), history, science, and philosophy, with half-a-dozen names under each subject, the two lists would not bear comparison. We may also infer . . . that if men are capable of a decided pre-eminence over women in many subjects, the average of mental power in man must be above that of woman. . . . [Men have had] to defend their females, as well as their young, from enemies of all kinds, and to hunt for their joint subsistence. But to avoid enemies or to attack them with success, to capture wild animals, and to fashion weapons, requires the aid of the higher mental faculties, namely, observation, reason, invention, or imagination. These various faculties will thus have been continually put to the test and selected during manhood.[19]

"Thus," the discussion ends, "man has ultimately become superior to woman" and it is a good thing that men pass on their characteristics to their daughters as well as to their sons, "otherwise it is probable that man would have become as superior in mental endowment to woman, as the peacock is in ornamental plumage to the peahen."

So here it is in a nutshell: men's mental and physical qualities were constantly improved through competition for women and hunting, while women's minds would have become vestigial if it were not for the fortunate circumstance that in each generation daughters inherit brains from their fathers.

Another example of Darwin's acceptance of the conventional mores of his time is his interpretation of the evolution of marriage and monogamy:

... it seems probable that the habit of marriage, in any strict sense of the word, has been gradually developed; and that almost promiscuous or very loose intercourse was once very common throughout the world. Nevertheless, from the strength of the feeling of jealousy all through the animal kingdom, as well as from the analogy of lower animals ... I cannot believe that absolutely promiscuous intercourse prevailed in times past. . . .[20]

Note the moralistic tone; and how does Darwin know that strong feelings of jealousy exist "all through the animal kingdom?" For comparison, it is interesting to look at Engels, who working largely from the same early anthropological sources as Darwin, had this to say:

As our whole presentation has shown, the progress which manifests itself in these successive forms [from group marriage to pairing marriage to what he refers to as "monogamy supplemented by adultery and prostitution"] is connected with the peculiarity that women, but not men, are increasingly deprived of the sexual freedom of group marriage. In fact, for men group marriage actually still exists even to this day. What for the woman is a crime entailing grave legal and social consequences is considered honorable in a man or, at the worse, a slight moral blemish which he cheerfully bears. . . . Monogamy arose from the concentration of considerable weath in the hands of a single individual — a man — and from the need to bequeath this wealth to the children of that man and of no other. For this purpose, the monogamy of the woman was required, not that of the man, so this monogamy of the woman did not in any way interfere with open or concealed polygamy on the part of the man.[21]

Clearly, Engels did not accept the Victorian code of behavior as our natural biological heritage.

SOCIOBIOLOGY: A NEW SCIENTIFIC SEXISM

The theory of sexual selection went into a decline during the first half of this century, as efforts to verify some of Darwin's examples showed that many of the features he had thought were related to success in mating could not be legitimately regarded in that way. But it has lately regained its respectability, and contemporary discussions of reproductive fitness often cite examples of sexual selection.[22] Therefore, before we go on to discuss human evolution, it is helpful to look at contemporary views of sexual selection and sex roles among animals (and even plants).

Let us start with a lowly alga that one might think impossible to stereotype by sex. Wolfgang Wickler, an ethologist at the University of Munich, writes in his book on sexual behavior patterns (a topic which Konrad Lorenz tells us in the Introduction is crucial in deciding which sexual behaviors to consider healthy and which diseased):

> Even among very simple organisms such as algae, which have threadlike rows of cells one behind the other, one can observe that during copulation the cells of one thread act as males with regard to the cells of a second thread, but as females with regard to the cells of a third thread. The mark of male behavior is that the cell actively crawls or swims over to the other; the female cell remains passive.[23]

The circle is simple to construct: one starts with the Victorian stereotype of the active male and the passive female, then looks at animals, algae, bacteria, people, and calls all passive behavior feminine, active or goal-oriented behavior masculine. And it works! The Victorian stereotype is biologically determined: even algae behave that way.

But let us see what Wickler has to say about Rocky Mountain Bighorn sheep, in which the sexes cannot be distinguished on sight. He finds it "curious":

> that between the extremes of rams over eight years old and lambs less than a year old one finds every possible transition in age, but no other differences whatever; the bodily form, the structure of the horns, and the color of the coat are the same for both sexes.

Now note: ". . . the typical female behavior is absent from this pattern." Typical of what? Obviously not of Bighorn sheep. In fact we are told that "even the males often cannot recognize a female," indeed, "the females are only of interest to the males during rutting season." How does he know that the males do *not* recognize the females? Maybe these sheep are so weird that most of the time they relate to a female as though she were just another sheep, and whistle at her (my free translation of "taking an interest") only when it is a question of mating. But let us get at last to how the *females* behave. That is astonishing, for it turns out:

that *both* sexes play two roles, either that of the male or that of the young male. Outside the rutting season the females behave like young males, during the rutting season like aggressive older males. (Wickler's italics)

In fact:

There is a line of development leading from the lamb to the high ranking ram, and the female animals (♀) behave exactly as though they were in fact males (♂) whose development was retarded. . . . We can say that the only fully developed mountain sheep are the powerful rams. . . .

At last the androcentric paradigm is out in the open: females are always measured against the standard of the male. Sometimes they are like young males, sometimes like older ones; but never do they reach what Wickler calls "the final stage of fully mature physical structure and behavior possible to this species." That, in his view, is reserved for the rams.

Wickler bases this discussion on observations by Valerius Geist, whose book, *Mountain Sheep*, contains many examples of how androcentric biases can color observations as well as interpretations and restrict the imagination to stereotypes. One of the most interesting is the following:

Matched rams, usually strangers, begin to treat each other like females and clash until one acts like a female. This is the loser in the fight. The rams confront each other with displays, kick each other, threat jump, and clash till one turns and accepts the kicks, displays, and occasional mounts of the larger without aggressive displays. The loser is not chased away. The point of the fight is not to kill, maim, or even drive the rival off, but to treat him like a female.[24]

This description would be quite different if the interaction were interpreted as something other than a fight, say as a homosexual encounter, a game, or a ritual dance. The fact is that it contains none of the elements that we commonly associate with fighting. Yet because Geist casts it into the imagery of heterosexuality and aggression, it becomes perplexing.

There would be no reason to discuss these examples if their treatments of sex differences or of male/female behavior were exceptional. But they are in the mainstream of contemporary sociobiology, ethology, and evolutionary biology.

A book that has become a standard reference is George Williams's *Sex and Evolution*.[25] It abounds in blatantly biased statements that describe as "careful" and "enlightened" research reports that support the androcentric paradigm, and as questionable or erroneous those that contradict it. Masculinity and femininity are discussed with reference to the behavior of pipefish and seahorses; and cichlids and catfish are judged downright abnormal because both sexes guard the young. For present purposes it is sufficient to discuss a few points that are raised in the chapter entitled 'Why Are Males Masculine and Females Feminine and, Occasionally, Vice-Versa?'

The very title gives one pause, for if the words masculine and feminine do not mean of, or pertaining, respectively, to males and females, what *do* they mean — particularly in a scientific context? So let us read.

On the first page we find:

Males of the more familiar higher animals take less of an interest in the young. In courtship they take a more active role, are less discriminating in choice of mates, more inclined toward promiscuity and polygamy, and more contentious among themselves.

We are back with Darwin. The data are flimsy as ever, but doesn't it sound like a description of the families on your block?

The important question is who are these "more familiar higher animals?" Is their behavior typical, or are we familiar with them because, for over a century, androcentric biologists have paid disproportionate attention to animals whose behavior resembles those human social traits that they would like to interpret as biologically determined and hence out of our control?

Williams' generalization quoted above gives rise to the paradox that becomes his chief theoretical problem:

Why, if each individual is maximizing its own genetic survival should the female be less anxious to have her eggs fertilized than a male is to fertilize them, and why should the young be of greater interest to one than to the other?

Let me translate this sentence for the benefit of those unfamiliar with current evolutionary theory. The first point is that an individual's *fitness* is measured by the number of her or his offspring that survive to reproductive age. The phrase, "the survival of the fittest," therefore signifies the fact that evolutionary history is the sum of the stories of those who leave the greatest numbers of descendants. What is meant by each individual "maximizing its own genetic survival" is that every one tries to leave as many viable offspring as possible. (Note the implication of conscious intent. Such intent is not exhibited by the increasing number of humans who intentionally *limit* the numbers of their offspring. Nor is one, of course, justified in ascribing it to other animals.)

One might therefore think that in animals in which each parent contributes half of each offspring's genes, females and males would exert themselves equally to maximize the number of offspring. However, we know that according to the patriarchal paradigm, males are active in courtship, whereas females wait passively. This is what Williams means by females being "less anxious" to procreate than males. And of course we also know that "normally" females have a disproportionate share in the care of their young.

So why these asymmetries? The explanation: "The *essential* difference between the sexes is that females produce large immobile gametes and males produce small mobile ones" (my italics). This is what determines their "different optimal strategies." So if you have wondered why men are promiscuous and women faithfully stay home and care for the babies, the reason is that males "can quickly replace wasted gametes and be ready for another mate," whereas females "can not so readily replace a mass of yolky eggs or find a substitute father for an expected litter." Therefore females must "show a much greater degree of caution" in the choice of a mate than males.

E. O. Wilson says that same thing somewhat differently:

One gamete, the egg, is relatively very large and sessile; the other, the sperm, is small and motile. . . . The egg possesses the yolk required to launch the embryo into an advanced state of development. Because it represents a considerable energetic investment on the part of the mother the embryo is often sequestered and protected, and sometimes its care is extended into the postnatal period. *This is the reason why* parental care is *normally* provided by the female. . . .[26] (my italics)

Though these descriptions fit only some of the animal species that reproduce sexually, and are rapidly ceasing to fit human domestic arrangements in many portions of the globe,[27] they do fit the patriarchal model of the household. Clearly, androcentric biology is busy as ever trying to provide biological "reasons" for a particular set of human social arrangements.

The ethnocentrism of this individualistic, capitalistic model of evolutionary biology and sociobiology with its emphasis on competition and "investments," is discussed by Sahlins in his monograph, *The Use and Abuse of Biology*.[5] He gives many examples from other cultures to show how these theories reflect a narrow bias that disqualifies them from masquerading as descriptions of universals in biology. But, like other male critics, Sahlins fails to notice the obvious androcentrism.

About thirty years ago, Ruth Herschberger wrote a delightfully funny book called *Adam's Rib*,[28] in which she spoofed the then current androcentric myths regarding sex differences. When it was reissued in 1970, the book was not out of date. In the chapter entitled "Society Writes Biology," she juxtaposes the then (and now) current patriarchal scenario of the daunt-less voyage of the active, agile sperm toward the passively receptive, sessile egg to an improvised "matriarchal" account. In it the large, competent egg plays the central role and we can feel only pity for the many millions of miniscule, fragile sperm most of which are too feeble to make it to fertilization.

This brings me to a question that always puzzles me when I read about the female's larger energetic investment in her egg than the male's in his sperm: there is an enormous disproportion in the *numbers* of eggs and sperms that participate in the act of fertilization. Does it really take more "energy" to generate the one or relatively few eggs than the large excess of sperms required to achieve fertilization? In humans the disproportion is enormous. In her life time, an average woman produces about four hundred eggs, of which in present-day Western countries, she will "invest" only in about 2.2.[29] Meanwhile the average man generates several billions of sperms to secure those same 2.2 investments!

Needless to say, I have no idea how much "energy" is involved in pro-ducing, equipping and ejaculating a sperm cell along with the other necessary components of the ejaculum that enable it to fertilize an egg, nor how much is involved in releasing an egg from the ovary, reabsorbing it in the oviduct if unfertilized (a partial dividend on the investment), or incubating 2.2 of them to birth. But neither do those who propound the existence and importance of women's disproportionate energetic investments. Furthermore, I attach no significance to these questions, since I do not believe that the details of our economic and social arrangements reflect our evolutionary history. I am only trying to show how feeble is the "evidence" that is being put forward to argue the evolutionary basis (hence *naturalness*) of woman's role as homemaker.

The recent resurrection of the theory of sexual selection and the ascrip-tion of asymmetry to the "parental investments" of males and females are probably not unrelated to the rebirth of the women's movement. We should remember that Darwin's theory of sexual selection was put forward in the midst of the first wave of feminism.[30] It seems that when women threaten to enter as equals into the world of affairs, androcentric scientists rally to point out that our *natural* place is in the home.

THE EVOLUTION OF MAN

Darwin's sexual stereotypes are doing well also in the contemporary literature on human evolution. This is a field in which facts are few and specimens are separated often by hundreds of thousands of years, so that maximum leeway exists for investigator bias. Almost all the investigators have been men; it should therefore come as no surprise that what has emerged is the familiar picture of Man the Toolmaker. This extends so far that when skull fragments estimated to be 250,000 years old turned up among the stone tools in the gravel beds of the Thames at Swanscombe and paleontologists decided that they are problably those of a female, we read that "The Swanscombe woman, or her husband, was a maker of hand axes. . . ."[31] (Imagine the reverse: The Swanscombe man, or his wife, was a maker of axes. . . .) The implication is that if there were tools, the Swanscombe *woman* could not have made them. But we now know that even apes make tools. Why not women?

Actually, the idea that the making and use of tools were the main driving forces in evolution has been modified since paleontological finds and field observations have shown that apes both use and fashion tools. Now the emphasis is on the human use of tools as weapons for hunting. This brings us to the myth of Man the Hunter, who had to invent not only tools, but also the social organization that allowed him to hunt big animals. He also had to roam great distances and learn to cope with many and varied circumstances. We are told that this entire constellation of factors stimultated the astonishing and relatively rapid development of his brain that came to distinguish Man from his ape cousins. For example, Kenneth Oakley writes:

Men who made tools of the standard type . . . must have been capable of forming in their minds images of the ends to which they laboured. Human culture in all its diversity is the outcome of this capacity for conceptual thinking, but the leading factors in its development are tradition coupled with invention. The primitive hunter made an implement in a particular fashion largely because as a child he watched his father at work or because he copied the work of a hunter in a neighbouring tribe. The standard handaxe was not conceived by any one individual *ab initio*, but was the result of exceptional individuals in successive generations not only copying but occasionally improving on the work of their predecessors. As a result of the co-operative hunting, migrations and rudimentary forms of barter, the traditions of different groups of primitive hunters sometimes became blended.[32]

It seems a remarkable feat of clairvoyance to see in such detail what happened some 250,000 years in pre-history, complete with the little boy and his little stone chipping set just like daddy's big one.

It is hard to know what reality lurks behind the reconstructions of Man Evolving. Since the time when we and the apes diverged some fifteen million years ago, the main features of human evolution that one can read from the paleontological finds are the upright stance, reduction in the size of the teeth, and increase in brain size. But finds are few and far between both in space and in time until we reach the Neanderthals some 70,000 to 40,000 years ago — a jaw or skull, teeth, pelvic bones, and often only fragments of them.[33] From such bits of evidence as these come the pictures and statues we have all seen of that line of increasingly straight and upright, and decreasingly hairy and ape-like men marching in single file behind *Homo sapiens*, carrying their clubs, stones, or axes; or that other one of a group of beetle-browed and bearded hunters bending over the large slain animal they have brought into camp, while over on the side long-haired, broad-bottomed females nurse infants at their pendulous breasts.

Impelled, I suppose, by recent feminist critiques of the evolution of Man the Hunter, a few male anthropologists have begun to take note of Woman the Gatherer, and the stereotyping goes on as before. For example Howells, who acknowledges these criticisms as just, nonetheless assumes "the classic division of labor between the sexes" and states as fact that stone age men roamed great distances "on behalf of the whole economic group, while the women were restricted to within the radius of a fraction of a day's walk from camp." Needless to say, he does not *know* any of this.

One can equally well assume that the responsibilities of providing food and nurturing young were widely dispersed through the group that needed to cooperate and devise many and varied strategies for survival. Nor is it obvious why tasks needed to have been differentiated by sex. It makes sense that the gatherers would have known how to hunt the animals they came across; that the hunters gathered when there was nothing to catch, and that men and women did some of each, though both of them probably did a great deal more gathering than hunting. After all, the important thing was to get the day's food, not to define sex roles. Bearing and tending the young have not necessitated a sedentary way of life among nomadic peoples right to the present, and both gathering and hunting probably required movement over large areas in order to find sufficient food. Hewing close to home probably accompanied the transition to cultivation, which introduced the necessity to stay put for planting, though of course not longer than required to harvest. Without fertilizers and crop rotation, frequent moves were probably essential parts of early farming.

Being sedentary ourselves, we tend to assume that our foreparents heaved a great sigh of relief when they invented agriculture and could at last stop roaming. But there is no reason to believe this. Hunter/gatherers and other people who move with their food still exist. And what has been called the agricultural "revolution" probably took considerably longer than all of recorded history. During this time, presumably some people settled down while others remained nomadic, and some did some of each, depending on place and season.

We have developed a fantastically limited and stereotypic picture of ways of life that evolved over many tens of thousands of years, and no doubt varied in lots of ways that we do not even imagine. It is true that by historic times, which are virtually now in the scale of our evolutionary history, there were agricultural settlements, including a few towns that numbered hundreds and even thousands of inhabitants. By that time labor was to some extent divided by sex, though anthropologists have shown that right to the present, the division can be different in different places. There are economic and social reasons for the various delineations of sex roles. We presume too much when we try to read them in the scant record of our distant prehistoric past.

Nor are we going to learn them by observing our nearest living relatives among the apes and monkeys, as some biologists and anthropologists are trying to do. For one thing, different species of primates vary widely in the extent to which the sexes differ in both their anatomy and their social behavior, so that one can find examples of almost any kind of behavior one is looking for by picking the appropriate animal. For another, most scientists find it convenient to forget that present-day apes and monkeys have had as long an evolutionary history as we have had, since the time we and they went our separate ways many millions of years ago. There is no theoretical reason why their behavior should tell us more about our ancestry than our behavior tells us about theirs. It is only anthropocentrism that can lead someone to magine that "A possible preadaptation to human ranging for food is the behavior of the large apes, whose groups move more freely and widely compared to gibbons and monkeys, and whose social units are looser."[34] But just as in the androcentric paradigm men evolved while women cheered from the bleachers, so in the anthropocentric one, humans evolved while the apes watched from the trees. This view leaves out not only the fact that the apes have been evolving away from us for as long a time as we from them, but that certain aspects of their evolution may have been a response to our own. So, for example, the evolution of human habits may have put a serious crimp into the evolution of the great apes and forced them to stay in the trees or to hurry back into them.

The current literature on human evolution says very little about the role of language, and sometimes even associates the evolution of language with tool use and hunting — two purportedly "masculine" characteristics. But this is very unlikely because the evolution of language probably went with biological changes, such as occurred in the structure of the face, larynx, and brain, all slow processes. Tool use and hunting, on the other hand, are cultural characteristics that can evolve much more quickly. It is likely that the more elaborate use of tools, and the social arrangements that go with hunting and gathering, developed in part as a consequence of the expanded human repertory of capacities and needs that derive from our ability to communicate through language.

It is likely that the evolution of speech has been one of the most powerful forces directing our biological, cultural, and social evolution, and it is surprising that its significance has largely been ignored by biologists. But, of course, it does not fit into the androcentric paradigm. No one has ever claimed that women can not talk; so if men are the vanguard of evolution, humans must have evolved through the stereotypically male behaviors of competition, tool use, and hunting.

HOW TO LEARN OUR HISTORY? SOME FEMINIST STRATEGIES

How *did* we evolve? Most people now believe that we became who we are by a historical process, but, clearly, we do not know its course, and must use more imagination than fact to reconstruct it. The mythology of science asserts that with many different scientists all asking their own questions and evaluating the answers independently, whatever personal bias creeps into their individual answers is cancelled out when the large picture is put together. This might conceivably be so if scientists were women and men from all sorts of different cultural and social backgrounds who came to science with very different ideologies and interests. But since, in fact, they have been predominantly university-trained white males from privileged social backgrounds, the bias has been narrow and the product often reveals more about the investigator than about the subject being researched.

Since women have not figured in the paradigm of evolution, we need to rethink our evolutionary history. There are various ways to do this:

(1) We can construct one or several estrocentric (female-centered) theories. This is Elaine Morgan's approach in her account of *The Descent of Woman* and Evelyn Reed's in *Woman's Evolution.*[35] Except as a way of parodying the male myths, I find it unsatisfactory because it locks the authors into many of the same unwarranted suppositions that underlie those very myths.

For example, both accept the view that our behavior is biologically determined, that what we do is a result of what we were or did millions of years ago. This assumption is unwarranted given the enormous range of human adaptability and the rapid rate of human social and cultural evolution. Of course, there is a place for myth-making and I dream of a long poem that sings women's origins and tells how we felt and what we did; but I do not think that carefully constructed "scientific" mirror images do much to counter the male myths. Present-day women do not know what prehistoric hunter/gatherer women were up to any more than a male paleontologist like Kenneth Oakley knows what the little toolmaker learned from his dad.

(2) Women can sift carefully the few available facts by paring away the mythology and getting as close to the raw data as possible. And we can try to see what, if any, picture emerges that could lead us to questions that perhaps have *not* been asked and that should, and could, be answered. One problem with this approach is that many of the data no longer exist. Every excavation removes the objects from their locale and all we have left is the researchers' descriptions of what they saw. Since we are concerned about unconscious biases, that is worrisome.

(3) Rather than invent our own myths, we can concentrate, as a beginning, on exposing and analyzing the male myths that hide our overwhelming ignorance, "for when a subject is highly controversial – and any question about sex is that – one cannot hope to tell the truth."[36] Women anthropologists have begun to do this. New books are being written, such as *The Female of the Species*[37] and *Toward an Anthropology of Women*,[38] books that expose the Victorian stereotype that runs through the literature of human evolution, and pull together relevant anthropological studies. More important, women who recognize an androcentric myth when they see one and who are able to think beyond it, must do the necessary work in the field, in the laboratories, and in the libraries, and come up with ways of seeing the facts and of interpreting them.[39]

None of this is easy, because women scientists tend to hail from the same socially privileged families and be educated in the same elite universities as our male colleagues. But since we are marginal to the mainstream, we may find it easier than they to watch ourselves push the bus in which we are riding.

As we rethink our history, our social roles, and our options, it is important that we be ever wary of the wide areas of congruence between what are obviously ethno- and androcentric assumptions and what we have been taught are the scientifically proven facts of our biology. Darwin was right when he wrote that "False facts are highly injurious to the progress of science, for they often endure long. ...."[40] Androcentric science is full of "false facts" that have endured all too long and that serve the interests of those who interpret as women's biological heritage the sexual and social stereotypes we reject. To see our alternatives is essential if we are to acquire the space in which to explore who we are, where we have come from, and where we want to go.

Harvard University

NOTES

* Reprinted by permission from *Women Look at Biology Looking at Women*, ed. by Ruth Hubbard, Mary Sue Henifin and Barbara Fried, with the collaboration of Vicki Druss and Susan Leigh Star, Cambridge, Mass.: Schenkman Publishing Co., 1979. I want to thank Gar Allen, Rita Arditti, Steve Gould and my colleagues in the editorial group that has prepared that book for their helpful criticisms of an earlier version of this manuscript.

1 For a discussion of this process, *see* Thomas S. Kuhn, *The Structure of Scientific Revolutions*, 2nd ed. (University of Chicago Press, 1970).

2 Berger and Luckmann have characterized this process as "trying to push a bus in which one is riding." [Peter Berger and Thomas Luckmann, *The Social Construction of Reality* (Garden City: Doubleday & Co., 1966), p. 12.]. I would say that, worse yet, it is like trying to look out of the rear window to *watch* oneself push the bus in which one rides.

3 Loren Eiseley, *Darwin's Century* (Garden City: Doubleday & Co., Anchor Books Edition, 1961), p. 24.

4 William Irvine, *Apes, Angels, and Victorians* (New York: McGraw-Hill, 1972), p. 98.

5 Quoted in Marshall Sahlins, *The Use and Abuse of Biology* (Ann Arbor: University of Michigan Press, 1976), pp. 101–102.

6 Francis Darwin, ed., *The Autobiography of Charles Darwin* (New York: Dover Publications, 1958), pp. 42–43.

7 *Ibid.*, pp. 200–201.

8 Richard Hofstadter, *Social Darwinism in American Thought* (Boston: Beacon Press, 1955), p. 45.

9 Though not himself a publicist for social Darwinism like Spencer, there can be no doubt that Darwin accepted its ideology. For example, near the end of *The Descent of Man* he writes: "There should be open competition for all men; and the most able should not be prevented by laws or customs from succeeding best and rearing the largest number of offspring." Marvin Harris has argued that Darwinism, in fact, should be

known as biological Spencerism, rather than Spencerism as social Darwinism. For a discussion of the issue, *pro* and *con*, see Marvin Harris, *The Rise of Anthropological Theory: A History of Theories of Culture* (New York: Thomas Y. Crowell, 1968), Ch. 5: Spencerism; and responses by Derek Freeman and others in *Current Anthropology* 15 (1974), 211–237.

[10] Antoinette Brown Blackwell, *The Sexes Throughout Nature* (New York: G. P. Putnam's Sons, 1975; reprinted Westport, Conn.: Hyperion Press, 1978). Excerpts in which Blackwell argues against Darwin and Spencer have been reprinted in Alice S. Rossi, ed., *The Feminist Papers* (New York: Bantam Books, 1974), pp. 356–377.

[11] Eliza Burt Gamble, *The Evolution of Woman: An Inquiry into the Dogma of her Inferiority to Man* (New York: G. P. Putnam's Sons, 1894).

[12] Charles Darwin, *The Origin of Species and the Descent of Man* (New York: Modern Library Edition), p. 69.

[13] Elaine Morgan, *The Descent of Woman* (New York: Bantam Books, 1973), pp. 3–4.

[14] Darwin, *Origin of Species . . .* , p. 567.

[15] *Ibid.*, p. 580.

[16] *Ibid.*, p. 582.

[17] *Ibid.*, p. 867.

[18] *Ibid.*, p. 873.

[19] *Ibid.*, pp. 873–874.

[20] *Ibid.*, p. 895.

[21] Frederick Engels, *The Origin of the Family, Private Property and the State*, E. B. Leacock, ed. (New York: International Publishers, 1972), p. 138.

[22] One of the most explicit contemporary examples of this literature is E. O. Wilson's *Sociobiology: The New Synthesis* (Cambridge: Harvard University Press, 1975); *see* especially chapters 1, 14–16 and 27.

[23] Wolfgang Wickler, *The Sexual Code: The Social Behavior of Animals and Men* (Garden City: Doubleday, Anchor Books, 1973), p. 23.

[24] Valerius Geist, *Mountain Sheep* (Chicago: University of Chicago Press, 1971), p. 190.

[25] George C. Williams, *Sex and Evolution* (Princeton: Princeton University Press, 1975).

[26] Edward O. Wilson, *Sociobiology: The New Synthesis* (Cambridge: Harvard University Press, Belknap Press, 1975), pp. 316–317. Wilson and others claim that the growth of a mammalian fetus inside its mother's womb represents an energetic "investment" on her part, but it is not clear to me why they believe that. Presumably the mother eats and metabolizes, and some of the food she eats goes into building the growing embryo. Why does that represent an investment of *her* energies? I can see that the embryo of an undernourished woman perhaps requires such an investment – in which case what one would have to do is see that the mother gets enough to eat. But what "energy" does a properly nourished woman "invest" in her embryo (or, indeed, in her egg)? It would seem that the notion of pregnancy as "investment" derives from the interpretation of pregnancy as a debilitating disease.

[27] For example, at present in the United States, 24 percent of households are headed by women and 46 percent of women work outside the home. The fraction of women who

work away from home while raising children is considerably larger in several European countries and in China.

28 Ruth Herschberger, *Adam's Rib* (1948; reprinted ed., New York: Harper and Row, 1970).

29 Furthermore, a woman's eggs are laid down while she is an embryo, hence at the expense of her mother's "metabolic investment." This raises the question whether grandmothers devote more time to grandchildren they have by their daughters than to those they have by their sons. I hope sociobiologists will look into this.

30 Nineteenth-century feminism is often dated from the publication in 1792 of Mary Wollstonecraft's (1759–1797) *A Vindication of the Rights of Woman*; it continued right through Darwin's century. Darwin was well into his work at the time of the Seneca Falls Declaration (1848), which begins with the interesting words: "When, in the course of human events, it becomes necessary for one portion of the family of man to assume among the people of the earth a position different from that which they have hitherto occupied, but one to which the *laws of nature and of nature's God* entitle them . . ." (my italics). And John Stuart Mill (1806–1873) published his essay on *The Subjection of Women* in 1869, ten years after Darwin's *Origin of Species* and two years before the *Descent of Man and Selection in Relation to Sex*.

31 William Howells, *Evolution of the Genus* Homo (Reading: Addison-Wesley Publishing Co., 1973), p. 88.

32 Kenneth P. Oakley, *Man the Toolmaker* (London: British Museum, 1972), p. 81.

33 There are also occasional more perfect skeletons, such as that of *Homo erectus* at Choukoutien, commonly known as Peking Man, who was in fact a woman.

34 Howells, p. 133.

35 Evelyn Reed, *Woman's Evolution* (New York: Pathfinder Press, 1975).

36 Virginia Woolf, *A Room of One's Own* (1945; reprinted ed., Penguin Books, 1970), p. 6.

37 M. Kay Martin and Barbara Voorhis, *Female of the Species* (New York: Columbia University Press, 1975).

38 Rayna R. Reiter, ed., *Toward an Anthropology of Women* (New York: Monthly Review Press, 1975).

39 This is what Sarah Blaffer Hardy and Nancy Tanner have done. *See* Sarah Blaffer Hardy, *The Woman That Never Evolved* (Cambridge, MA: Harvard University Press, 1981); and Nancy Makepeace Tanner, *On Becoming Human* (Cambridge: Cambridge University Press, 1981).

40 Darwin, *Origin of Species* . . . , p. 909.

MICHAEL GROSS AND MARY BETH AVERILL

EVOLUTION AND PATRIARCHAL MYTHS OF
SCARCITY AND COMPETITION

Nature, as depicted in biological science, is a man's world. For researchers inevitably project the visions their imaginations, and the attitudes their life experiences make available,[1] and most biologists have been men. A feminist task is to reconsider patriarchal images: to understand them as reflections of a male mentality; to consider whether they even answer any questions feminists wish to ask; and to remake the image of nature in metaphors conformable to women's reality.

Here we focus on two related themes in the patriarchal image of nature – scarcity and competition – showing how they entered evolutionary theory and how they are used currently in evolutionary and ecological thought. We then take note, briefly, of general difficulties with biological theories derived from the conventional assumptions of patriarchal "objective" science, and suggest, tentatively, ingredients which might be integrated into a study of nature from a feminist perspective. We conclude with the suggestion that this study begin with the substitution, for scarcity and competition, of the opposite characteristics: that nature may be better understood in terms of plenitude and cooperation.

One of the "liberties" – in the best expansive sense of the word – we believe feminist scholarship may take is the freedom to risk intellectually, to sketch incomplete projects, and thereby to inspire a collective quest. So we have not surveyed a field in "review article" style, nor have we provided line-by-line analyses of primary research. Instead, we look at the image of nature conveyed by selected (but representative and respected) biologists in order to encourage others to criticize and refine our interpretation, to develop and extend such an examination and reconstruction.

We shall discuss in detail here concepts expressed by Eugene P. Odum (Professor of Ecology and Director of the Institute of Ecology at the University of Georgia) and Robert Ricklefs (Associate Professor of Biology at the University of Pennsylvania), both of whose textbooks are used in graduate courses across the country. We shall also comment on recent remarks by Jared Diamond (Professor of Physiology, University of California, Los Angeles) who, as a relative newcomer to ecology but a physiologist of repute, recently argued forcibly for an increased consideration of competition in ecological

71

Sandra Harding and Merrill B. Hintikka (eds.), Discovering Reality, 71–95.
Copyright © 1983 *by D. Reidel Publishing Company.*

research. However, themes of struggle and competition run through so much of the literature on evolution that many other works might have been chosen here without changing the essential points, although evolutionary theory is not monolithic. Below, for example, we will refer to some of the problems raised by important biologists such as T. Dobzhansky, G. E. Hutchinson, and R. Lewontin,[2] but those doubts do not suffice to disturb the scientific community's commitment to competition and scarcity as underlying ideas in evolutionary theory.

I. "ARE YOU THERE, NATURE? THIS IS MAN CALLING."

"Evolution" is a story of progress, of "improvement," of expansion, invasion,[3] and colonization. Its episodes and events express the familiar sorts of processes and characteristics which men think promote progress and create history: competition, struggle, domination, hierarchy, even cooperation — but only as a competitive strategy. A number of other characteristics and kinds of process do not appear, among them nurturance, tolerance, intention and awareness, benignity, collectivism. Altruism does in fact agitate debates in evolutionary biology, but mainly because scientists struggle to explain it away by reducing it to unintentional cooperation resulting from instinct, or competitive advantage, or expectation of future reciprocation.[4]

Competition is a core concept of evolutionary theory. Evolution and the mechanism of "natural selection" are notions which answer the questions "How did the current assortment of living forms come into existence?" and "What explains the close fit between the characteristics of living beings and the environments they inhabit (i.e., the phenomenon of adaptation)?" The second question, and indirectly the first, are answered by "natural selection," roughly expressed in the Spencerian phrase "survival of the fittest." Survival of the fittest only matters because the individuals which constitute a species differ slightly among themselves, and some of the differences are inheritable. Unpredictable variations in specific characteristics arise continually — both through the recombination of genetic characteristics during sexual reproduction and by the introduction of altogether new variations by mutation. Both mutation and recombination are (supposedly) random or chance processes, so there is no obvious reason why they should lead to a closeness of fit between the attributes of the organism and the character of its environment, or to changes which suit it to an environment which is itself undergoing change. A process of competition culls those random variants so that the best adapted individuals, the "fittest," make the greatest numerical contribution to the population of the next generation, and thus to the composition of its gene pool. Competition, the essential process of natural selection, gradually eliminates the less fit.

Now we turn to the introduction into evolutionary thought of the competitive notion of "survival of the fittest" and its attendant concept of scarcity. We shall then examine briefly the current status in evolutionary and ecological thought of the principle of competition.

II. FROM CREATIONISM TO CAPITALISM

Charles Darwin, a scientific beneficiary of British imperialism, was able to spend some five years voyaging around the world as ship's naturalist aboard the *Beagle*. He left England holding the assumption then prevalent among scientific circles there that each species was independently created by a singular and specific act of God. He returned all but convinced that species had in fact come into existence through a naturalistic and mechanistic process of historical emergence, evolution.[5]

Although Darwin was convinced that species had come into existence through a process of "descent with modification," he found it difficult to show how the characteristics of an environment engendered adaptive characteristics in its inhabitants. In the several years (1835—1838) between his return to England and his invention of the theory of natural selection, he labored over a number of approaches to the problem. For instance, he frequently wondered whether somehow the environment acts directly on the womb, or some other aspect of the reproductive process, to modify descendants. He considered but dismissed Lamarck's assumption that abilities or capacities developed during the life history of an individual were passed on to the next generation by heredity (inheritance of acquired characteristics). And he queried animal breeders and plant hybridizers for their knowledge of heredity and the impact of the environment, but could find no agent in nature which he could assimilate to the self-conscious choices made in the development of domesticated animals and plant hybrids.[6] Darwin's problem was that he could see no way to explain how the environment might act so as to produce adaptive characteristics in successive generations of plants or animals. When he looked into the laws of inheritance he fould that novelties emerged unpredictably and were not necessarily adaptive, and that the patterns of hereditary transmission of characteristics were inchoate.[7]

When he read (1838) Thomas Malthus's *An Essay on the Principle of Population* (1798) he found a solution, as would A. R. Wallace some twenty years later. Malthus showed Darwin that if one assumed scarcity of resources, especially food, a competition would ensue which affected the composition of successive generations. Here Darwin saw a source of adaptation and an engine of evolutionary progress: in the course of such a competition, those best suited to an environment would be able to produce relatively more offspring; by inheritance their characteristics would predominate in the next generation. The randomness and confusion of the reproductive process would no longer matter because competition would insure a logic to its endproduct by the elimination of unfit parents before they could reproduce, or the survival of a relatively smaller number of their less fit offspring.[8]

Malthus had paired two laws, so-called, of nature as the basis of his principle of population: populations expanded by way of generation or reproduction at a geometric rate of increase (in proportion to the numerical progression 1, 2, 4, 8, 16, ...) while food supply can increase only at an arithmetic rate (1, 2, 3, 4, 5, ...) given that there is a limited amount of land which can be made to support agriculture. In this, Malthus thought he was applying to human populations a principle widely acknowledged about the rest of the natural world:

The race of plants and the race of animals shrink under this great restrictive law. And the race of man cannot, by any efforts of reason, escape from it.[9]

At the core of Malthus' attitude is the association of nature both with hyperfecundity and with food scarcity, ideas which can also be seen as contradictory (since one kind of organism's "overproduction" may be another kind of organism's food supply). The consequence of these principles is that there can never be enough to go around, a vision of the human situation based on putative laws of nature which Darwin would project on all of nature as a fundamental principle. Malthus had a more limited and specific political purpose, however: he applied the principles of population to argue against the perfectability of mankind in general and the hope in particular of improving the lot of the poor by way of the English "poor laws," the equivalent of our state welfare for the disabled and indigent. He argued that such generosity allowed the unfit — equated with the poor — to reproduce, indeed to reproduce faster than the upper classes which showed "moral restraint." He predicted that as a consequence humanity would deteriorate. As social policy we hear early intimations of eugenics, of the "culture of poverty," of forced sterilization, of the genocidal currents in "population bomb" arguments.[10]

Thereby a typical patriarchal theme of male control of reproductive choices for the sake of abstract political-economic goals combined with the capitalistic defense of middle class accumulation, expansion, and domination. These beliefs were sustained by another typical attribute of patriarchal thought: objectification of rather than identification with the "other," in this instance the members of the "poorer classes." But while the exigencies of survival were a source of dismay for Malthus who saw in scarcity and competition the decline of the English bourgeoisie and aristocracy under the provisions of the poor law, for Darwin they were positive in their consequences for plant and animal populations, where the hand of liberalism did not reach out to preserve the unfit.[11]

III. AT WAR WITH NATURE

Darwin carried Malthus' social and political views – in the form of a biological theory about nature's inherent imbalance – into his theory of evolution.[12] The weaker seedlings are crowded out, choked off; slower deer fall to predation; while taller giraffes get the bigger leaves the shorter see their offspring falter from undernourishment, weaken and fall ill:

from the war of nature, from famine and death, the most exalted object of which we are capable of conceiving, namely the production of higher animals, directly follows.[13]

Struggle, "in a large and metaphorical sense" was the way of nature, its most essential fact – even though Darwin self-consciously extended its meaning far beyond the obvious instances, from the war "of" or in nature to a war of life *with* nature:

a plant on the edge of a desert is said to struggle for life against the drought, though more properly it should be said to be dependent upon the moisture. A plant which annually produces a thousand seeds may more truly be said to struggle with the plants of the same and other kinds which already clothe the ground ... As the mistletoe is disseminated by birds, its existence depends on them; and it may methodically be said to struggle with other fruit-bearing plants, in tempting the birds to devour and thus disseminate its seeds.[14]

Darwin thus employed struggle rhetorically "for convenience sake," casting every significant interaction in nature in the language of competition within and among the species, and the struggle between organism and its environment. At his first reading of Malthus, the image sprang immediately to mind:

One may say there is a force like a hundred thousand wedges trying to force every kind of adapted structure into the gaps in the œconomy of nature, or rather forming gaps by thrusting out weaker ones.[15]

The same image of nature as battleground and passive victim, and life as essentially a competitive struggle with a limited number of places at the top, captured the imagination of Darwin's contemporaries[16] and has had recurrent appeal.

We wish to suggest that this alienated perception of nature which emphasizes its parsimony may derive largely from male socialization to strive against others and to manipulate nature in the world of work; and it may little correspond with women's traditional experience, in western history, in the realm of family and home, where the main emphasis is upon relationship.[17]

IV. COMPETITION IS FOR THE BIRDS

Images of struggle remain deeply embedded in evolutionary and ecological thought. Research projects which assume competition as a basis for population shifts — the plethora of laboratory and field studies, and the esoterica of mathematical population biology — simply do not raise questions about the fundamental principles of scarcity and competition which underly them. The methodological sophistication of population studies seems almost to obscure the circularity of the way competition is assumed as the source of the shifts then used as evidence of competition. As we shall see, while the laboratory situation is effective because it allows researchers to control and limit variables which in nature are myriad, that very limitation reduces the generalizability of laboratory studies to nature.

A final difficulty with the usefulness of both laboratory and field studies is rarely acknowledged. Evolution takes a long time, but careers in science, relatively speaking, do not. (Particularly under the existing pressure to publish positive findings rapidly in order to advance within the guild, time is limited.[18]) Consequently, even well-documented evidence of population shifts in nature do not constitute evidence of evolutionary change — which means change in the genetic make-up of a species, not a shift in its habitat.

The central concept of ecology to which competition is relevant is the "niche" — loosely, the situation in which an organism lives, which may include food source, nesting site, light, water supply. In other words, it is defined by a combination of spatial, temporal, nutritional, metabolic and behavioral characteristics. Two main collections of research supposedly document the competitive basis for niche formation: (1) field observations that species having nearly identical resource requirements (water, light, food, nest site, etc.) are distinguished by some characteristic such that they do not overlap; (2) laboratory observations showing that when two species of animals which eat the same food are contained in the same system, eventually one species dies out (competitive exclusion). Both lines of evidence, however, do not demonstrate competition *per se* but only its supposed outcome, one species per niche.[19] The principle of one species per niche is problematic and potentially circular because if one is convinced that cohabiting species must differ one can probably find evidence − after all, "different species" are different.[20] Conversely, observed differences will be rationalized with the assumption that they must have emerged as ways to avoid coming into direct competition. One can make endless distinctions: "niche differences are not simply a matter of differences in habitat or food, but also consist of differences in techniques for finding the same food in the same habitat."[21]

In Robert Ricklef's definition of competition it subtly becomes pervasive. Competition is "the use of a resource (food, water, light, space) by an organism which thereby reduces the availability of a resource to others . . . a resource consumed by one individual can no longer be used by another."[22] If any action which reduces the availability of a resource to others is competitive, then almost any behavior can be so construed and life itself becomes, virtually by definition, competitive:

organisms that *potentially* may use the same resources are called competitors . . . . When a fox captures a rabbit, there is one less rabbit in the prey population for other foxes, or for bobcats, hawks, and others that also prey on rabbits.[23]

By way of further illustration of the role of competition, Ricklefs observes that:

The slowing of population growth as a population approaches the carrying capacity of its environment results from competition between individuals in the population . . . . At high population densities intense intraspecific competition reduces resources below the level that will sustain further population growth, and thus regulates the size of the population.[24]

Competition is an invisible hand here: the event is demographic; the supposed cause is competition.[25] At one point, competition refers to eating, breathing, or harvesting sunlight; here it means the slowing of a population's rate of growth. However, in at least some cases where the reasons for the slowing of growth rate in a population have been investigated in detail, there turn out to be density dependent feedback mechanisms which lower fertility. For instance, laboratory populations of rodents show a variety of reproductive abnormalities as density rises,[26] but to call these events competition would be farfetched. Cessation of yeast cell growth (either in isolated or in mixed ("competitive") populations) seems to depend also on blocking reproduction — on the accumulation of the metabolic waste product alcohol which kills young buds.[27] The term "competition" comes to have so many meanings that it loses scientific precision; yet its wide use reveals how important it is to biologists to see competition as the underlying cause of a diversity of events in nature.

In understanding why "we frequently observe many ecologically similar species co-existing in nature, clearly using the same resources . . . ." while, "by contrast, closely related species rarely coexist in the laboratory," Ricklefs does not question the relevance of the laboratory model but notes:

Whenever ecologists have examined groups of similar species in the same habitat they have found small but significant differences in size or foraging behavior that enables species to use slightly different resources and avoid intense competition.[28]

In other words, whether or not competition is observed in nature, it remains the invisible hand guiding the emergence of differences between cohabiting types ("character displacement") which the theory predicts one should find as a consequence of competition. Competition must be imminent even if the observed behavior is at best an effort to avoid it. Methodologically and psychologically, there may be no way to avoid being anthorpomorphic. But seeing competition avoidance as the motive underlying non-competitive behavior simply projects it even where it manifestly is not.

Is behavior observed in nature which might justify the expression "competition"? Consider Eugene Odum's comments about territoriality, "any active mechanism that spaces individuals or groups apart from one another."[29]

In most territorial behavior actual fighting over boundaries is held to a minimum. Owners advertise their land or location in space by song or displays and potential intruders generally avoid entering an established domain.[30]

Here, "active antagonism" boils down to singing or performing a kind of dance. So much for intraspecific competition. On the question of interspecific competition, similarly, Jared Diamond dismisses "fights" on the basis of a lack of evidence (they are likely to be intermittent, he explains, and brief, lasting only until territories are staked out), and also on principle. Fights are disadvantageous because "dangerous to the winner as well as to the loser" (since they waste time and make both participants vulnerable to predators). Instead, "one species harvests resources more efficiently and lowers resources to the point where it can still survive but its competitor cannot." But field biologists "have rarely attempted to measure resource levels directly."[31] And when they do we might expect the same leap from correlations to a competitive causal mechanism which we find in the use of demographic shifts to argue for the causal agency of competition.

Moreover, since efficient harvesting really means an organism should go on about its own business as best it can ("potential competitors" notwithstanding), why should this be called competition at all? More than a rhetorical question, this gets to the heart of the matter: going on as efficiently as possible about one's business, in the patriarchal mentality, does not mean doing well for its own sake but striving to excel specifically at the expense of one's colleagues (read: competitors). We would suggest that underlying motives include (a) fear that others' success somehow diminishes one's own (underlain of course by the assumption of scarce resources — for instance, limited quantities of praise and recognition), and (b) anxious and transient satisfaction at the failure or relative losses of one's competitors.[32]

Diamond, in examining evidence for interspecific competition (and calling for more confirmatory research) uses a business analogy for efficient harvesting in nature which further illustrates these attitudes. Competition between Hertz and Avis, he remarks, does not manifest in battles between personnel at adjacent car rental counters; rather "the mechanism of competition consists of trying harder for customers so as to starve out the rival's resource base, and not of fighting."[33] The imputed purpose — "to starve out the rival's resource base" — assumes that businesses operate on the model of inherently scarce resources, as do theoretical biologists. The authentic motive behind business activity may be to expand the scope and range of one's business — for instance to encourage new customers to rent cars — irrespective of whether it harms one's competitors. Indeed, starving a rival may injure one's own business even in the capitalist market system because it lowers the visibility of one's own product or service.

So resilient and intractable are the images of scarcity, of others bearing down, of satisfaction deriving from the relative failure of others, that images of competition are hard to pare away from evolutionary or ecological theory. As G. E. Hutchinson has written,[34]

although animal communities appear qualitatively to be constructed as if competition were regulating their structure, even in the best studied cases there are nearly always difficulties and unexplored possibilities.

Among such difficulties are observations that species are not always found in unoccupied niches which are apparently suitable for their habits.

These difficulties suggest that if competition is determinative it either acts intermittently, as in abnormally dry seasons . . . or it is a more subtle process than has been supposed . . . .

Unfortunately there is no end to the possible erection of hypothesis fitted to particular cases that will bring them within the rubric of increasingly subtle forms of competition.

Replying to the criticism that "the conclusion of competition is an inference rather than a direct demonstration" because other factors may cause population shifts, Jared Diamond asks, "what scientific conclusion is immune to the objection that there might be a different undetected and unspecified explanation?"[35] Diamond believes this affirms the usefulness of competition as a fundamental principle; we think the search for a different explanation is precisely the goal for feminists.

If there is one class of organisms that most often have been observed to squabble in natural settings, it is birds, which are especially noted for contesting territories which then become nesting sites and often also food sources.[36] One might predict that as a result studies have investigated which sorts of birds turn out to be the best competitors, and how their success affects the genetic composition of the next generation. But the time scale for such a study may be so short as to make its connection to evolutionary change difficult or impossible. Moreover, we can anticipate the probable results based on what *is* known about primate and human dominance hierarchies. Contrary to the assumption that a "dominant" male is dominant in any category, say the best hunter also controls the largest territory and inseminates the most females, researchers find that the individual who ranks at the top of a hunter hierarchy may not be the most successful at controlling territory or inseminating females or achieving at any other measure of dominance used.[37] Similarly, for instance, with territorial behavior in birds: the winner may not be, evolutionarily, "the fittest," except by definition. Even in the rare cases where competition can be established and linked to reproduction, still its relationship with evolutionary progress or change is unproven.

V. DO YOU BELIEVE IN MAGIC?

Most of us came of age in an intellectual environment where evolution is taken for granted and where opposition to it is identified with the regressive or parochial views of religious fundamentalists, and we find it hard to think about natural history any other way. Of course it was not always so. In its modern form, evolution is an invention of nineteenth century Victorian society. So if in the early 1800's one wondered where the diversity of species came from (and distrusted French materialists), one turned for the answer to Scripture and found there not a story of progressive development through millenia of history, but an essentially static account of God's expert engineering of diverse living forms, each of which he created once and for all, set on earth, and admonished to "be fruitful and multiply."[38] Until the late nineteenth century two attitudes remained generally widely accepted: that nature is essentially benign and peaceable, and that God had created each species separately and purposefully.

But in evolutionary theory, the orderliness of distinct species belies the chaos of reproduction: the irrational, random processes of variation – recombination and mutation. Competition is the process which imposes order on this chaos, by selecting the ("fittest") individuals to constitute a species. Competitive struggle was implanted into the image of nature concurrent with the abrupt and disruptive advent of modern industrial capitalism and, even more to the point, the impact of early feminist political action. And the role of competition in evolutionary theory seems to reflect one response to political developments. As men were engaging in political struggle with women, so, in the realm of evolutionary thought, men installed competition as the force which imposes order on the chaos they perceived in the process of reproduction, which they associated with women.[39] Evolution and natural selection, as products of nineteenth century thought, coincide with other reflections of men's anxiety about women, most plainly displayed in their preoccupation with her reproductive ability: her uncontrolled sexuality, her ("pathological") reproductive physiology, even her (hysterical) psychology.[40] The nineteenth century medicalization of women's reproductive capacities, as an attempt to control and contain women's fecundity, parallels the emphasis on domination and competition in nature as the main restraints over unbridled chaos in the orderly evolution of species. Feminists can criticize the association of reproduction with disorder from an awareness that – apart from patriarchal interference – reproduction is a most orderly process. Such an awareness suggests, as we shall point out below, a very different view of the role in evolution of female reproductive behavior.

Evolutionary theory then not only appears as a cultural product of the early days of industrial capitalism but also expresses patriarchal concern with the "problem" of disorder in the reproductive process, and further reveals a preoccupation with its control. Such an understanding of the roots of evolutionary theory encourages the dissection of its various elements, and a conscious evaluation of which parts are compatible with feminist values and which are not.[41] Evolution, for instance, assumes that historical change is progressive and unidirectional rather than, say, cyclical or non-directional. The prevailing ideology of science allows it to suppose no conscious directing agent, but only material causes such as random chemical aberrations (mutations) and naturalistic processes (such as differential reproduction). Particular assumptions of evolutionary thought such as the one we have emphasized — that scarcity is inevitable and in turn demands competition which is expressed in dominance relationships that make for evolutionary "progress" — reflect patriarchal culture. Evolutionary thought also partakes of other, general characteristics of modern science. Like other scientific theories, the mechanism proposed for evolution must meet criteria of simplicity and comprehensiveness: the further the impact of a single principle like competition can be generalized, the better. Fields which bear on evolution such as geology and ecology are institutionally embedded in a framework of resource exploitation and technological manipulation[42] — their results are linked to petroleum mining, or to pest elimination; to the relations between crowding and "social pathology," or to the difficulties of (Third World) population control.[43] The sciences — as the paradigm of modern academic disciplines — maintain the self-serving if misleading pretense of "dispassionate objectivity," an attitude which promotes a sense of separation between self and other, observer and observed, scientist and nature.

Perhaps evolutionary theory as an interpretation of natural history can draw inspiration from feminist approaches to human history. Much of patriarchal history has been a history of the patriarchy: its wars, its politics; a progressive journey toward professionalized, urbanized, bureaucratic capitalistic or socialistic states. But feminists, looking afresh at history,[44] see that the events of interest and the developments of importance must be re-evaluated. Epochs defined by wars or political hegemonies give way to historical time defined by women's concerns: such superimposed changes as shifts in domestic arrangements and household technology, transformations in family structure or child-rearing practices, changes in the role of women in the domestic and non-household labor forces; and also, perhaps more importantly, the rediscovery of strains of women's culture suppressed or ignored by patriarchy. Even on the level of historiography, one might speculate that a history defined by events — historical products, as it were, such as treaties, electoral results, legislative outcomes — might give way to a concern with process and continuity. Perhaps we can re-think evolution, and cease seeing it as the story of an increasing capacity to manipulate nature, or the progressive development of increased specialization. Instead, one might emphasize the successive emergence of new forms of opportunity for existence, or the continual diversification of new modes of being, or new patterns of harmonious coexistence.

In linking evolution to a feminist sense of continuity and to women's consciousness, we can identify elements which might be incorporated into the story. Although it is beyond the scope of a paper such as this to attempt a fully-articulated alternative theory, here we shall suggest some directions in which to begin speculating.

VI. WOMEN, EVOLUTION, AND ETHICS

If we grant that the future of a species is intimately tied to its biological reproduction, still a women's perspective sheds an altogether different light on the specific factors which guide the process. Elizabeth Fisher[45] has demonstrated how women's experience and the findings of women researchers combine to suggest an altogether different account than the one popularized by biologist apologists for patriarchy.[46] Among other lines of evidence suggesting the central importance of women in the processes leading to the evolution of humans, she cites a diversity of examples showing that, among higher primates, females choose their mates, and that these may not be the most vigorous or aggressive males at all. Insofar as females direct the genetic development of the species, they seem not to select for bellicosity but for wisdom. In terms of the development of human social organization, she suggests that the mother-child bond and social networks among mothers, not the band of male hunters, are the crucial units of proto-human social organization.

Fisher's work points to one key approach which may be generalizable. In organisms where reproduction involves social interactions, females are likely to play a role equal to or greater than that of males in determining the genetic constitution of the next generation, insofar as they determine the male to whose sperm they will expose their eggs. However, models which emphasize the role of females in reproduction must not become supports for the male equation (reduction) of women as being essentially reproductive agents.[47] The point is to call attention to processes and phenomena important to women's reality but undervalued or overlooked in patriarchal thought.

But even if the female parent were the agency of change in the genetic composition of the next generation, how might that change have evolutionary consequences? The traditional model – so tenaciously a part of the culture that it is difficult to imagine change occurring any other way – would assert that a genetic change has evolutionary impact only if it gives offspring a competitive advantage in the eternal, incessant competition for scarce resources; that of all the random and unpredictable variants, succeeding generations are composed only of those which, under constant pressure, show themselves to be the most "fit." But genetic change may not be purely random – we don't know what order the female or male might impose on such "random" processes as mutation and recombination because the question is not asked; the reigning dogma is that they are unpredictable and inchoate. Besides questioning whether variation really is random, we need also to re-examine the image of natural scarcity which supports the competitive process that purportedly brings order by permitting only certain variants to gain a foothold or by giving them reproductive superiority.

If there are many places in nature, if opportunity is rife, not restricted, what keeps processes of variation (if indeed they are random) from blurring lines between species into indistinctness? What process promotes the apparent closeness of fit or adaptation between a species and its habitat? The customary explanation is that small variations coupled with constant selection constantly heightens the degree of adaptation, while simultaneously keeping species lines distinct by eliminating all but the fittest individuals. But species lines may not be so distinct as the methods and assumptions of taxonomists and other biologists imply; indeed the assumed distinctness may reflect primarily a scientific need to impose some sort of categorical order on nature. And the closeness of adaptive fit may not be so close as biologists suppose, guided as they are by an urge to see the extant types as "winners" rather than just there in a particular habitat. Indeed, the notion of adaptive closeness of fit imposes a static concept on a fundamentally dynamic process: environments are constantly undergoing change, and both species' characteristics and population distributions respond, leading to further change.[48] The reconstruction or invention of models for evolutionary change is then a rich opportunity for feminists.[49]

There is, furthermore, the Malthusian "dilemma," that in many species so many more individuals are born than manifestly survive. Malthus, Darwin and their followers assumed that competition – the stronger starving out or crowding out or shading out the weaker – kept the numbers down, a suggestion paralleled by Diamond's notion, quoted above, of starving out a rival's resource base. But it may be equally likely that population regulation is a matter of fertility control in some cases, and is a consequence not of scarcity but of predation in others.

There is, however, no reason to expect that predation will fall on the less "fit" or vigorous, especially where very large numbers of offspring fall prey in every generation. With predation (or disease) there may be a large random element involved in determining which individuals perish. There are of course characteristics which seem to be specific adaptations for avoiding predation (Batesian and Müllerian mimicry − looking like an unpalatable type − and cryptic patterning − not looking like a meal at all). But the kind of research necessary to untangle causal mechanisms is very difficult if not impossible. In the absence of compelling findings there seems no good reason to attribute such characteristics to a "competition" to develop protective attributes; indeed in the absence of better evidence on predator-prey patterns there is no reason for thinking only in terms of selective mechanisms except a lack of imagination.

When we turned the notion of competition upside down, we exposed the guiding logic of female choice of a male mate. Likewise, when we turned the concept of scarcity upside down we found plenitude and opportunity as the condition for innovation. Why not see nature as bounteous, rather than parsimonious, and admit that opportunity and cooperation are more likely to abet novelty, innovation, and creation than are struggle and competition? Evolution in this perspective can be seen not as a constant struggle for occupation and control of territory but as a successive opening of opportunities, each new mode of biological organization providing a new opportunity for still more diverse forms of life − new sources of food, new habitats, new means of dispersal.

If a feminist theory of evolution is to be responsible to value considerations, then we must look carefully at the issue of scarcity. In patriarchal culture, an awareness of scarcity remained virtually the sole brake on unbridled exploitation of natural resources (and it has hardly been effective).[50] It is essentially selfish and pragmatic − fully in tune with the spirit of capital accumulation. If we find in nature not scarcity but bounty, why conserve? Here the bases of an ethic may come from the elimination of the subject-object dualism characteristic of scientific thinking, and removal of the very fear or abhorrence of nature which in the past inspired the coupled attitudes of awe and an urge to control (characteristic of the technocratic extravagances of patriarchal culture). An ethic of conservation may emerge from the very contrast we delineated between struggle, exploitation, and competition, on one hand, and cooperation on the other. Loving nature does not mean trying to change, deface, rape, or despoil it; it means appreciating it for its own sake rather than for what can be wrenched from it; it means opening ourselves to experiencing it, learning from it, and concerning ourselves primarily with *maintaining its integrity*. From that will follow conserving values such as minimal use of non-renewable resources, minimal tampering with the environment, and careful attention to the myriad interactions that result from any invasive act.

School of Natural Science, Hampshire College (MG)
The Hammonasset School, Madison, Conn. (MBA)

NOTES

[1] Polanyi, Michael, *Personal Knowledge* (Chicago: University of Chicago Press, 1958) and Peter Berger and Thomas Luckmann, *The Social Construction of Reality* (Garden City: Doubleday, 1966).

[2] Dobzhansky is discussed in note 49; Lewontin in note 20, and Hutchinson cited on p. 79.

[3] Eugene P. Odum, typically, uses the expression in passing in *Fundamentals of Ecology* (New York: Saunders, 1971), p. 241. It becomes a major element in an extended militaristic metaphor developed in a recent popularization by Harvard Assistant Professor Robert Cook in *Natural History* magazine, 'Reproduction by duplication,' 89 (1980), 91: "[W]hile oaks and maples produce tall trunks and overtop other individuals, a spreading clone subverts from within – or rather beneath . . . .

"This is how grasses gradually acquire turf, and the process has been aptly compared to guerilla warfare. At its leading edge, a large [system of runners], extensively interconnected, presents an array of advanced raiders, each provisioned by an elaborate logistical network occupying already conquered terrain. The tactical advantage of this competitive strategy depends upon the physiological capacity and duration of the supply lines . . . ."

Cook goes on to discuss the evolution of such plants in similarly combative language (pp. 92–93): "a long, protracted combat sets in," "many die at the front . . . others advance supplied from the rear"; one plant "slowly retreats," while another is "quickly invaded and conquered," and "they continue to skirmish, each well entrenched and none able to oust the others." It is difficult for any reader to remember that the underlying events all this refers to are cell divisions or distensions.

[4] R. L. Trivers, 'The evolution of reciprocal altruism,' in *The Sociobiology Debate*, ed. Arthur L. Kaplan (New York: Harper & Row, 1978). See also the bibliography compiled by Alan Miller, *The Genetic Imperative: Fact and Fantasy in Sociobiology* (Toronto: Pink Triangle Press, 1979), for additional sources.

[5] Darwin had been especially impressed by the general resemblance but reduced size and slightly altered character of living reptilian species and fossil dinosaurs. He was also struck by the remarkable modifications of species which fitted them for particular habitats (for instance the specially shaped beaks of groups of finches each largely restricted in its habits to one or another island in the Galapagos archipelago, and all of them resembling a mainland type from which he deduced they had originated). In

addition, he observed and collected an immense variety of previously unknown species, and saw some of the diversity of human societies (through the racist eyes of the era, which made him view native tribesmen as being closer to primates than to "civilized" man). Finally, he became imbued with a feeling for gradual change in nature over long periods of time from his acquaintance with Charles Lyell's uniformitarian geological theories.

A good introduction to Darwin's work and life is the biography by Gavin de Beer, *Charles Darwin* (Garden City: Doubleday/Anchor, 1963). H. E. Gruber and P. H. Barrett, *Darwin on Man* (New York: Dutton, 1974) is helpful on Darwin's intellectual development during the period when he formulated his theory of natural selection. Michael Ruse synthesizes the most recent scholarship on the scientific, philosophical, religious and social context of *The Origin of Species* in *The Darwinian Revolution – Science Red in Tooth and Claw* (Chicago: University of Chicago Press, 1979).

6 Peter Vorzimmer, 'Darwin's questions on the breeding of animals,' *Journal of the History of Biology* 2 (1969): 269–281.

7 George Grinell, 'The rise and fall of Darwin's first theory of transmutation,' *Journal of the History of Biology* 7 (1974): 272. Mendel's findings on the mathematical regularities of characteristics across generations of sweet pea hybrids impressed neither Darwin nor his contemporaries. See De Beer, *Darwin*, pp. 170–171.

8 Two of several useful studies of Darwin's debt to Malthus are: Peter Vorzimmer, 'Darwin, Malthus, and the theory of natural selection,' *Journal of the History of Ideas* 30 (1969): 527–542 and Peter J. Bowler, 'Malthus, Darwin and the concept of struggle,' *Journal of the History of Ideas* 37 (1976): 631–650.

9 T. R. Malthus, *An Essay on the Principle of Population* [1798] (Middlesex, England: Penguin, 1970), p. 72.

10 Looking backward we know that Malthus was wrong in his interpretation. He had not identified a situation controlled simply or even fundamentally by laws of nature, but one resulting from political and economic motives and choices – in particular the early effects of rapid urbanization and capital accumulation. Likely as not, an increase in food supply permitted the population rise and even encouraged it, although the debate about whether "over" population related mainly to food supply, historically, continues. See, for instance, Thomas McKeown, *The Modern Rise of Population* (New York: Academic Press, 1976) and another view in Etienne van de Walle's review, 'Accounting for population growth,' *Science* 197 (1977): 652–653.

11 Darwin did follow Malthus in worrying over the human consequences of interference with "nature" (*Descent of Man* I [1871], p. 168): "We civilised men . . . do our utmost to check the process of elimination; we build asylums for the imbecile, the maimed and the sick; we institute poor laws; and our medical men exert their utmost skill to save the life of every one to the last moment . . . . Thus the weak members of civilised society propagate their kind. No one who has attended to the breeding of domestic animals will doubt that this must be highly injurious to the race of man."

12 Discussions of the relationship between Darwin's ideas and the Malthusian-Spencerian climate of thought in England include Ruse, *Darwinian Revolution*, pp. 150–155, R. M. Young, 'Malthus and the evolutionists: the common context of biological and social theory,' *Past and Present* 43 (1969): 109–145, Derek Freeman, 'The evolutionary theories of Charles Darwin and Herbert Spencer,' *Current Anthropology* 15 (1974): 211–237, and Sandra Herbert's remarks on these sources in 'The place of man in the

development of Darwin's theory of transmutation, II,' *Journal of the History of Biology*
10 (1977): 155–227 on pp. 195–196.
[13] Charles Darwin, *The Origin of Species* [second edition, 1860] (New York: New
American Library, 1958), p. 450.
[14] *Ibid.*, p. 75.
[15] 'Darwin's notebooks on the Transmutation of Species,' Part VI, *Bulletin of the
British Museum (Natural History), Historical Series* 3 (1967): 142 [MS. p. 135].
[16] Anarchist thinker Peter Kropotkin realized the limited scope of the competitive
vision when he wrote in *Mutual Aid* (1902) that during episodes of struggle "the whole
portion of the species, which is affected by the calamity, comes out of the ordeal so
impoverished in vigour and health that *no progressive evolution of the species can be
based upon such periods of keen competition.*"
  Moreover, he put his finger on the narrowness of conception, the fixation upon
and fetishization of a relatively infrequent occurrence: "how false is the view of those
who speak of the animal world as if nothing were to be seen in it but lions and hyenas
plunging their bleeding teeth into the flesh of their victims. One might as well imagine
that the whole of human life is nothing but a succession of war massacres. Association
and mutual aid are the rule with mammals."
  Still, Kropotkin's biological ideas, which were largely ignored, would not have
disturbed the essentially competitive model anyway, for he still placed mutual aid into a
competitive framework as the strategy a species uses to gain the upper hand in relation
to other species.
[17] See Nancy Chodorow, *The Reproduction of Mothering* (Berkeley: University of
California Press, 1978), pp. 167, 169, 179, 187, 189. In many social-economic situations,
for instance in the workplace, the conviction of scarcity (of people, money, markets,
etc.) is an artificial construct men seem to impose in order to establish a competitive
and hierarchical situation (competitive against one another and against the milieu which
they define as "external"). Accustomed to the rules of such milieux (or, as Chodorow or
Dorothy Dinnerstein might suggest, blocking memories of childhood powerlessness
in an institution which is its antithesis) they recreate competitive rules and hierarchical
structures again and again.
[18] Eleanor Vander Haegen and Michael Gross, 'Feminist science: a vision for the future,'
in *Toward a Feminist Analysis: Proceedings of the Women & Society Symposium*,
eds. Buff Lindan and Carey Kaplan (Winooski, VT.: The St. Michael's College Press,
1981), pp. 21–36.
[19] Jared Diamond cites the observations on chickadee population shifts by a Russian
ornithologist who "*surmised* that the retreat" of one species resulted from its poorer
adaptation "in the face of competition" with an expanding species from a neighboring
region. "Here," Diamond writes, "is a case where the development of niche segregation
(and the refinement of reproductive isolation) was *actually observed.*" ('Niche shifts
and the rediscovery of interspecific competition,' *American Scientist* 66 (1978): 322–
331 on p. 325, our emphasis) But *was* competition the underlying cause? Or was it an
instance of a crucial change in micro-climate, or parasites, or . . . ?
  Some evidence shows that two closely related species living in the same region differ
more markedly in certain structures than the same species living in different localities.
(*Ibid.*, p. 326) The assumption is that competition exacerbates differences. But again all

we see are the demographic, behavioral, or structural consequences which constitute the "evidence" for competition.

[20] Richard Lewontin, 'Adaptation,' *Scientific American* **239** (3): 213–229, September, 1978 is eloquent about the difficulties in adaptation and niche theory: such features as circularity, untestability, and unavoidable simplification of complex physiological or genetic interrelations in an organism.

[21] J. Diamond, 'Niche shifts', p. 324.

[22] Robert E. Ricklefs, *The Economy of Nature* (Portland, Oregon: Chiron Press, 1976), p. 266.

[23] *Ibid.*

[24] *Ibid.*

[25] Odum, *Fundamentals*, p. 217 injects competition in a similarly misleading way when he describes Gause's laboratory study of two species of *Paramecium* grown together, in which, after sixteen days, only one species survived. In this "'classic' example of competitive exclusion," Odum reports that "neither organism attacked the other or secreted harmful substances." The survivor species "simply had a more rapid growth rate (higher intrinsic rate of increase) and thus 'out-competed'" the other species. The quotation marks around "out-competed" are important because this is a case where competition is again the assumed invisible hand behind the phenomenon, in this case, "a more rapid growth rate."

Also, in reporting the research, Odum does not tell the full story. Depending upon culture conditions, food may be the limiting factor, or the accumulation of waste products may be, and the conditions will determine the outcome, the "superior competitor." Nor does "higher *intrinsic* rate of increase" as a cause of competitive victory convey the spirit of Gause's remark that "the superiority of one species over another in competition did not simply reflect the properties of those species taken independently, but *was often essentially modified by their process of interaction.*" (G. F. Gause, *The Struggle for Existence* [New York: Williams and Wilkins, 1934], p. 112, emphasis in original)

[26] J. J. Christian et al., "The role of endocrines in the self-regulation of mammalian populations," *Recent Progress in Hormone Research* **21** (1965): 501–578.

[27] Gause, *Struggle*, pp. 74–75, 89.

[28] Ricklefs, *Economy*, p. 269.

[29] Odum, *Fundamentals*, p. 209.

[30] *Ibid.*, p. 210.

[31] Diamond, 'Niche shifts,' p. 330.

[32] The literature of social psychology, in "operationalizing" a concept like competition, scarcely explores the underlying emotions or attitudes. But it does document amply that competition is a male characteristic. Of 18 studies cited in *Psychology Abstracts* published between 1974 and 1978 concerning competition with respect to sex differences, 11 find (elementary school and college age) males are more competitive than females. All the rest but one find no difference. (The exception is a study in which female dyads were found to be more competitive than male dyads among a college age sample.) Another four studies show that boys are more self-serving in distributing rewards gained during competition than are girls. (See especially: John McGuire and Margaret Hanratty Thomas, 'Effects of sex, competence, and competition on sharing

behavior in children,' *Journal of Personality and Social Psychology* 32 [1975]: 490–494, and Mark A. Barnett *et al.*, 'Children's reward allocation after competition: sex differences and the effect of task structure,' *Journal of Genetic Psychology* 133 [1978]: 149–150.) On the other hand, as might be expected, there is some evidence that the sex of the experimenter intersects with the results obtained in such studies. (Vincent Skotko et al., 'Sex differences as artifact in the prisoner's dilemma game,' *Journal of Conflict Resolution* 18 [1974]: 707–713.)

An especially interesting finding is that in conditions of artificial crowding, male college students assume a competitive posture while females behave cooperatively. (Yakov Epstein and Robert Karlin, 'Effects of acute experimental crowding,' *Journal of Applied Social Psychology* 5 [1975]: 34–53.) Finally, even though American males seem on the whole to be more competitive than females, Indian males are even more competitive than American males of the same age, according to several cross-cultural studies. This may depend, according to one research team, on "a 'view of the world' that is based on scarcity and limited resources." (Daniel Druckman *et al.*, 'Cultural differences in bargaining behavior: India, Argentina, and the United States,' *Journal of Conflict Resolution* 20 [1976]: 413–452.)

For an optimistic vision of cooperative possibilities see Dee G. Appley and Alvin E. Winder, "An evolving definition of collaboration and some implications for the world of work," *The Journal of Applied Behavioral Science* 13 (1977): 279–291.

33 Diamond, 'Niche shifts,' *op. cit.*, p. 329.

34 G. E. Hutchinson, 'Summary,' *Cold Spring Harbor Symposia on Quantitative Biology* 22 (1957): 415–427 on p. 419.

35 Diamond, 'Niche shifts,' *op. cit.*, p. 327.

36 V. C. Wynne-Edwards, 'Population control in animals,' *Scientific American* (August, 1964): 68–74, argues specifically that territoriality is a way to insure that food scarcity never becomes the limiting factor on population size in the birds he studied – since the territory staked out for a nest is more than adequate to feed the family nesting in it.

37 Gina Bari Kolata, 'Primate behavior: sex and the dominant male,' *Science* 191 (1976): 55–58 provides an entry into the main findings of recent research. See also Lila Leibowitz, *Females, Males, Families* (Bound Brook, CT.: Duxbury Press, 1978) and Suzanne Chevalier-Skolnikoff and Frank E. Poirer (eds.) *Primate Bio-Social Development* (New York: Garland, 1977), especially Gershon Berkson, 'The social ecology of defects in primates,' (pp. 189–204) and Linda M. Fedigan and Laurence Fedigan, 'The social development of a handicapped infant in a free-living troop of Japanese monkeys,' (pp. 205–222) which show that monkeys in the wild and in laboratory settings care for handicapped members of the group ("fitness" and "selective advantage" notwithstanding), and also the review article by Richard C. Savin-Williams and Daniel G. Freedman, 'Bio-social approach to human development,' (pp. 563–601) which reports (but plays down or tries to explain away) inconsistencies among various measures of dominance applied to the same group of children, disagreements between sociometric and behavioral ratings of dominance, and complexly intransitive dominance relationships (rather than linear hierarchies) among those "at the top."

38 Eighteenth and early nineteenth century naturalists did worry about some of the specifics of the story: What about fossils of animals which manifestly no longer exist? When or how often did the Deluge come? But essentially the picture was one that had first been painted by Plato: the "Good" (i.e. God) conceived of the entire plan of

creation, including all living forms, which follow a hierarchical chain from the slime up through man to the angels. Succeeding centuries of thinkers complicated the story by replacing a single hierarchical sequence with a group of hierarchies for particular categories – plants, invertebrates, vertebrates. (Arthur O. Lovejoy, *The Great Chain of Being* [Cambridge, MA.: Harvard University Press, 1964])

Eighteenth century thinkers did recognize some competitive interactions in nature – usually in a rather aristocratic form, in terms of a chivalrous competition among stags for a female, for instance. But even where they did recognize competition, they did not see it leading to change but instead to promoting stability: eliminating the damaged or weaker members of a species kept it true to type.

39 We are grateful to Janice Raymond for her suggestion that we take account of the mid-nineteenth century women's rights struggle in relation to theories emphasizing competition, and her insight that men's responses reflected fear of women's associations and mistrust of all forms of women's creativity.

40 Historical perspective is provided by Barbara Ehrenreich and Deirdre English, *For Her Own Good* (Garden City, N.J.: Doubleday-Anchor, 1978) and G. J. Barker-Benfield, *The Horrors of the Half-Known Life* (New York: Harper & Row, 1976). Adrienne Rich, *Of Lies, Secrets and Silence* (New York: W. W. Norton, 1979) succinctly describes the current situation (p. 270): "A male-dominated technological establishment and a male-dominated population control network view both the planet and women's bodies as resources to be seized, exploited, milked, excavated, and controlled. Somehow, in the nightmare image of an earth overrun with starving people because feckless, antisocial women refuse to stop breeding, we can perceive contempt for women, for the children of women, and for the earth herself."

Susan Griffin explores these ideas on a more metaphorical level in *Woman and Nature* (New York: Harper & Row, 1978); Dorothy Dinnerstein sees the same connections through psychoanalytic theory in *The Mermaid and the Minotaur* (New York: Harper & Row, 1977); and the ethical and political dimensions are explored in H. B. Holmes, B. B. Hoskins and M. Gross (eds.) *Birth Control and Controlling Birth: Women-Centered Perspectives* (Clifton, N.J.: Humana, 1981). On hysteria, see Phyllis Chesler, *Women and Madness* (New York: Doubleday, 1972).

41 Others who have contributed so far to this analysis include Griffin, *Woman and Nature*, Donna Haraway, 'Animal sociology and a natural economy of the body politic, parts I and II,' *Signs* (Autumn, 1978): 21–60; Ruth Hubbard, 'Have only men evolved?' in this volume; and Evelyn Reed, *Sexism in Science* (New York: Pathfinder, 1978).

Discussing the history of theories of animal behavior, Donna Haraway observes ('The biological enterprise: sex, mind and profit from human engineering to sociobiology,' *Radical History Review* 20 [1979]: 206–237 on pp. 232–233): "Nature, including human nature, has been theorized and constructed on the basis of scarcity and competition. Moreover, our nature has been theorized and developed through the construction of life science in and for capitalism and patriarchy. That is part of the maintenance of scarcity in the specific form of appropriation of abundance for private and not common good. It is also part of the maintenance of domination in the form of escalating logics and technologies of command-control systems fundamental to patriarchy. To the extent that these practices inform our theorizing of nature, we are still ignorant and *must* engage in the practice of science. It is a matter for struggle. I do not know what life science would be like if the historical structure of our lives minimized domination. I do know that the history of biology convinces me that basic knowledge would reflect and reproduce the world, just as it has participated in maintaining an old one."

[42] For instance, postdoctoral research by one co-author of this paper (MBA) seemed theoretical in scope (microbial populations of Douglas fir trees) but was financially supported because of its later intended use in a large scale computer program to predict the effects on a forest ecosystem of clear-cutting.

[43] A few representative items in this controversial literature include: Alice T. Day and Lincoln H. Day, 'Cross-national comparisons of population density,' *Science* 181 (1973): 1016–1023; Patricia Draper, 'Crowding among hunter-gatherers: the !Kung bushmen,' *Science* 182 (1973): 301–303; and Omer R. Galle *et al.*, 'Population density and social pathology: what are the relations for man?' *Science* 176 (1972): 23–30.

[44] Sharon Shepela, 'Feminism as the defining concept for feminist disciplines,' paper presented at 'Women and Society: Past, Present and Future – A Symposium,' (St. Michael's College, Winooski, VT., March, 1979). Also see Mary Daly's brilliant and wide-ranging radical feminist analysis of conventional patriarchal disciplines in *Gyn/Ecology* (Boston: Beacon, 1979).

[45] Elizabeth Fisher, *Women's Creation* (Garden City, N.J.: Anchor/Doubleday, 1979). Also, Leibowitz, *Females, males, families*; Reed, *Sexism in Science*; Hubbard, 'Have only men evolved?' and Haraway, 'Animal sociology.'

[46] The traditional account of human evolution – now well-disseminated by works of "pop ecology" by Tiger, Morris etc. – saw hunting as a proto-typically male activity, and a crucial factor in the evolution of proto-humans. As those accounts have it, the critical step in human evolution was the invention of tools, meaning weapons. The more aggressive males could, as the best hunters, support the most offspring and, as the more dominant, mate with the greatest number of females; consequently the most assertive males directed the course of human evolution while females occupied a rather passive role as childbearers. Gathering, still the major food source in traditional societies, has generally been women's work, although men may or may not participate; this, Fisher suggests, places women at the center of the economic order and makes hunting a less important luxury. She suggests that while tools were probably important in evolution, the first tools were likely to have been carrying baskets developed for food gathering and readily adapted also for carrying infants and young children.

[47] The entire notion of uncontrolled fecundity, associated with women, in the process of reproduction, needs further qualification. The historical evidence on human societies suggests that women have long used a number of family limitation practices from herbal abortifacients to infanticide. Social and dietary patterns seem also to have played a role in regulating fertility. Women's fecundity has therefore never been "out of control" except when catastrophically disrupted by colonialism, by the unbalanced exportation of western medicine without the corresponding pattern of economic and social organization. See Susan George, *How the Other Half Dies* (Montclair, N.J.: Allanheld, Osmun, 1977), and F. M. Lappe and Joseph Collins, *Food First: Beyond the Myth of Scarcity* (Boston: Houghton-Mifflin, 1977). Similarly, as noted earlier, even under the artificial conditions of laboratory investigations, rodent populations are self-limiting with respect to population density. In short, we should probably look elsewhere than at unbridled fecundity constrained only by competition to find the order underlying evolutionary change.

[48] Field studies of "adaptive radiation" – the process through which "a group of

organisms with a common ancestor evolves and speciates to fill many of the adaptive zones in the environment" − exemplify the difficulty with using a concept of rigidly-defined species in relation to a historical process of evolution. In Elizabeth C. Dudley's study of this process in relation to plant variations at diverse altitudes ('Adaptive radiation in the Melastomataceae along an altitudinal gradient in Peru,' *Biotropica* 10 (1978): 134−143) "species" reduces, in terms of the practical collection of data, to a collection of measurements of such factors as leaf length and width, shape, surface, texture, petiole and internode lengths, etc. Some of these vary with respect to altitude. But instead of considering the evolution of distinct species − which is what the theory is about − this research correlates variations in plant characters with continuous variations in factors like altitude, climate, soil type. Genotype − the genetic make-up which constitutes the basis for evolutionary change − is not studied directly but only indirectly in terms of phenotype − the actual expression in plant morphology of the interaction between genes and environment. A further reason why genotype is not being investigated is related: the phenotype which a given genotype may produce varies greatly in response to environmental factors ("phenotypic plasticity").

But even careful studies of phenotypic plasticity are not very informative about genotype, which is what distinct species are defined by. For instance, when four samples of a species of columbine growing in four diverse habitats in western Massachusetts are studied carefully it remains difficult to assess whether one of the populations "either did not have genes to allow plasticity . . . or . . . if present, they were unable to express significant differences in the environments present".

The same author further concludes that "none of the characteristics studied were under strong genetic control" and that significantly varying characteristics (with respect to environment) are "under strong environmental control and have a high degree of plasticity." In particular, "no consistent pattern emerged in the expression of the various characteristics suggesting that different sets of genes were operating on different characteristics in the different environments." It becomes essentially an arbitrary imposition to apply such categorical concepts as species, race, variety on the immense variability of such an organism in its several environments. (Germain LaRoche, 'An experimental study of population differences in leaf morphology of *Aquilegia canadensis* L. (Ranunculaceae),' *American Midland Naturalist* 100 (1978): 341−349. The study reveals typical methodological problems: misapplication of statistical methods, and a failure to use a procedure which might avoid experimental bias. These methodological difficulties do not disturb the validity of our observation of the arbitrariness of taxonomic categories.)

As for adaptation, a study of two bumblebee species' foraging behavior on two species of flower shows that any attempt to study adaptive relationships must be enormously complicated. Although the purpose of this study (David W. Inouye, 'Resource partitioning in bumblebees: experimental studies of foraging behavior,' *Ecology* 59 (1978): 672−678) is to test the hypothesis that "if bumblebees are indeed competing for food resources, there should be observable changes in foraging behavior (i.e. a niche shift)," it illustrates that the simple relationship between the length of a bumblebee proboscis and the depth of the corolla tube from which it gathers nectar is very complex. For the effective depth of nectar in the corolla is not determined simply by corolla length but also by the frequency and legnth of bumblebee visits, time of day, shifts in division of labor in the hive, age of the bees, time of season, and probably also such climatic variables as temperature and rainfall. In short, a static relationship of bumblebee and flower morphologies belies the actual subtlety of the processes of bumblebee and flower behavior and consequently questions arise about the supposed examples of adaptive "closeness of fit." (The paper, by the way, does indeed demonstrate that shorter proboscis bees will forage on a long corolla flower when long proboscis bees are removed, and – unsurprisingly – concludes that "the results of my study imply the action of competition between bumblebees in montane environments." Then, typically, it redefines competition to mean consumption: "interference competition was never observed during the study, and I have never observed aggressive interactions between bumblebee species in the East River Valley. Competition for nectar probably occurs through direct depletion of resources." [pp. 676–677])

49 Although it would be presumptuous to suggest here how to construct an alternative evolutionary theory, two biological mechanisms deserve some consideration: pre-adaptation and isolation. Isolation may play an important role in the formation of new species without requiring that competitive pressure be involved. "Pre-adaptation" shifts the emphasis in the development of new adaptations from differential reproduction to the production of the innovation in the first place. T. Dobzhansky discusses evolution in relation to these factors and manages to avoid the term "competition." (Yet he retains the concept in his ideas about fitness: "gene constellations that fit the environment survive better and reproduce more often than those that fit less well." [*Genetics of the Evolutionary Process* (1970), p. 431]) Pre-adaptation – the fortuitous acquisition of a characteristic which happens to suit the variant for a slightly different environment or somewhat alters its behavior – may itself serve as an isolating mechanism; if so, the need to invoke competition to account for the differentiation of species is further diminished. Moreover, any pre-adaptation which significantly alters form, geographical range, or behavior may shift the variant slightly outside the accustomed range of its major predator(s), thereby further promoting a prompt increase in its population.

50 Lynn White remarks ('Historical Roots of our ecological crisis,' *Science* 155 [1967]: 1203–1207): "by destroying pagan animism Christianity made it possible to exploit nature in a mood of indifference to the feelings of natural objects." Here we can make a psychological connection to patriarchal structures of thought and feeling. Deriving from the social structure of the family – with the father as toiler in the world and the mother as primary care-giver in the family – is the psychological contrast between the instrumental, goal-directed, manipulative tendency of men, and the interpersonal, empathic, relational style of women. (Chodorow, *Mothering* [see note 17 above]) William Leiss (*The Domination of Nature* [Boston: Beacon Press, 1974], p. 34) qualifies White's statement with the observation that "Christian doctrine sought to restrain man's earthly ambitions by holding him accountable for his conduct to a higher authority," but notes further (p. 35) that as the conflict of religion and science in the eighteenth and nineteenth century shed the sense of man's subordination to God, it maintained and extended the vision of man as "lord of nature." (Leiss also notes two other sources fostering an attitude of control and domination: the ambivalent notions of fear and loss of control alongside desire for benefits which attach to instruments of manipulation (pp. 27–29); and the intense interest in nature and its operations as embodied in Renaissance magic. Again there are strong sexual identifications in the relation of woman and nature as entities to be "penetrated" for their "secrets" in the attitudes of the Renaissance alchemists. [See Sally Allen and Joanna Hubbs, 'Outrunning Atalanta – an investigation of the feminine image in alchemical transformation,' *Signs* 6 (1980), 210–229.])

Thus an exploitative attitude toward nature conformable to patriarchal psychology was intrinsic to Christian doctrine and, as Judeo-Christian belief gave way to exclusive faith in science and technology, that attitude of mastery and domination persisted and strengthened, now uninhibited by any wider ethical framework from religion.

ANN PALMERI

CHARLOTTE PERKINS GILMAN:
FORERUNNER OF A FEMINIST SOCIAL SCIENCE

Charlotte Perkins Gilman considered herself a social scientist and a feminist theorist. In Gilman's eyes, doing social science and doing feminist theory were not two separate enterprises, they were one. But modern historians have dismissed her claim to being a social scientist and have resurrected her solely as an important feminist from the past. There are at least two reasons for this dismissal. First, the social evolutionary theory on which Gilman based her claims for social reform was discarded long ago by social scientists because of its neo-Lamarckian reasoning. Secondly, since her feminism was an essential part of Gilman's scientific argument, her social theory seems to be discounted on this basis alone. In other words, feminism, since it is a form of moral reasoning, has no place in social scientific research. Although historians of science are now engaged in assessing theories in the context of their time, no real consideration has been given to the role of moral reasoning in social scientific research. This is particularly important for those engaged in trying to reconstruct what a feminist social science might be.

I cannot give a full account of the role of moral reasoning in the social sciences here, but, I suggest that through a look at Gilman, we can begin to comprehend more fully the claim of present feminists that adopting a view of women as fully human and as actors in history leads to a more well-founded social science. While a commitment to the equality of women and men by no means commits any social scientist to a particular theory, the acceptance of this moral assumption does preclude the consideration and acceptance of certain sorts of theories.

While feminists have argued that there is a role for such moral reasoning in the social sciences, they have not given a philosophical account of social scientific reasoning in order to justify their claim. One of the first steps in such a justification is to study historical cases which provide *evidence for* the acceptance of the legitimate role of moral reasoning in social science. I offer the social theories of Charlotte Perkins Gilman as one such case. This case offers us evidence that moral reasoning is often used as a *preliminary* justification for pursuing a line of research, often different from the previously acceptable kinds of research. Consequently, moral reasoning may be invoked

97

Sandra Harding and Merrill B. Hintikka (eds.), Discovering Reality, 97–119.
Copyright © 1983 *by D. Reidel Publishing Company.*

as an appeal to a background of moral principles from which a new line of research is to be made convincing. It is no accident, then, that our present day appeals for a feminist social science are primarily appeals to moral principles from which it is hoped (and often promised) acceptable social scientific research might come.

Charlotte Perkins Gilman (1860–1935) has been praised as "the greatest theoretician the women's movement ever produced"[1] and "the most intellectually gifted of them all"[2] ("about the only one, in fact, the American movement ever produced"[3]) and yet is still disregarded because she borrowed from Lester Ward many of the ideas that buttressed her claims about the female half of the species. What is remarkable about these evaluations of Gilman is their failure to give a thorough study of Gilman's work. Instead, these historians seem satisfied to give an account of her views on the home and motherhood that do not offer full comprehension of the biological and moral principles upon which they are based. The most notable analysis (and probably the origin of the prevailing view on Gilman) is by Carl Degler,[4] who treats Gilman's use of the science of her day as some kind of self-deluded appeal to false evidence.

Convinced herself of the power of science, and especially Darwinism, she cast her study in *pseudo-scientific* terms. Her favorite device of comparing relations between the human sexes with those between animals gives a tone, if not a conviction, of universal validity to her arguments.

Later on he writes, ". . . her aim was not to prove her point by evidence, so much as it was to *shock* her readers into seeing the relations between the sexes from a new point of view."[5] Yet Degler also tells us that *Human Work*, an application of evolutionary theory to human society, was the book Gilman thought her best and most important.[6] None of these historians, if they read this book at all, make any attempt to connect her analysis of human evolution to her social critique of women's position. Since Degler reduces to a reform tract all of Gilman's efforts to found her social analysis upon the science of her day,[7] he undermines, I think, his claim that she was a theoretician at all. Furthermore, such a reduction reveals a failure to delineate the actual role of moral and political reasoning in social scientific thought.

One of the main reasons, it seems, that historians have failed to give Gilman a more thorough treatment stems from their views of social scientific reasoning itself. By suggesting that Darwinian or evolutionary thinking turned out to be an improper mode of reasoning when applied to the social realm and functions as a mere cover-up for those advocating various social policies (from "laissez-faire" to "socialism"), these historians seem to suggest that any moral reasoning in the social sciences (and perhaps that any science used in moral reasoning) is inappropriate. Moreover, this view implies that any moral argument found in scientific reasoning necessarily disqualifies it. I shall argue that it is not moral argument that calls into question certain lines of scientific reasoning but evidence of *improper* moral argument. Moral ideals may have an important place in the social sciences, they shape, mold, and fashion what count at a particular time as legitimate ways of arguing. While we easily recognize the illegitimate nature of Spencer's "laissez-faire" argument and comfortably dismiss his grandiose theorizing as "pseudo-science," the case against the other social evolutionists is not so clear. And neither is the line between "science"and "pseudo-science." The plausibility of Ward's and Gilman's reasoning depends in part upon how their moral ideals shaped their reasoning and made their "explanations" and "accounts" a more legitimate line of inquiry than other modes of evolutionary argument. Of course, such ideals are no guarantee of a plausible line of social research. But in some cases moral claims *enhance* a scientific argument despite its later demise on other grounds. And Gilman's case might show that given the evolutionary model current at the time, a more plausible account of the evolution of sexual relations *depended upon* the insertion of a more plausible *moral reasoning*.

I have two purposes in writing this paper: (1) To give Charlotte Perkins Gilman's views of social evolution a sufficient rendering so we may assess her as the foremost early feminist theoretician, and (2) to give some insight into the kind of feminist reasoning that characterized these early social thinkers.

I

Degler's dismissal of Gilman's evolutionist arguments reflects a general skepticism of all evolutionary thinking of the late nineteenth century. We habitually think of this period as one in which moral posturing gave rise to the profound abuse of a newly-minted biological theory of Darwin. Yet the doctrine of social evolutionism of the late nineteenth century was a revival in new dress of a much older theory that rebloomed in the Darwinian aura. The subsequent rejection of the framework of social evolutionism should not diminish our interest in the influence of this kind of thought in the moral development and insights of its proponents. The main thesis of this doctrine, inherited from the eighteenth century, is that there is a sequence of social forms "which followed inevitably from the uniformity of the laws of nature and of human nature unimpeded by local or accidental circumstance."[8] That social life evolved in a way analogous to biological life became a powerful, reigning idea and the first articulation of a social science.

Gilman's mentor, Lester Frank Ward (1842–1913), was entrenched in the prevailing debate on the character and nature of this social evolution. Called the founder of American sociology, Ward was mainly known as a critic of Herbert Spencer's views. These criticisms stemmed from Ward's deep aversion to Spencer's doctrine of "laissez-faire" and "the survival of the fittest." In response to such doctrines, Ward developed his own "dynamic sociology" which amounted to a theory of the development of human consciousness. His odd mixture of Darwinian and neo-Lamarckian principles (which were heartily endorsed by Gilman) comes to the forefront in his analysis of the orgin of partriarchy or what he calls his "gynaecocentric" theory. In order to unravel the character of Gilman's thinking we not only have to understand this major influence on her argument, but also just what problems these arguments and explanations hoped to answer.

Giving an account of the evolution of social structures was a vital task for some nineteenth century scientists. Spencer had suggested that the *ultimate* explanation for social systems is to be found in a determination of the laws of pre-social man; all phenomena are to be explained in terms of physical causes. The stages of society, going from the simple to more complex forms of organization, are just following a natural law applicable to all species of animals. It is from this thesis on natural law Spencer thought his "laissez-faire" doctrine followed.[9]

In treating the human mind as a *natural* product of evolution ("its achievements are to be classed and studied along with other natural phenomena"),[10] Ward follows Spencer. But in delineating the mind as a *separate* object of study, in recognizing mind as a *new power*[11] in the world, Ward argues "it is only to a limited extent and in the most general way that we can apply the same canons to the organic as to the inorganic world." All human institutions are a result of invention, human practical art. To study them we must invent the "artificial method." "If nature's process is rightly named natural selection, man's process is artificial selection."[12] Art operates in protection of the weak *against* natural forces. The development of just institutions, the feelings of morality, all are aimed at "resisting the law of nature." Paradoxically, Ward claims that in the understanding of the human mind would be the true understanding of nature.

When nature comes to be regarded as passive and man as active, instead of the reverse as now, when human action is recognized as the most important of all forms of action, and when the power of the human intellect over vital, psychic, and social phenomena is practically conceded, then and then only, can man justify claims to have risen out of the animal and fully to have entered the human stage of development.[13]

The distinction between natural development and human action was a distinction between the causes of motion, one *genetic* and the other *telic*. Ward, once called the "American Aristotle," undermines the "laissez-faire" doctrine by arguing that nature has no design, no purpose; he proposes instead that it is humans, because of the powers of their minds, that generate this sense of *purpose*.

By making this sharp distinction between the *genetic* and the *telic*, Ward made it impossible to construct simple-minded analogies from biological causes to social evolution. It is no surprise, then, that Ward is known as the creator of American sociology. For following his arguments, the principles of social evolution must take purposeful human behavior into account. Human making is the human condition and any science must recognize this essential characteristic. Thus the "laissez-faire" argument that the "survival of the fittest" is best served by the non-intervention of human art, simply does not make sense, since all human behavior *is* intervention. Ward suggested this argument through a complex intertwining of scientific and moral argument. The issue is not whether human beings would intervene or not intervene, but *how*, whether according to plan, based on certain principles, or whether haphazardly. In discovering natural law, Ward argued, humans were able to marshall the forces of nature for their own purposes. If we were to gain knowledge of social welfare, we would have the power of producing happiness. We can do this with knowledge of *social* laws. So it follows, Ward continues, that

The *special* problem of sociology is to control these forces, to remove throughout its vast domain all those which obstruct the natural course of the feelings, to increase and intensify those which are favorable to that course, and to guard against any form of stimulation whose reaction will count more strongly against the general sum of human happiness than the stimulus itself counts in its favor.[14]

Although this sentiment echoes the utilitarian arguments of the early nineteenth century, for Ward, a simple calculus will not do. The evolutionary framework requires a knowledge of the history of the human species, in short, an accurate account of the development of the human mind.

Another important feature of Ward's thought, apart from his criticism of

Spencer, is his "gynaecocentric" theory. This theory is inspired by Darwin's account of the evolution of sex roles through the principle of sexual selection. Unlike Ward, Darwin uses the principle of sexual selection to explain the necessity of patriarchy. As Darwin defines it, the principle of sexual selection is "the advantage which certain individuals have over others of the same sex and species solely in respect to reproduction."[15] In other words, the male or female picks its mate on the basis of who is most likely to help reproduce more and better adapted offspring. The problem, of course, which Darwin duly notes, is "in understanding how it is that the males which conquer other males, or those which prove the most attractive to the females, leave a greater number of offspring to inherit their superiority than their beaten and less attractive rivals."[16] Now superiority, as Darwin understood it, meant adaptation to the environment. What the principle of sexual selection suggests is that there is a correlation between well-developed secondary sexual characteristics and fitness as a parent. So, the choice of a mate on the basis of how well the mating dance is done or on how bright the plumage is must somehow correspond with the likelihood of the potential mate to produce viable offspring. Darwin's principle of sexual selection is important for understanding monogamous species and, in particular, for explaining the peculiar characteristics of males and females of the human species.

In certain species, females make the selection of a mate; while in others, males make the selection. Darwin suggests that the male power of selection for humans was wrested from the female through the development of male superiority.[17] In doing so, Darwin makes two assertions: (1) that superior intelligence has been selected − by the principle of natural and sexual selection − and (2) that males have developed superior intelligence to females. The first assertion he explains by applying his usual principle of natural selection − that superior intelligence was better adapted to the environment and that such intelligence in the early period of human existence was selected for by the female sex. Yet, Darwin claims in the second assertion, the power of selection was wrested from the female sex because the development of the male's superior intelligence to the female. "Man is more powerful in body and mind than woman, and in the savage state he keeps her in a far more abject state of bondage than does the male of any other animal; therefore, it is not surprising that he should have gained the power of selection."[18] Darwin seemed to think that the selection of intelligence was primarily a selection for male intelligence, that a male's adaption to the environment, moreover, required that it be superior to female intelligence. While, of course, there was the general claim that the species of homo sapiens demonstrated that intelligence was a superior trait in the sense that it was well adapted to the environment that homo sapiens lived in, Darwin wished also to argue that the discrepancy in male and female intelligence was a further adaption well-suited to the environment and that such a discrepancy was in part a result of sexual selection – the choice of more intelligent mates.[19] The origin of patriarchy is derived from the power of males in selecting a mate.

From the vague suggestion by Darwin of an early period of sexual selection by women,[20] Lester Ward argues that female selection was probably a long stage in human history. To believe in this early stage of "gynaecocracy" is "the logical and inevitable conclusion that must follow the admission of the animal origin of man."[21] The introduction of the notion of "gynaecocracy," Ward suggests, ends the androcentric world view of "male superiority." Instead, he argues that "androcracy" came not as a result of male superiority in intelligence but in a recognition of paternity. With this recognition, Ward reasons, came a recognition of equal authority over progeny and a realization that men and women are unequal in strength. Before this time, there was no reason to acknowledge such an inequality. Physical strength could be translated into superior power of man over woman as a recognition of woman's *economic* value to man. The primitive androcracy consisted of polygamous marriages and celibate men; the women, children, and celibate men all were enslaved to the patriarch. This enslavement which continued through the development of monogamy, Ward suggests, blinds us to the androcentrism revealed in the notion of male superiority and the propriety of the male subjection of women.

The male and female differences that the androcentric view suggests are innate, Ward argues, are the result of the long subjection of women that have exaggerated their differences from men. Further, in the leisure class especially, Ward notes, males have selected women for their beauty and this selection has tended "to dwarf her stature, sap her strength, contract her brain, and enfeeble her mind."[22] Not using your physical muscles to exercise, your brain to solve problems, according to this view, makes these faculties *atrophy*.[23]

Not only do the principles of natural selection and sexual selection work to accentuate the sexual differences but, according to Darwin and Ward, certain *habits* may be *acquired*, and, it seems, *inherited*. This neo-Lamarckian claim was hotly debated at the time,[24] and for social evolutionists was an important feature of their theories. It would be easy to mock its assertion in these days in light of the acceptance of the Darwinian principle of natural selection, but for these social theorists it provided an important means of understanding what we would now call "cultural transmission." Darwin himself cautions that the transmission of such habits does not always prevail. "It must be borne in mind that the tendency in characters acquired by either sex late in life, to be transmitted to the same sex at the same age, and of early acquired characters to be transmitted to both sexes are rules which, though general, do not always hold."[25] Darwin does not specify the cases where the inheritance does not hold but the fact that Darwin and others thought that *any* acquired habits are transmitted is extremely significant in understanding how social evolution was supposed to occur.[26] In the case of sexually acquired habits, Ward wanted to show that although androcracy was a product of genetic evolution, the complete subjection of women and the extraordinary sexual differences we witness are habits of a "culture" that have been transmitted and, though not easily, are transformable through the development of different social habits. While this interpretation of a neo-Lamarckian principle reflects certain moral and political reasoning that was clearly aimed at undermining the Spencerian "laissez-faire" doctrine, what Ward proposed was not totally the result of wishful political reformist thinking.

The currency of Lamarckian thought around 1900 showed the first inklings of making sense of the notion of "culture" and trying to make such a notion conform to acceptable scientific principles.[27] Since Ward's main thesis was that the principles of social science might be quite distinct, the inheritance of acquired characteristics gave him further ammunition in providing a mechanism for man's mental evolution to higher levels. Furthermore, and not without great consequence for the nineteenth century, neo-Lamarckianism was invoked to explain the differences between the "races" or, as we now would put it, between "cultures." Just why were some "races" in different stages of development than others was one of the central questions to be answered. Neo-Lamarckianism, George Stocking suggests, provided the social evolutionists with the means of keeping in tune with the newly discovered evolutionary principles of biology while providing a principle with which to understand the evolution of humans in society.[28] If the adaptations of parents could be transmitted to their offspring, then differing social characteristics, or cultural habits, between those of the same species would be accounted for more easily.[29] In essence, what the neo-Lamarckians presented was an "environmentalist" thesis recast in evolutionary terms.[30] While the neo-Lamarckian point of view was used to explain various concepts of race, the main point here is that, whatever the political motive, the acceptance of such a principle was embraced partially as a way to establish specific evolutionary principles for understanding humans' mental evolution, and thus to delineate a separable area of inquiry, a "social science." In addition, Ward uses a neo-Lamarckian argument to suggest that evolution must mean progress, that the history of humankind is a history of learning, and, that androcracy *need not* always prevail. He suggests, but by no mean guarantees, that a future stage might be *androgynocracy* – a state in which sexual differences will be minimalized. What the "inheritance of acquired characteristics" provided for Ward, especially dramatized through Darwin's hypothesis of use and disuse, was a *moral* argument for the altering of human habits, in short, the *transformation* of human "culture" by planned human intervention corresponding to articulated social goals.

As I have indicated, the neo-Lamarckian principle which admits the inheritance of the direct effects of the environment, was sanctioned by Darwin himself;[31] it provided an important and needed mechanism for explaining cultural evolution. The demise of this principle among the neo-Darwinians around the turn of the century did not destroy the effort to articulate the notion of "culture" and the effect of the environment on human behavior, what it did destroy instead is the connection of this nascent social science with Darwinian evolutionary principles.[32]

Giving up neo-Lamarckianism meant the destruction of Ward's gynaeco-centric theory which has the virtue, on the one hand, of recognizing that the differences in the sexes are the result of biological evolution without, on the other hand, giving up the powerful environmentalist argument [33] that such differences are surmountable by human design over a shorter period of time.[34] For Ward, therefore, neo-Lamarckianism cast a powerful spell and the power of this spell, although it was extinguished by 1915 in most respectable biological circles, haunted Ward to the end of his life.

Despite the demise of Ward's views, we can appreciate the value of linking the gynaecocentric theory to social evolution. The notion of the female as the perfect, original form out of which the male sexual characteristics developed was based on Ward's acceptance of three basic principles: (1) the "laissez-faire" doctrine was inadequate to explain human mental evolution, (2) the evolution of sexual differences was a result of natural selection, sexual selection, and the inheritance of acquired habits, and (3) the subjection of women is transformable by planned human intervention in accordance with natural and social law.

II

In her review of Ward's *Pure Sociology*, Charlotte Perkins Gilman suggests

that the presentation of woman as the race-type – the original type of life
– will clear "all our dark and tangled problems of unhappiness, sin, and
disease, as between man and woman."[35] While her enthusiasm for Ward's
theory is admiration beyond its worth, this comment marks the deep effect
Ward's theory had on Gilman's imagination. In fact, her two major books,
Women and Economics (1898) and *Human Work* (1904) are extensions and
modifications of Ward's theory. Yet it is also clear that Gilman deepened
Ward's theory in various ways, particularly, and most importantly, in her
analysis of the economic value of women. Ward only mentions this feature
while Gilman expands this analysis in various directions, first, to show how
women's present state is connected to the evolution of sex differences and,
second, to show a way to reform this economic relation that is in accord
with biological and *socialist* principles. While no doubt Gilman is more
hopeful than even her own theory would allow her to be, her critical analysis
of the economic relation of the sexes is telling. First, I will lay out Gilman's
economic foundations.

In her book *Human Work*, Gilman attacks the well accepted Want Theory
(or economic version of "the survival of the fittest"), the thesis that "man
works to gratify wants, and that if his wants are otherwise gratified he will
not work."[36] Gilman believed like all social evolutionists of her day that the
law of development meant that certain social stages must occur in sequence,
and, in addition, that these stages led ultimately from "self-supporting
individualism" (an egocentric system) to a "collectively supporting socialism"
(a socio-centric system). Her argument against the Want theory is simple:
the theory of evolution requires only that individuals try to preserve the
race (the species), not oneself. Work becomes contemptible because it is not
work for the joy of working and serving others, it is viewed instead as a
way of harming others by competing with them. Gilman was soundly convinced that a *proper* interpretation of evolutionary theory discounted the
Want theory of work, as well as the corresponding notion of Supply and
Demand. Work, properly viewed, is a natural human activity which the wage-
labor system has perverted into slavery. Pain and degradation are not *essential
conditions* but are the result of mistaken human action, a failure of humans
to see what is to their benefit. The human species must, in order to fulfill
its organic needs and to survive as a species, construct a society where work
is most efficiently and pleasantly performed. Work, Gilman defined, as the
social expenditure of energy. "The course of evolution," she writes, "has
been to develop more and more complicated instruments for the transmission
of energy."[37] How this energy is best transmitted becomes the foundation

for her claim that women are not only not being fulfilled but also fail to serve the society in the best way. Thus the burden of Gilman's argument lies in her claim that at one and the same time we are *controverting* the laws of evolution and the ethical laws of humanity in our present economic system.

Instead of the drudgery exacted in the wage-labor system, Gilman suggests that

Normal conditions of human work require, first, that the worker shall be well nourished physically and socially, well educated to his fullest height of ability, and well placed in work he likes best and does best. ... A worker, so placed is in no way overtaxing his own energy, but is merely giving expression of social energy, and finds in that process exhaustless joy.[38]

Normal conditions, of course, are not the conditions in which we presently exist, and yet, Gilman claims, we can see, despite the cultural opposition, how we long for such normal conditions.

So irresistable is our growth in this direction that even under all our artificial hindrances, against the combined resistance of religion, tradition, supersitition, habit, custom, education, and condition, still the normal child does want to work, tries to work, and in some cases bursts through the whole cordon of opposition and does the work he is made for... [39]

Even the specialization, the division of labor, and powerful economic development in human society is "ruthlessly degrading and defrauding" the worker. Gilman is arguing that it is the *system* of work, not work itself, that should be condemned.[40]

The dilemma Gilman faces is to show how the present conditions developed as a natural outcome of organic forces while still proposing a stage in the future which satisfies real human needs. Her solution seems to be, as with Ward, that our previous failures to satisfy ourselves occurred because we were ignorant. Consequently, the main way out of our present situation is through education, universally prescribed. "The workman should have such education as shall give him for a background the full knowledge of social evolution; and the special place of his own trade in that evolution ... "[41] It is the human being as maker and doer, as "worker," that is the proper focus of such a study of human evolution. These basic socialist principles are at the heart of Gilman's evolutionary theory of the sexes.

III

In her most famous and influential work *Women and Economics*, Gilman suggests that the economic dependence of women on men is an "unnatural" condition because it does not fulfill the needs of women or of the society as a whole. The economic development of specialization has occurred only through the progressive development and specialization of the male. "This is not owing to a lack of the essential human faculties necessary to such achievements, nor to any inherent disability of sex, but to the present condition of women, forbidding the development of this degree of economic ability."[42] What needs to be explained, of course, is how women came to be in this position.

In a partly humorous yet telling analogy Gilman compares women with horses. Like horses, women are domestic slaves.

> The horse works it is true; but what he gets to eat depends on the power and the will of his master. His living comes through another. He is economically dependent. . . . The labor of women in the house, certainly, enables men to produce more wealth than they otherwise could; and in this way women are economic factors in society. But so are horses.[43]

Household labor (and other labor for low wages a woman might do) is economically profitable. While this conclusion is not remarkable to our ears, especially after the revival of the economic analysis of the role of women, Gilman pioneered such an analysis along evolutionary lines.

At the outset, Gilman is willing to grant the first premise of the argument for male superiority,

> the female of the genius homo is supported by the male whereas in other species of animals male and female alike graze and browse, hunt and kill, climb, swim, dig, run, and fly for their livings, in our species the female does not seek her own living in the specific activities of our race, but is fed by the male.[44]

What Gilman questions, however, is the necessity and universality of this claim. In order to demonstrate her case, she must show that it could be, at least in the future, otherwise. Female dependence, it is often argued, is a result of motherhood. Yet Gilman disagrees, it is not motherhood that binds a woman to her subservient status but the other work that is required of her.

> It is not motherhood that keeps the housewife on her feet from dawn til dark; it is house service, not child service . . . In spite of her supposed segregation to maternal duties, the human female, the world over, works at extra-maternal duties for hours enough to provide her with an independent living, and then is denied independence on the ground that motherhood prevents her working.[45]

With this line of argument Gilman is asserting the possible transformation of sex roles on the basis that such a transformation would be natural while still demonstrating the development of present sexual inequality through the basic principles of evolutionary theory. Gilman's first strategy is to separate the duties of the household – a cultural development-from motherhood – the natural exercising of a certain biological function. The distinction between the genetic and the telic, proposed by Ward, is put to good use by Gilman's analysis of the economic relation of the sexes.[46] By separating duties of the household as a cultural phenomenon, Gilman can then go on to suggest that continuance of the particular economic relation of the sexes found in our stage of development would somehow be *abnormal*, that is, not along the evolutionary lines that will help the species survive.

The gradual development of the masculine and feminine organs and functions cannot of course be denied as a natural development, but Gilman suggests, most provocatively, an "unnatural feature by which our race holds an unenviable distinction consists mainly in this; – a morbid excess in the exercise of this function."[47] The morbid action is the excessive indulgence in the sex attraction beyond the original needs of the organism. The immediately acting cause of this excessive attraction, Gilman proposes, is the wide differentiation between the sexes.[48] The secondary characteristics which, as Darwin indicated, are signs of who will be the better parent and thus function only for reproduction of the species can be "personally unfavorable." When such secondary characteristics are overdeveloped they harm personal development by making the individual conspicuous and render her an easy mark for enemies. Overdevelopment of sexual characteristics, then, can ultimately undermine species development as well. This is precisely what has happened, Gilman proposes, to the human race. "Our excessive sex-distinction, manifesting the characteristics of sex to an abnormal degree, has given rise to a degree of attraction which demands a degree of indulgence that directly injures motherhood and fatherhood."[49] The checks to excessive sex-distinction, Gilman suggests, lie in the basic principle of evolutionary theory, the principle of natural selection. If the sex distinction grows excessive, Gilman argues, then the differences might threaten the survival of the species. "The force of natural selection, demanding and producing identical race qualities, acts as a check on sexual selection, with its production of different sex-qualities."[50] The conclusion which follows from this, Gilman thinks, is obvious. "When, then, it can be shown that sex-distinction in the human race is so excessive as not only to affect injuriously its own purposes, but to check and pervert the progress of the race, it becomes a matter for most serious consideration."[51] So the whole argument of *Women and Economics* is to show how the economic dependence of the female upon the male is a development injurious to the human species.

The major feature of this abnormal development is the dependence of the female on the male as a source for food. The male so "modifies" the female's environment that the sex-attraction primarily used for attracting a mate for reproduction becomes the basis for her individual survival. Because of this dependence, women had developed into "the weaker sex" and a cult has been made of their "femininity," meaning their feeble clumsiness. The dependence on the male for her very life has caused the female to pour everything into this relationship, into "love," so that it overrides all other human faculties she might possess. The exercise of *human* faculties primarily became the province of the male. The active capacity of women for obtaining their own food has shriveled; "the human female was cut off from the direct action of natural selection."[52] The development of skill, courage, and endurance became sex-linked — the development of the species came through the male, the female developing whatever characteristics it took to hold on to him. "With the growth of civilization, we have gradually crystallized into law the visible necessity for feeding the helpless female, ... "[53] Civilization, Gilman argues, has been primarily responsible for the impediment of the human faculties of the female sex. Gilman does not want to deny that male and female children inherit characteristics from both parents[54] yet the sex differentiation that occurs early in childhood, carried through with lack of opportunity and active repression, continually keeps such faculties from developing in the female. "Man is the human creature. Woman has been checked, starved, aborted in human growth, and the swelling forces of race-development have been driven back in each generation to work in through sex-functions alone."[55]

The familiarity of the effects a civilization has upon us is a problem. These practices, being commonplace, bring with them an air of "naturalness," "inevitability," and "normality." In order to make her case, Gilman tries to penetrate through such familiar practices and demonstrate that they are a result of abnormal sex-development. In doing so, she is challenging the sanctity of religious belief, law, and custom. Her favorite institution for attack is marriage. Monogamous relationships are natural for the human species, Gilman suggests, but the way marriage is presently practiced is a violation of its true expression and function. When economic independence for women is proposed, critics claim that it would violate this sacred institution. This reveals, Gilman retorts, the underbelly of this present practice.

Sex-relations are for sale. Love and devotion become commodities for buying, "No wonder that men turn with loathing from the kind of commodity they have made."[56] Yet we approve of this relation in marriage while "condemning it unsparingly out of marriage."[57]

The abnormality of marriage as practiced can be criticized, Gilman argues, on good economic grounds. While the origin of the present marriage relation may have been the result of benefits gained at an earlier time from the specialization of certain functions, the increase in specialization and organization has made such a relation obsolete. Our individual impulses which were developed must now be overthrown by social need.[58] Although many individual women survive and benefit economically from the present marriage relation, in the long run women are deeply damaged by such a relation and so is the rest of society. The severance of sex from economics is, therefore, vital. While the popular mind tends to think this severance destroys marriage, Gilman argues that we would thereby make marriage what it truly is – a sexual relation, not an economic one. So Gilman envisions that

a pure, lasting, monogamous sex-union can exist without bribe or purchase, without the manacles of economic dependence, and that men and women so united in sex-relation will still be free to combine with others in economic relation, we shall not regard devotion to humanity as an unnatural sacrifice, nor collective prosperity as a thing to fear.[60]

This brings us to Gilman's contribution to feminist theory, and only now can we understand what that contribution is. Since economic production is an expression of human energy, in order that women be fully human, Gilman concludes that their abilities to produce must be allowed to develop fully. The power of the women's movement is found in the mutual recognition by women of their real capacities, their development, along side the labor movement, toward collectivity, toward what Gilman calls "the social spirit." "The economic independence of women" will make possible "a higher sex life than has ever yet been known."[61]

To my mind, the central and important feature of Gilman's thought and the source of her biting critique of contemporary conditions comes from the next step she thinks evolution must take beyond women's economic independence, the vision of androgyny. This, for her, means a *lessening* of the sexual differences that are culturally reinforced, a lessening of sexual attraction based on physical allurement, instead, she hopes for a world based on "another love." And to this ideal we now turn.

IV

The most important aim of social evolutionist thought was to provide a mechanism for change that was comparable to Darwin's principle of natural selection. For social evolutionists, like Spencer and Sumner, it was easy — any alleged discovery of a biological principle would satisfy them. Ward and Gilman, in maintaining the separate forces of the genetic and the telic, needed and were able to isolate a separate uniquely human mechanism of change. Gilman's problem in particular was to show why the economic dependence of women upon men was outmoded and how the movement out of this stage might occur. Gilman's suggestion in *Human Work* that human evolution shows a progression from the ego-centric to the socio-centric point of view, clearly, is an argument that the complexity of function which has come with the human adaptation to the environment *requires* a sociocentric point of view. This progressive development, Gilman suggests, already exists in a seminal-ovian form and puts the outmoded nature of the present sex relations into high relief. By citing the women's movement as one indication of this change in consciousness, Gilman demonstrates several important things: (1) while the economic dependence of women on men is outmoded some clearly have a vested social interest in maintaining it and (2) while men have been the leaders so far in changing our civilization this change will come mostly through the development of the powers of women through the increasing social collectivity of work.[62]

The advent of women into the labor force, by necessity, has been accompanied with a growing sense and need of economic independence. But another need has developed as well, the recognition of reproduction as the central focus of human society, and, consequently, mothering as the most important *human* work. Present day mothers cannot really perform the motherly function of educating their children because they do not possess the knowledge to teach. Continually staying at home to perform drudgery undermines mothering.

We are now, Gilman thinks, well beyond the point where humans should be ruled by "instinct." Motherhood is not incompatible with the work of women but rather is enhanced by it. But the raising of the human species is *ultimately* the work of both sexes. In predicting this great future, Gilman writes

We see the mother, the race type, manifesting new faculties, transmitting her faculties to her young, and devising more and more efficacious means to promote that great process. We see the father, reaching race equality at length, contributing more and more of service to the young.[63]

We cannot deny the facts of biological reproduction and that monogamy was a product of this development; yet how we raise children now might be a different question and one of the greatest importance to our civilization. What we need now is "city-mothers" responsible not for their own children but for all children, mothers responsible for what Gilman calls "child-culture."[64] It is not enough to have reproductive control of our own bodies, what women need is social control of related institutions. This kind of control, Gilman argues, requires as wide and extended an education for women as it does for men. As a result,

The child will get a far more just and healthful idea of human relations when he finds himself lifted and led on by a mother whose life has a purpose of its own, than when he finds himself encompassed and overwhelmed by a mother who has no other object or interest than himself.[65]

This mother's new power over reproductive institutions was not so she could control just her own destiny, but the destiny of society itself; mothers would make community in a highly organized society possible. As long as they were responsible for the having and rearing of the children they should be in control of those institutions and others as well. Since Gilman thought developing a "child-culture" the highest human art and science and would considerably alter all "human work," it would be a mistake to narrow our view of her argument.[66]

It was not that Gilman believed in an innate difference of the male and female brain but the modification by sex — especially in the male (of combativeness, for instance,) — must be undermined. "After a few centuries of full human usefulness on the part of the women, we shall have not only new achievements to measure but new standards of measurement."[67] In her fantasy *Herland*, she describes the power of this new "female culture," what a two thousand year inherited experience of mothering could attain for the fully human sense of cooperation.[68] Motherhood is no "cult" in Gilman's argument, but the valuing of reproduction in social life and social theory.

Gilman did not lack faith in women, nor did she reduce women to mother, rather she simply did not trust male power and authority to promote these important values. The female control over reproduction is not only required for women themselves but for the preservation and development of both sexes into full humanity.

The moral power of Gilman's argument, then, is in presenting present evils and in outlining a future moral ideal. The power of women as mothers is required as a guideline for reconstruction of our institutions in accordance with the "evolutionary goal" of survival of the species. So the predominance of "female values" is ultimately in service of an androgynous ideal. "We can make no safe assumption as to what, if any, distinction there will be in the free human work of men and women, until we have seen generation after generation grow up under absolutely equal conditions."[69] The evolutionary thesis that women are the first sex was, for Gilman, primarily a moral, not a factual assertion; our survival depends upon our recognition of motherhood as a primary function of social life for all human beings. Human beings, Gilman believed, have the capacity for warmth, love, and caring – this potential, she also maintained, was compatible and enhanced by the technological and scientific break-throughs of the nineteenth and twentieth centuries.

V

Present day feminists would object to many of Gilman's proposals. Most anthropologists have dismissed the notion of an early matriarchy, some feminists object to any use of "androgyny" as a moral ideal, and still others might object to placement of "mothering" as the central function of a human society. While these are important controversies to be discussed by present feminists, they can not undermine Gilman's important contribution to feminist thought.

The moral ideal of "androgyny" and "mothering" only makes sense in light of other features of Gilman's political philosophy. Her arguments for a new notion of the economic value of women entailed features of a widely accepted line of thought about the functioning of the individual in a social organism. The cooperative, highly organized, almost bureaucratic features of her thought were shared not just by American progressives, but such widely diverse figures as F. H. Bradley, Emile Durkheim, Karl Marx, Sidney Webb, Max Weber and Lenin.[70] The special feature of this thought is its rejection of individualism; the true social life of humans must be attained not in an ordering by contractual obligation of self-interested individuals, but in returning to a sense of "community" within a highly public, bureaucratic, specialized organization.

Although the idea of environmentalism had been around since Hobbes and the Enlightenment, the individualistic nature of Hobbes's thought and the liberals who followed could never fully establish a "science" of culture. Their individualistic articulation of a "political society" inhibited a notion of a communally shared "culture" or "society." Accordingly, feminist thought until this time is highly individualistic and organized around the notion that women ought to be recognized as rational creatures and, consequently, should be accorded all the rights and duties of any citizen. With the advent of the notion of a "culture" or "society" as a highly organized, complex communal structure, the liberation of women could be talked about in a new way. Gilman's gift to feminist thought was in adapting metaphorically this political framework to a theory of social evolution and poltical reform.

Instead of ignoring women's biological and cultural history, Gilman argues that we must take account of it and demonstrate how women's *special* features as "mothers," biological and cultural reproducers of the species, make them the "prime movers" in altering evolution towards our moral ideal. So instead of saying women are "rational" just like males, Gilman instead appeals to us to take a look at the material conditions under which men and women have lived.

Yet the materialist view that Gilman proposes is by no means a reductionist view *nor* a view which completely severs biological processes from cultural processes.[71] Gilman's power as a theorist, then, is found in her relentless assertion that the progress of human civilization depended upon our interest in the reproduction of mothering and the mothering of reproduction.

What are we to make of the historians' charge that Gilman and the Social Darwinists engaged in "pseudo-science" in order to promote their own political aims? Is it just wishful thinking to distinguish Ward and Gilman as the true social scientists because their moral and political ideas are more amenable to our own? Ward and Gilman both argued that their scientific claims were more plausible *because* their political commitments were more well-founded. As we have seen, while such political and moral commitments cannot guarantee a line of reasoning will lead to success, improper moral reasoning can often lead to bad scientific reasoning, as the racial theories of the nineteenth century give ample evidence. While Ward and Gilman had little of what we would now call "scientific" evidence, it would be a mistake to dismiss the neo-Lamarckian line of argument as pseudo-science.

Ward and Gilman, as well as present feminists, do not think the issue is whether moral reasoning is appropriate or inappropriate in the social sciences. In fact, Ward and Gilman, by skillfully using such moral and political reasoning, undermined the assumptions of the other social evolutionists while presenting what was, at the time, a reasonable alternative theory. That Ward and Gilman recognized so early what the invasion of an unacceptable moral theory could do to social scientific thinking should give us pause.

Although we cannot accept an alternative social theory *just because* it has a different, more acceptable moral view, in this case, the moral-political ideal of androgyny made the claim that the abnormal exaggeration of sex was damaging to the human species an interesting and plausible hypothesis.

Ward and Gilman were the first to ask questions of the social evolutionists and the first to understand the patriarchal assumptions involved in their theories. The reemergence of evolutionary questions in anthropological thinking, while most often specifically directed at the adaptation of particular cultures, nevertheless reflects the power of this line of thought that Ward and Gilman foresaw. The androgynous ideal functions as a commitment to recognize evidence of female subordination and to gain full understanding of its causes, while still taking women's equality as a self-evident moral assumption. The feminist commitment, as present feminists agree, suggests questions and evidence of a wholly different sort than previous, androcentric theories would suggest.[72]

Because this is so, what needs to be explained is transformed. Questions such as how women in various and myriad ways express their power and influence in unequal roles could not be asked without such a moral commitment. Moral commitments, then, suggest basic assumptions from which certain *lines* of reasoning become acceptable. The explicit nature of these moral commitments made by feminists in Gilman's time and ours only demonstrates the predominance of androcentric theories. These commitments may be very costly for some social scientists to give up.

What is important to emphasize once again is that the feminist moral arguments act as a principle of exclusion — certain lines of reasoning ought not to be pursued not because they are morally unacceptable (although this is also true) but because they are scientifically fruitless. Any theory which presumes women's innate inferiority or disregards their role is ill-founded *scientifically*. We have specific scientific reasons to doubt its plausibility.[73]

Gilman's social theory, then, is a forerunner to our learning to doubt a sexist social theory and to construct a feminist social science in its place.

Hobart and William Smith Colleges

NOTES

* This paper is dedicated to Janet Braun-Reinitz and Richard Reinitz. I also wish to thank the Women's Studies Program Committee at Hobart and William Smith Colleges, the Mid Western Division of SWIP, and, especially Susanne McNally.

[1] Judith Nies, *Seven Women: Portraits from the American Radical Tradition* (New York: Penguin Books, 1977), p. 127).

[2] William O'Neill, *Everyone Was Brave: A History of Feminism in America* (New York: Quadrangle Books, 1971), p. 130.

[3] O'Neill, p. 39.

[4] Carl N. Degler, 'Charlotte Perkins Gilman on the Theory and Practice of Feminism,' *American Quarterly* 8 (1956), 21–39 and his Introduction to *Women and Economics* (New York: Harper Torchbooks, 1966 [1898]), pp. vi–xxxv.

[5] Degler, Introduction, pp. xxix–xxx. Italics mine.

[6] Degler, Introduction, p. xxix.

[7] Degler, Introduction, pp. xxx–xxxi.

[8] George W. Stocking, Jr., *Race, Culture and Evolution: Essays in the History of Anthropology* (New York: The Free Press, 1968), p. 114.

[9] See J. W. Burrows, *Evolution and Society: A Study of Victorian Social Theory* (Cambridge: Cambridge University Press, 1966).

[10] Henry Steele Commager, ed., *Lester Ward and the Welfare State* (Indianapolis: Bobbs-Merrill, 1967), p. 70.

[11] Commager, p. 75.

[12] Commager, p. 79.

[13] Commager, p. 84.

[14] Lester Ward, *Dynamic Sociology*, Vol. I (New York: Appleton, 1883), p. 69. Italics mine.

[15] Charles Darwin, *The Descent of Man* (New York: Modern Library, n.d.), p. 568.

[16] Darwin, p. 572.

[17] Darwin, p. 901.

[18] Darwin, p. 901.

[19] While Stephen Gould suggests in *Ever Since Darwin* (New York: W. W. Norton, 1976) that Darwin's version of evolution, "descent with modification," eschews any notion of "higher" and "lower," (Gould, p. 36) Darwin clearly violates his own recommendation with respect to the evolution of the sexes. The superior intelligence of the male does not mean here "better adapted" but some more ordinary social sense of superior. "The chief distinction in the intellectual powers is shown by man's attaining to a higher eminence, in whatever he takes up, than can woman – whether requiring deep thought, reason, or imagination, or merely the use of the senses and hand." (Darwin, p. 873).

[20] Darwin, p. 901.

[21] Lester Ward, *Pure Sociology* (New York: Macmillan, 1903), p. 940.

[22] Ward, *Pure Sociology*, p. 372.

[23] Darwin, p. 568. The principle of atrophy, Darwin suggests, applies to all animals.

[24] For an account of the extent and depth of neo-Lamarckian thought see Edward J. Pfeifer, 'The Genesis of American Neo-Lamarckianism' in *Isis* 56 (1965), 156–167.

[25] Darwin, p. 874.

[26] See Pfeifer.

[27] E. B. Tylor was the first to use the concept of "culture" in 1871. Stocking suggests, however, that Tylor's usage has only the vaguest resemblance to the modern pluralistic concept and there is little sense of the notion of cultural transmission.

[28] See especially Stocking's essay 'Lamarckianism in American Social Science 1890–1915,' pp. 234–270. See note 8.

[29] Of course, the hierarchical structure and progressive development was a component of this argument but not a necessary component, and, therefore, we will set the issue of racism aside for the purposes of this essay.

[30] See Pfeifer.

[31] Darwin, p. 919. The role of the transmission of acquired habits is not at all clear in Darwin's writings. He did think it played some role in human evolution.

[32] Stocking, p. 268.

[33] See Stocking's essay 'Franz Boas and the Culture Concept in Historical Perspective,' pp. 195–234.

[34] Pfeifer gives an account of the uniformitarianism vs. catastrophy theories in the light of neo-Lamarckianism.

[35] Charlotte Perkins Gilman, *The Forerunner* 1, No. 12 (1910), p. 26.

[36] Gilman, *Human Work* (New York: McClure, Phillips & Co., 1904), p. 65.

[37] *Human Work*, p. 187.

[38] *Human Work*, pp. 187–8.

[39] *Human Work*, p. 195.

[40] *Human Work*, pp. 285–6.

[41] *Human Work*, p. 257.

[42] *Women and Economics*, p. 9.

[43] *Women and Economics*, pp. 7, 13.

[44] *Women and Economics*, p. 18.

[45] *Women and Economics*, pp. 20–1.

[46] *Women and Economics*, p. 23.

[47] *Women and Economics*, p. 30.

[48] *Women and Economics*, pp. 31–2.

[49] *Women and Economics*, p. 33.

[50] *Women and Economics*, p. 36.

[51] *Women and Economics*, p. 37.

[52] *Women and Economics*, p. 62.

[53] *Women and Economics*, p. 63.

[54] *Women and Economics*, pp. 69–70.

[55] *Women and Economics*, p. 75.

[56] *Women and Economics*, p. 98.

[57] *Women and Economics*, p. 97.

[58] *Women and Economics*, p. 106.

[59] *Women and Economics*, p. 107.

[60] *Women and Economics*, pp. 115–116.

[61] *Women and Economics*, p. 143.

[62] Contrary to O'Neill's claim, Gilman's feminism is saturated with her Socialism.

[63] Gilman, *His Religion and Hers* (Westport, Conn.: Hyperion Press, 1976, rep. 1923), pp. 240–1.

[64] Gilman, *Concerning Children* (Boston: Small, Maynard, 1900).

[65] *Concerning Children*, p. 197.

[66] Linda Gordon in *Woman's Body, Woman's Right* (Baltimore: Penguin Books, 1976) tends to lump Gilman with those who promoted a "cult of motherhood" and suggests that evolutionary arguments at this period undermined feminism. I hope this paper shows that Gilman's version of the evolutionary arguments, at the very least, minimized such an undermining. (See Gordon's Section Two: 'Toward Women's Power').

[67] Gilman, *The Forerunner* 3, No. 9 (1912), p. 249.

[68] Gilman, *Herland* (New York: Pantheon Books, 1979).

[69] Gilman, *The Man-Man World or Our Androcentric Culture* (New York: Charlton Co., 1911), p. 250.

[70] Sheldon Wolin, *Politics and Vision* (Boston: Little, Brown, 1961), chp. 10.

[71] *Herland*, p. 78. Gilman is perfectly aware of the acceptance of Weissman's views in the scientific community and seems more willing to accept them than Ward.

[72] Reiter, Rayna, R., *Toward an Anthropology of Women* (New York: Monthly Review Press, 1975) and Rosaldo M. and Lamphere, L., eds., *Women, Culture, and Society* (Standford: Standford University Press, 1974).

[73] Dudley Shapere, 'The Character of Scientific Change,' in T. Nickles, ed., *Scientific Discovery, Logic, and Rationality* (Dordrecht: D. Reidel, 1980), pp. 61–116.

LOUISE MARCIL-LACOSTE

THE TRIVIALIZATION OF THE NOTION OF EQUALITY

The general claims of this paper are that an epistemological analysis of feminism is culturally necessary and that reactions to feminism are epistemologically determined. The issues to be discussed may be summarized by the following questions: in what sense is it accurate to say that feminist writings are a repetition of men's writings; even supposing that they are, for what reason should this be given as so powerful an objection to feminist writings; and finally in what sense is it necessary for women to go on "validly" repeating men? As I hope will be shown, the answers to these questions are much more complex than one would gather from the argument that the definition of women's identity has always been made by the concept of resemblance to the oppressor.

To begin with a simple presentation of this topic, let us consider a dual set of commonplace reactions toward the "plethora" of feminist writings. A first reaction is that feminist writings are quite tedious because they are repetitive. In this context, the fact that feminist writings would presumably say the same thing over and over again is not taken as a *prima facie* clue to the universality and perpetuity of the issues involved in feminism, thereby to its seriousness. Rather, it is taken as a proof of futility.[1] Another reaction is that feminist writings are not only quite varied but deeply contradictory. The fact that feminist writings would presumably not say the same thing is not taken as a *prima facie* clue to the complexity of the issues involved in feminism, thereby to its heuristic interest. Rather, it is taken as a proof of self-denying invalidity.[2]

The present analysis will focus on the sense in which these apparently disconnected reactions are epistemologically related. In the first case, it is held that we know that feminism is against sexism. In the second case, it is held that we know that a concern to avoid sexism amounts to a failure to provide conclusive arguments for an acceptable alternative. In both cases, it is assumed that nothing really and significantly new is to be expected from feminist writings.

But if we analyse this view and particularly the notion of novelty with which it is connected, we realize that it rests on a very puzzling assumption: either feminist writings are ways of repeating men, or else what they say is of

121

Sandra Harding and Merrill B. Hintikka (eds.), Discovering Reality, 121–137.
Copyright © 1983 by D. Reidel Publishing Company.

no primary importance. Some illustrations of this charge may be derived from conflicting ways of defining what a feminist writing is. When feminist writings are defined as those whose author is a woman, the novelty of the fact that women write books is easily reducible to the fact that they do now what historically only males did: they become authors.[3] Feminist writings may also be defined as those whose topic is mainly about women, no matter who the author is and regardless of the stand taken on women's liberation. However complicated this issue, it is easy to interpret it as a repetition of males' writings on a live issue which is in vogue. A more subtle way of making the repetition charge may be derived from defining feminist writings as those in which the woman issues are not merely a topic, but the issues in which one has a stake. The charge here could be that women borrow from scientific models, presumably defined by males, whatever they deem worthy of application to the case of women. Finally, against feminist writings defined as those which explicitly promote women's liberation, the charge may be that, though radical, such writings are repetitive because they refer to already known utopias or models of reform or revolution. If greater room is made for novelty, such writings may be given as examples of what Hegel has called "the inverted world". They would propose the same models as males but would rather have women at the top. However dialectically construed, this description is easily reducible to the fact that women come to replicate whatever evils were already committed by men.[4]

Considering the above examples of the repetition charge, one may be inclined to dissolve them in yet another and more general pattern of repetition. One may argue that it is a feature of all publications to be highly repetitive and that the cases of genuine novelty are quite rare. As Whitehead has said: All philosophy is a footnote to Plato. However, we shall here consider the repetition charge from the point of view of its frequency and specificity when it is applied to feminist writings. A first thing to notice is the systematic reduction of the notion of novelty which the above examples imply. Novelty is indeed reduced to the author, whatever the content of the writing. The content is reduced to a topic as obsolete as it is vague. The live issue is reduced to a borrowing of models and, finally, the radicalness of proposals is reduced to a mimic of males' evils. All these reductions amount to saying that somehow feminist writings are on a par with any other writing, but for one feature. Specifically, they are allowed to share whatever *uninteresting* features our ways of practicing knowledge may have. This, one may submit, is our epistemological "repos du guerrier", or the epistemological version of a form of social sexism where women are recognized every right to share with males everything that fails.[5]

In the attempt to assess the descriptive validity of the repetition charge, the first difficulty therefore is to overcome an *a priori* reduction of the possible novelty of feminist writings. This reduction, however, is not as easy to counter by reference to the content of feminist writings as one might imagine. The reason for this is that when faced with the question: what is new in such writings? the answer would seem to lie in secondary, *ad hoc*, empirical information. For example, in a recent book which has been applauded for its novelty, T. Boslooper and M. Hayes attempt to show how women have been excluded or discouraged from sports, this being a paradigm of *The Femininity Game*.[6] In making a case for the "true" nature of women as strong, competitive, aggressive, and in thereby denouncing the physical and psychological biasses on women, the authors provide interesting statistical data and historical corrections to some myths like the myths of the Amazones and of Atalanta. However, one may argue, this book does not seem to add much, say, to Simone de Beauvoir's *Le deuxième sexe*. One may further argue that Simone de Beauvoir did not make much of a discovery herself in showing that many prohibitions (e.g. not to climb trees, not to run) had philosophical preconceptions at their origin, as well as philosophical implications for the status of women.[7]

In other words, at least from a certain way of looking at feminist writings, it would seem that the novelty they introduced is local or incidental, and that their addition is incremental in a rather non-significant way. At best, such writings would provide quantitative growth on second order issues of knowledge.[8] As with French grammar, where an "expletive" expression merely fills the gap in a sentence without being necessary to its meaning, feminist writings would fill a gap in the general sentence of our culture without being necessary to its meaning. They would thus have the "book-keeping" feature that I. Levi ascribes to local induction: a reiterated way of using a rule to answer the same question over and over again, adding still more information to the question.[9]

The information, however, cannot be explained by means of the Surprise Thesis of information.[10] Indeed, when a good case for equality is found in feminist writings, the purport of the conclusion would seem to be an invitation to imitate men. Thus, Boslooper and Hayes conclude their analysis in saying that "In order to win the war and end the Femininity game, women will have to play, for the time being at least, a man's game". Similarly, Simone de Beauvoir concludes that women should quit the "immanent" role ascribed to them and rather endorse the "transcendent" role ascribed to men.[11]

This description of the conclusions reached in feminist writings may appear simplistic. It leaves aside whatever the arguments for equality might have been and whatever qualifications were added to the conclusion. For example, the conclusion of Boslooper and Hayes is qualified by saying that women should work within, but not for, the existing system. Specifically, they must not conceive victory in males' terms. One might be at a loss to determine what kind of a game women would play if in playing a male's game they refuse victory in males' terms. But what needs attention here is that except for such undetermined possibilities, the claim for equality carries the notion of imitation in what may be termed the "main" contribution of feminist writings and that it is in relation to *such* contributions that feminist writings are ascribed an expletive status. It would thus seem necessary to probe the repetition charge further in raising the question whether, except for side issues, feminist writings are described as repetitive for lack of awareness of the novelty involved in their treatment of "broad" issues.

A revealing way of answering this question may be derived from feminist writings known as "women studies" or "women scholarship", where the analysis of the women issues is done through different models provided by sociology, economics, political science, anthropology, history, philosophy, etc. The specific question here is whether such writings could be said to offer a new form of "rationality" or new models, methods of inquiry, systems, etc.[12] The answer to this question may be summarized in saying that in general, such writings are "downstream" from already given forms of rationality rather than sources from which altogether new forms of rationality flowed. Typical of such writings is the fact that rationality itself is at stake, rather than directly and explicitly challenged in its most general sense. Usually, one form or another of rationality is applied to the question of women itself seen as a case of the former.[13] This relationship holds, I think, for writings denouncing the "imperialism" of reason, logic, or rationality taken as rationalism. There, the question of women is held to present a good case of, or a good argument for, revolutionary claims about anarchy, permanent revolution, ecstatic experience, radical freedom, the value of the irrational, etc.

Paradigmatic of this kind of relationship between rationality and femininity is the example of recurrent debates over the notion of reason as opposed to those of feeling or intuition when such notions are applied to women.[14] In feminist writings, attempts will be made to show that it is erratic to define women's reason by means of feeling or intuition; or else, it will be argued that feeling or intuition, ascribed to women, should be given a higher epistemological status than it has in rationalistic theories; or, finally it will be maintained that the distinction itself is groundless or unapplicable.[15] These three arguments are not identical, yet they all imply that the issue involved in feminism is defined on the basis of one form or another of rationality, itself defined regardless of femininity.

To describe in more precise terms the type of critical analysis which is done in feminist writings of an academic type, one may use the metaphor of the river again and say that it involves a "downstream/up-stream" movement. One form or another of rationality is applied to the women question and then presuppositions, implications, criteria, or corrections to the models are delineated. This description may be confirmed by considering the writings of the great theorists of feminism. Thus, one may say that *Le deuxième sexe* by Simone de Beauvoir is "downstream" existentialist ontology; that B. Friedan's *The Feminine Mystique* is "downstream" *gestalttheorie* as a correction to "functionalism"; that *The Dialectic of Sex* by S. Firestone is a "downstream" Marxist universe, as corrected by a Freudian-Reichian universe; that K. Millett's *Sexual Politics* is "downstream" the Weberian model of *Herrschaft*; that L. Irigaray's *Speculum de l'autre femme* is "downstream" Lacanian psychoanalysis; etc.[16]

It should be noticed that in reference to a given model, not all feminist writings consider the question of applicability in the same way. Significant here is the way in which the negative dimension of the women's situation is related to the model of argumentation. Thus, de Beauvoir wants to show how ontological categories such as "being for oneself" and "being for another", with the particular inauthenticity of the latter, *explain* the very consistency of the universal but immoral application of the category of "being for another" to women. Friedan wants to show *why* liberal ideals of equality have not been applied to women: the obstacles are summed up in the "feminine mystique". Firestone wants to show how Marxist and Freudian models *could* be applied without sexism. Millet wants to show *that* male power is consistently applied to women even in writings that would seem to be mere fictions. Irigaray wants to show *that* philosophy itself is a phallocratic application of the males' inability to consider women as different. Etc.

In a similar manner, one should recognize that the attitude toward the model varies. For example, Friedan and Millet do not attempt to correct their respective model. Friedan values the ideals of the American revolutionists and hopes to promote their application to women. Millet holds that Weber's model of *Herrschaft* is adequate in order to describe the situation of women as it really is, though it is not a morally justified model of human relationships. On the other hand, while de Beauvoir, Firestone, Irigaray provide corrections to their respective models, these are used as a basis for reconstruction once the women condition has been rigorously analysed.

But however different such references to a model and its applicability to the case of women can be, we may safely conclude that feminist writings of an academic type refer more or less creatively to already given forms of rationality, rather than offer new forms of rationality. However, in analysing the "downstream/up-stream" mode of inquiry found in feminist writings, one detects the presence of three basic epistemological categories (historicity, materiality, values) by which they can be seen as, at least, announcing new forms of rationality.

Historicity is introduced by the fact that feminist writings do critically refer to already given forms of rationality. This implies that our ways of reasoning have historical attributes which must be incorporated in our models of valid thought, and that epistemological models must be seen as including what E. W. Adams has called "consequence beliefs" or the entire field of practical influence beliefs may have in terms of consequences.[17] Materiality is introduced in feminist writings by the fact that they introduce the women condition as the issue, with identifiable though complex concrete factors. This indicates that the most important field of analysis is neither formal, nor empirical. Rather, it lies in the relationship between rules and facts, formal properties and quantifiers, empirical generalizations and "ultimate partition" or counter-examples, etc.[18] Finally, values are introduced in feminist writings by the fact that they have a stake in the women issue. Rather than persistently arguing that knowledge is value-laden, feminist writings attempt to show how, when, and why such and such a value plays such and such a role in the unfolding of such and such an epistemological model. This process includes the recognition that values are involved in the very choice among alternative theories, a normative act implying a conscious selection of a certain way of understanding and interpreting the world.[19]

In other words, feminist writings may be described as actually practicing what P. K. Feyerabend has described as "science at its best", the practice of research defined as an interaction between theories stated in explicit manner and older views which have crept into the frozen observation language.[20] One may also describe feminist writings as providing the basis for a Kuhnian puzzle-solving tradition in matters related to historical, material, and axiological attributes of models. The feminists' attempt at articulation between a constellation of beliefs and significant, if not shared examplars, would thus be accountable within the notion of the growth of knowledge.[21] In any case, in introducing historicity, materiality, and values as fundamental epistemological categories, feminist writings represent a forceful challenge to critical thought seen as a formal and meta-discourse.

An important consequence of this description of the novelty of feminist writings on "broad" issues is that it provides a basis for refuting the expletive view. It shows indeed that in missing the crucial import of such categories as historicity, materiality, and specific values, one misses the extent to which the feminists' analyses of the women question provide not merely cases of already given forms of rationality, but also and more importantly crucial elements of test-cases to the theories themselves. For example, the model of piecemeal knowledge-formation which is certainly used in many feminist writings provides important elements of a test-case of, say, Popperian epistemology. Questions related to feminism are certainly part of this "background knowledge" whose every bit is and must be open to critical analysis and falsification though, according to Popper, *only* in a piecemeal way.[22] For example again, feminists' analyses of material conditions and empirical situations could be treated as paradigmatic cases on which to construe the conditions of what E. W. Adams calls "partial rationality" or the special circumstances in which reasoning in real life which appears to be probabilistically unsound may be shown to have a sound pattern of inference.[23] Or else, issues involved in the analysis of women as "creatures of fate" ('Cinderella as a Winner') do appear as formidable puzzles to be solved in any attempt to evaluate in utilitarian terms the "lucky estimate" model of (reasonable) probability and the conditons of felicity in using a random process in relating long range and short range values.[24]

In other words, the question is: what prevents us from seeing that while models could and should be applied to the women question and thereby help to clarify the feminist issues, the analyses of the women condition in turn could and should be applied to models and help clarify epistemological issues? I submit that the main epistemological reason for this lies in the assumption that whatever *valid* claims, models, or criteria are used in feminist writings, they are doomed to be a repetition of men: they will be undistinguishable from what is validly said by men *qua* human beings. What needs emphasis here however is that the notion of epistemic validity in relation to conditions which must be true universally and specifically regardless of sex *determines* the notion of novelty applied to feminist writings and thereby the repetition charge.

The argument may be summed up as follows. Because validity is defined in sexually neutral terms, the condition of novelty applied to feminist writings is that a significant addition to our corpus of knowledge must include the production of an altogether new form of rationality, in the same manner that in political terms, a significant sexual revolution would require the creation of a new human nature; otherwise, it is said, women repeat males' models of revolution.[25] That this condition, however, is sexist may appear in analysing its self-defeating character. Because epistemic validity is defined in sexually neutral terms, any alternative to our ways of practicing knowledge must also hold regardless of sex. At the same time, any feminist alternative must be shown to have been defined in such a way that only women could have done so.[26] Thus, in order to be valid, feminist writings must enunciate their claims in due order; but when put in this sexually neutral order, it is no longer relevant to know what this order says about women, nor what the introduction of the issues of feminism says about it.

The tricky thing here is that for a feminist writing to offer a significantly new item in our corpus of knowledge, it has to search in the line of "incommensurable theories", while the request for admittance is one of formal verisimilitude.[27] Another tricky thing is that in order to provide significantly new items, feminist writings must focus on the broadest epistemological issues possible, while the actual content of these writings is systematically given a merely expletive function. Finally, in disconnecting broad issues and specific issues, one further makes their potential connection irrelevant as long as the actual puzzle-solving attempts do not lead to an altogether new logic.[28]

One way to confirm this interpretation of the origin of the repetition charge is to notice that it is systematically ambiguous. It covers any meaning to be given to the notion of repetition, such as saying the same thing over and over again, presenting a duplicate of something, offering a pattern uniformly repeated over a surface, providing a fresh supply of goods similar to those already received, offering a mimicry inverted or not, seeking again, causing to appear, demanding the restitution of something, etc. What is significant here is that, from the point of view of a sexually neutral notion of validity, all these meanings of "repeating" men are systematically identified by retreating into formalism.

Furthermore, implicit in the sexually neutral notion of validity is the idea that a statement is valid by virtue of its silence on sexual variables. Silence is thus here an epistemological imperative and plays the role of what I. Lakatos calls a "negative heuristic", the paths which must be avoided in research.[29]

It follows that sex-laden statements are ascribed a zero probability which implies that any reference to sexual differences is strictly irrelevant. A good example of this problem is the charge raised against writings attempting to show the extent to which former ontologies were sexist. As P. K. Feyerabend puts it: "when a faulty ontology is comprehensive . . . then every description inside the domain must be replaced by a different statement or by no statement at all". The feminists' attempt to correct faulty ontologies would thus be of the former kind, while the objection to such attempts especially in terms of lack of interest would be of the latter kind. By the imperative of silence, however, it is the "no-statement-at-all" alternative that must be chosen, a point which may explain why it is so easy to find cases of sexism in the history of ideas, while it is so difficult to design procedures in order to avoid justifying it. In turn, the imperative of silence explains why an expletive function is ascribed to feminist writings in terms of concreteness, because "with a view of science as a purely abstract notion, it is left to us to fill it now with this, and now with that concrete content".[30]

The most important point here is not that feminist writings do fill abstract concepts with concrete contents. The most important point is that by the imperative of silence, the delineation of well-designed procedures to see where the limits of abstract reasoning are situated in such matters is made impossible, let alone desirable. As a result, the claim that in repeating men, women would repeat men *qua* human beings or *qua* males is systematically unverifiable.[31] H. Törnebohm wrote that: "The immunity from refutation of once confirmed hypothesis in a piecemeal knowledge-formation brings about that this long trend is to be expected [i.e. no risk of refutation], provided that no attempts at systematization of accepted pieces of knowledge is made".[32] In our case, by the rule of silence, any reference to sexual variables is secondary and has only a piecemeal character; on the other hand, the attempt to produce a systematization is determined by the request to produce a new "feminine" logic which would hold regardless of sex; thus, no risk of refutation of the repetition charge can be present. Accordingly, the objection that "everything becomes sexist" with feminist writings is not only an unverifiable statement: it is also a form of intellectual sexism when it derives from a refusal to set up ways of deciding about this issue. In other words, the gist of the imperative of silence is not so much the claim that in a sexually neutral notion of validity males and females are indistinguishable. Rather, it is the claim that for all epistemologically valid purposes it must remain so.

The consequences of the imperative of silence, especially with respect to the limits it imposes on the growth of knowledge, are numerous. For our purpose, it will be sufficient to underline its role in two typical, but misguided, attempts to overcome the problem of knowledge defined in sexually neutral terms. These attempts are misguided to the extent that they are determined by the self-defeating scenario which the imperative of silence imposes on feminist studes in terms of novelty. The first attempt to solve this dilemma is to deny the possibility for any human being to pronounce any statement *qua* human being: one would thus judge *qua* male or female. Yet, unless one is willing to dismiss the universality and validity of, say, logical, mathematical, necessary, or analytical truths, unless one is willing to preclude the possibility of women's access to such truths, or the possibility of using them when discussing about feminism, this argument turns against itself. An example of the problem here involved may be found in recent debates over sexism in logic textbooks. It is certainly possible to substitute Xantippe for Socrates in the famous syllogism about humans being mortal. Such a substitution is surely culturally revealing as are recent attempts to de-masculinize expressions in ordinary and technical language.

Yet, as far as the argument goes with respect to the repetition charge, this substitution leaves the validity of the syllogism unaffected. This is exactly the point made *against* feminist writings by means of the repetition charge. On the other hand, as the feminists here argue that the so-called neutral components of knowledge are already sexist in that "men" actually means "males", then the substitution of Xantippe for Socrates must be reckoned as a repetition of males in an inverted form.[33] In both cases, however, it is the assumption that a significant addition to logic would require the production of a feminine sexually neutral logic that explains the futility attached to such attempts.

The second misguided attempt to counter the repetition charge is not to deny it but rather to appeal to neutral components of knowledge. The point is to dismiss as inadequate the switch from men to males or females. In this view, the feminists' appeal to neutral components of knowledge is no more a matter for them of repeating men than it is for males a matter of repeating women. The interest of this view lies in the postulate that if we are consistent with a sexually neutral notion of validity, then we all talk *qua* human beings.[34] The problem however is that the inadequacy of the switch from men to males or females does not imply irrelevance for any statement implying a reference to non-sexually neutral components of knowledge.[35] The fact that knowledge does include sexually neutral items is so far, here, from the solution that it is precisely the beginning of the problem.

Clearly, something in our account of neutrality itself is at stake when admitting both the existence of sexually neutral components of knowledge and the issues of sexism. One would have assumed that consistent with the neutrality thesis, the issues of pseudo-neutrality would be given the status of a "positive heuristic" or of the paths which must be pursued in research. But this is not the way that the feminists' questions about neutrality is understood. We rather have a case of what I. Levi described as the "infallibility" attitude toward items in a corpus of knowledge: unconcerned with the "pedigree" of this conviction about neutrality, one is far less interested in systematic counter-examples such as those present in certain forms of the repetition charge.[36]

Thus, noticing that Simone de Beauvoir's account of the woman's condition is "downstream" existentialist ontology, one assumes that for all philosophically important purposes, de Beauvoir repeated Jean-Paul Sartre. At best, she would have added a few illustrations to the Sartrian thesis and her writings would have an expletive function in the history of existentialism. Against this view, it has been argued that Sartre was sexist and that de Beauvoir has corrected the implications of his philosophy on this point: she was then repeating man. On the same basis, it has been argued that in referring to Sartrian ontology, de Beauvoir cannot escape sexism herself: she was then repeating the male.[37] Yet, we still lack a systematic study of the crucial issue here: the epistemological changes that de Beauvoir has introduced in the existentialist ontology because of her focus on ethics and its material conditions as revealed by the problem of sexism. In other words, suppose that the notions such as "being for oneself" or "being for another" and such existentialist notions as freedom or absurdity *are* sexually neutral items in a corpus of knowledge. Then the question is: what is being changed in these notions and in the model of explanation related to them once the male/female distinction is not denied and the issues of sexism are discussed? The next question is: what do such changes reveal about the neutrality thesis itself?

To be clear, the extent to which there is epistemological novelty in applying theories or models to a new case, here the case of women, is a matter of analysis. The extent to which there is novelty in the theory or the model itself once corrected according to this new application is also a matter of analysis.[38] But it is only in making such specific analyses that we shall be able to design procedures by which the neutrality thesis could be tested, thereby the claim that women repeat men could be falsifiable. For all these conditions to obtain, the "ultimate partition" of the imperative of silence must be rejected as a clear case of trivialization of "answerhood" in matters related to the neutrality thesis.[39] This trivialization arises because the class of well-formulated questions in issues related to feminism is restricted to one: whether there is and should be any neutral component of knowledge. The potential answers to this question are reduced to one positive answer. Accordingly, the *further* questions raised by feminist writings, especially concerning neutrality, can neither be closed, nor open questions. They cannot be closed questions because there is a clear restriction on the ranges of variables, the set is "humanhood". They cannot be open questions because the rule of silence prohibits the search of effective methods for enumerating or building up the potential answers to any given questions. It is, ultimately, a no-question problem.

To conclude on the questions raised in the beginning of this paper, it is in its most formal sense that it is accurate to say that feminist writings repeat men. But the sexually neutral formalism of this answer rests on the imperative of silence which, as a negative heuristic, makes the repetition charge strictly unverifiable and unfalsifiable. Nonetheless, the repetition charge is a more serious charge against feminist writings than it is against any other because the imperative of silence requires that the novelty to be sought in feminist writings correspond to the production of an altogether new "feminine" *and* sexually neutral logic, an altogether new epistemology and, by implication, a new human nature. It follows that the repetition charge against feminist writings is systematically misleading in that the male/female distinction is made irrelevant by the very appeal to what may be termed "epistemic equality". But since the ways we have hitherto appealed to sexually neutral components of knowledge have included as many ways as possible to justify sexism, and since, on the other hand, feminist writings call for important revisions in our ways of practicing knowledge, in particular in considering the notion of validity as including not only epistemic conditions, but also epistemic consequences and utilities, we conclude that it is necessary for women to go on "validly" repeating men. It is necessary to unfold the theoretical and practical conditions and implications of a sexually neutral notion of validity, if only to determine when and how neutrality becomes pseudo-neutrality.

As things turn out, it could be that the main reason why an epistemological analysis of feminism is culturally necessary lies at the cross-roads of the prescribed "undistinguishableness" of the notion of validity and the search for valid ways of analysing the issues of equality in moral, psychological, social, economic, political terms. It has been the contention of this paper that the logical force of the argument from repetition rests on an unacceptable assumption. Let us add: by means of the repetition charge, one further has a logical instrument for denying the issue of equality, either as a genuine or as a solvable problem. One also has a logical instrument for reducing the issue of equality to an already given, recognized and therefore trivial formal principle. But in order to find ways by which the appeal to epistemic equality will not degenerate into a trivial way of justifying any sort of inequality, we shall have to substitute for the imperative of silence a commitment to free knowledge itself from sexist assumptions. To proceed to this "revision of our credal state",[40] we shall need more than formal assertoric logic. We shall also need to delineate a logic of questions and a logic of discovery.

Université de Montréal

NOTES

[1] On this, see V. L. Bullough, *The Subordinate Sex, A History of Attitudes toward Women* (New York: Penguin Bk, 1974, esp. pp. 336ff) and M. Ellman, *Thinking about Women* (New York: Harvest Bk, 1968).

[2] On this, see G. G. Yates, *What Women Want, The Ideas of the Movement* (Cambridge: Harvard U.P., 1975) and J. Hole and E. Levine, *Rebirth of Feminism* (New York: NYT Quatrangle, 1975).

[3] This definition is tautological and nominalistic only in appearance. It covers H. Morgan's *Total Woman* (traditional), M. Cardinal's *Des mots pour le dire* (radical) and academic writings where no mention is made to the question of women (taken as a proof of intellectual emancipation). This definition raises the issue of whether a male might produce an adequate account of women and whether it is sufficient to be a woman to do so. This also raises the issue of "outsider/insider" in sociology of knowledge; on this see E. E. Almquist, 'Women in the Labour Force', *Signs* 3 (1977), 843–855, 871.

[4] For this type of reduction, see S. Lilar, *Le malentendu du deuxième sexe* (Paris: P.U.F., 1970), N. Mailer, *The Prisoner of Sex* (Boston: Little, Brown, 1971), R. R. Barber, *Liberating Feminism* (Boston: Delta Bk, 1976). Notice that in copying males' evils, women are less justified than males; on this, see H. Marcuse, *Counterrevolution and Revolt* (Boston: Beacon Press, 1972) and J.-P. Sartre, in 'Simone de Beauvoir interroge Jean-Paul Sartre', *L'ARC* 61, 3–12. On this question, see also H. R. Hays, *The Dangerous Sex: The Myth of Feminine Evil* (New York: Putnam's Sons, 1964), F. d'Eaubonne, *Histoire et actualité du féminisme* (Paris: Alain Moreau, 1972), D. Paulme, *La mère dévorante* (Paris: NRF Gallimard, 1976), C. Garside, 'Women and Persons' in M. Anderson (ed.), *Mother Was Not a Person* (Montréal: Content Publ. Comp., 1972, pp. 194–204).

[5] See C. Rochefort, *Le repos du guerrier* (Paris: Grasset, 1958), S. de Beauvoir, *La femme rompue* (Paris: Gallimard, 1965), H. Maure, *L'amour au féminin* (Paris: J'ai lu, 1973). Notice that to the possibility of sharing what fails is related the possibility of

sharing eveny dream about what life could be (as in J. J. Rousseau's *La nouvelle Héloïse*) as well as "compensatory" theses such as women's lesser intellect compensated by their greater moral qualities or women's exclusion from politics compensated by their "informal" power. On this, see A. B. Muzzey, *The Young Maiden* (Boston: W. Corsby, 4th ed., 1943), S. P. White, *A Moral History of Woman* (New York: Doubleday, 1937), C. J. Furness, *The Genteel Female: An Anthology* (New York: Knopf, 1931), M. R. Beard, *Woman as a Force in History* (New York: Macmillan, 1946), C. Garside, 'Good and Evil for Women', in J. Goldenberg and J. Romero, 'Women and Religion' (*Proceedings of the American Academy of Religion*, 1973, pp. 104–127).

6　Published in New York: Stein and Day, 1973. See also J. I. Roberts, *Beyond Intellectual Sexism, A New Woman; A New Reality* (New York: D. McKay Company, 1976) and S. Hammer (ed.), *Woman, Body and Culture* (New York: Harper & Row, 1975).

7　*Le deuxième sexe* (Paris: NRF Gallimard, 1949, esp. Vol. II, Part I).

8　On this, see H. Törnebohm, 'On Piecemeal Knowledge-Formation', in R. J. Bogdan (ed.) *Local Induction* (Dordrecht: D. Reidel, 1976, pp. 297–319, esp. p. 317). Attempting to explicitly assume this "piecemeal" condition are, *e.g.* the writings of D. Boucher and M. Gagnon, *Retailles, Complaintes politiques* (Montréal: L'Etincelle, 1977).

9　I. Levi, 'Acceptance Revisited' in Bogdan's *Local Induction* (*op. cit.*), pp. 1–73, esp. p. 43 and I. Levi, *Gambling with Truth* (New York: Knopf, 1967).

10　See G. Menges & E. Kofler, 'Cognitive decisions under partial information' in Bogdan's *Local Induction* (*op. cit.*), pp. 183–189, and references to Carnap, Hintikka, Popper, Levi, etc.

11　*Le deuxième sexe* (*op. cit.*), esp. general introduction and Vol. II, Part III; *The Femininity Game* (*op. cit.*), pp. 101–115; 183–184. See also C. Valabrègue, *La condition masculine* (Paris: Payot, 1968).

12　More detailed analysis of this question to be found in my 'Féminisme et rationalité', in *Rationalité aujourd'hui, Rationality Today* (Ottawa: Presses de l'Université d'Ottawa, 1979), pp. 475–484.

13　In a sense, there is one exception to this claim in the writings of L. Irigaray, *Speculum de l'autre femme* (Paris: Minuit, 1974) and *Ce sexe qui n'en est pas un* (Paris: Minuit, 1977): in denying the applicability of Freudian-Lacanian models to the case of women, Irigaray challenges the notion of rationality in its most general sense.

14　Recent debates in K. Stern, *The Flight from Woman* (New York: Farrar, Strauss & Giroux, 1965), J. Stoller, *Sex and Gender: On the Development of Masculinity and Femininity* (New York: Science House, 1968), A. S. Rossi, 'Sentiment and Intellect: The Story of John Stuart Mill and Harriet Taylor Mill', *Midway* 10 (1970), 29–51.

15　These typical arguments (1. counter-example; 2. hierarchy of criteria; 3. unapplicability of the model) may be related to typical levels of analysis in women's studies (1. discovery of new phenomena; 2. re-interpretation of traditional notions; 3. demonstration of the relevance of traditional theories). See M. B. Parlee on psychology in *Signs*, Vol. I, no. 1, Fall 1975, pp. 132–138 and K. Boals on political science in *Signs*, Vol. I, no. 1, Fall 1975, pp. 161–174. Notice that the chosen model of analysis may be taken as an open model; *e.g.* C. Broyelle, *La moitié du ciel, le mouvement de libération des femmes aujourd'hui en Chine*, Paris: Denoël-Gonthier, 1973.

16　S. de Beauvoir, *Le deuxième sexe* (*op. cit.*); B. Friedan, *The Feminine Mystique* (New York: Norton, 1963) and *It Changed My Life* (New York: Random House, 1975); S. Firestone, *The Dialectic of Sex; The case for Feminist Revolution* (New York: Bantam

Bk, 1970); K. Millett, *Sexual Politics* (London: Hart-Davis, 1971) and *Flying* (New York: Ballantine Bk, 1974); L. Irigaray, *Speculum de l'autre femme, op. cit.* One may also see M. Daly, *Beyond God the Father* (Boston: Beacon, 1973) as "downstream" P. Tillich's model, or H. Cixous, *La* (Paris: Gallimard, 1976) and C. Clément, *La jeune née* (Paris: Féminin futur, 1975) as "downstream" the "nouvelle écriture" or the philosophy of the Other. Trying to explicitly assume the "downstreamness" of the woman's condition and studies, see *e.g.* S. James, *Sex, Race and Class* (London: Race Today Publ., 1976) and M. Dalla Costa, *The Power of Women and The Subversion of the Community* (Bristol: Falling Wall Pr., 1978).

[17] E. W. Adams, *The Logic of Conditionals, An Application of Probability to Deductive Logic* (Dordrecht: Reidel, 1975, esp. pp. 87–90). The notion of consequence is not reduced to that of action. Notice that there are consequences to "negative" beliefs. Historicity is a critical category in as much as implying an awareness that no matter which form of rationality has been dominant in history, it has served to hide, justify, maintain, or not to disturb sexism. On this, see V. L. Bullough, *The Subordinate Sex: A History of Attitudes toward Women* (New York: Penguin Bk, 1974) and B. A. Carroll, *Liberating Women's History* (The University of Illinois Press, 1976).

[18] This may be presented as the epistemological version of the "in between" status of women. On this, see M. Strathern, *Women in Between, Female Roles in a Male World, Mount Hagen, New Guinea* (New York: Seminar Pr., 1975) and L. Irigaray, 'Des marchandises entre elles' in *Ce sexe qui n'en est pas un* (*op. cit.*). Notice that this claim is contrary to J. S. Mill's claim on women's *essential* counterpart-to-abstractness role in *The Subjection of Women* (The M.I.T. Press, 1970).

[19] On this, see S. G. Harding (ed.), *Can Theories be Refuted? Essays on the Duhem-Quine Thesis* (Dordrecht: Reidel, 1976, esp. p. xxi).

[20] In 'The Rationality of Science' (from 'Against Method') in Harding's *Can Theories Be Refuted?* (*op. cit.*) pp. 289–316, esp. p. 289. Notice Feyerabend's metaphor when arguing against the view that rationality is agreement with fixed rules: "Rather than choosing either a dragon or a pussycat as our companion . . . We can turn science from a stern and demanding mistress into an attractive and yielding courtesan who tries to anticipate every wish of her lover" (p. 311).

[21] T. Kuhn, *The Structure of Scientific Revolutions*, 2nd Edition, The University of Chicago Press, 1970. Kuhn's notion of paradigm could also be used as an explanatory model for the repetition charge: the reason why there has not been a "scientific revolution" in matters related to feminism would be that feminists' paradigms are "incommensurable" with the formalist view of the growth of knowledge. On the importance of "articulation" between models and examplars, see my "Paradigmes et bon sens", *Dialogue* 16 (1977), 629–652.

[22] K. R. Popper, *Conjectures and Refutations* (London: Routledge & Kegan Paul, 1963, pp. 238–239).

[23] E. W. Adams, *The Logic of Conditionals, op. cit.*, esp. p. 18.

[24] 'Cinderella as a Winner' is a chapter in T. Boslooper and M. Hayes, *The Femininity Game, op. cit.* On the "lucky estimate" model, see E. W. Adams, *The Logic of Conditionals, op. cit.*, pp. 82–87.

[25] See G. Greer, *The Female Eunuch* (New York: McGraw-Hill, 1971, esp. p. 8 and pp. 328–329) on "men are not free" as an argument why nobody should be; S. Rowbotham, *Women, Resistance and Revolution* (New York: Penguin Bk, 1972, esp. pp. 200–247);

S. G. Harding, 'Feminism: Reform or Revolution', *Philosophical Forum* 5, Nos. 1–2, Fall 1973/Winter 1974, pp. 271–285; and C. Pierce's review in *Signs* 2 (1976) 422–433; J. Russ, *The Female Man* (New York: Bantam Bk, 1975).
26 On this, see *e.g.*, P. M. Spacks, *The Female Imagination* (New York: Avon Bk, 1972), S. Paradis, *Femme fictive, femme réelle. Le personnage féminin dans le roman canadien-français* (Ottawa: Garneau, 1966), E. G. Davis, *The First Sex* (New York: Penguin Bk, 1971), Des femmes de Musidora, *Paroles . . . elles tournent!* (Paris: Editions des femmes, 1976). Notice that if we apply I. Levi's rule for the contraction/extension of a corpus of knowledge – that the risk of error in entertaining a hypothesis be probabilistically related to the addition of information which this hypothesis could provide – to feminist writings, the repetition charge eliminates the "risk of error" (no new information is to be expected) and further creates what may be termed "a risk of truth" (rehearsing unpalatable features of knowledge). See I. Levi, 'Acceptance Revisited', *op. cit.*, esp. p. 9.
27 I refer to P. K. Feyerabend's notion of "incommensurable theories", in 'The Rationality of Science', *op. cit.*
28 I refer to T. Kuhn's notion of puzzle-solving in *The Structure of Scientific Revolutions, op. ci.* This "dislocation" of broad and specific issues, thereby the neglect of the core of women's studies, may be taken as an epistemological version of what B.-C. de Coninck calls *La partagée* (Paris: Minuit, 1977).
29 I. Lakatos, 'Falsification and the Methodology of Scientific Research Programmes', in Harding's *Can Theories be Refuted?*, *op. cit.*, pp. 205–260, esp. p. 241.
30 P. K. Feyerabend, 'The Rationality of Science', *op. cit.*, pp. 299–301. The multiplicity of concrete contents thereby ascribed to feminist writings in terms of expletive novelties could be taken as an epistemological version of what A. Ceresa calls *La fille prodigue* (Paris: Editions des femmes, 1975).
31 This may be seen as a good epistemological case of what P. K. Feyerabend calls the "sociology of repression" in 'The Rationality of Science', *op. cit.*, pp. 314–315. As well put by L. Irigaray, we do not know exactly what a male's language is because the assumption is to define everything in sexually neutral terms; yet, we can say that there is no language but a male's language. See *Ce sexe qui n'en est pas un, op. cit.*, p. 138.
32 H. Törnebohm, 'On Piecemeal Knowledge-Formation', *op. cit.*, p. 317.
33 See J. Moulton, 'The Myth of the Neutral Man', summary in C. Pierce's article on philosophy in *Signs* 1 (1975), 487–503; J. Agassi's comment and C. Pierce's reply in *Signs* 2 (1976), pp. 512–514. On language, see K. One, 'Manglish', *Everywoman*, Nov. 1972, p. 12 and M. R. Key, *Male/Female Language* (Metuchen, N.J.: Scarecrow Pr., 1975).
34 A positive way of putting this thesis appears in the "androgynous" perspective. On this, see C. G. Heilbrun, *Toward a Recognition of Androgyny* (New York: Knopf, 1973); A. S. Rossi, 'Equality between the Sexes: An Immodest Proposal', *Daedalus* 93 (1964), 608–643; M. Daly, *Beyond God the Father* (*op. cit.*); G. G. Yates, *What Women Want, The Ideas of the Movement* (*op. cit.*). Another way appears in pleas for the couple as the model; see R. Gary, *Clair de femme* (Paris: NRF Gallimard, 1977); M. Champagne, *La violence au pouvoir* (Montréal: Editions du Jour, 1971). A puzzling suggestion appears in G. Deleuze and F. Guattari, *L'anti-Oedipe* (Paris: Minuit, 1972, esp. pp. 350–352): there is not one, not two, but "*n* sexes". This thesis is given as an interpretation of Marx's puzzling statement on a human and a non-human sexes, in *Critique de la philosophie de l'État de Hegel*.

35 This is my way of understanding the epistemological origin of the problem whether women studies are *on* or *of* women; see, *e.g.* M. Parlee in *Signs* 1 (1975), 119–139; M. T. Shuch Mednick, *Signs* 1, Part 1, 763–770. This is also my interpretation of the reason why the "starting-point" problem should be so difficult in feminist writings; see, *e.g.* H. Cixous, *Les commencements* (Paris: Grasset, 1970). This is finally my suggestion for reconstructing the problem of universalization from a feminist perspective; see, *e.g.* C. G. Allen, 'True sex polarity', and 'Sex-Identity and Personal Identity' in W. Shea and J. King Farlow (eds.), *Contemporary Issues in Political Philosophy* (New York: Science, History Publ., 1976, pp. 93–125).

36 I. Levi, 'Acceptance Revisited' and *Gambling with Truth, op. cit.*

37 See *L'ARC, Simone de Beauvoir et la lutte des femmes*, Vol. 61; R. D. Cottrell and J. O. Grace in *Signs* 1 (1976), 977–979; M. L. Collins and C. Pierce, 'Holes and Slimes: Sexism in Sartre's Psychoanalysis', *Philosophical Forum* 5, Nos. 1–2 (Fall 1973/Winter 1974), pp. 117ff; S. Lilar, *Le malentendu du deuxième sexe* (*op. cit.*); B. Friedan, *It Changed My Life* (*op. cit.*); H. Nahas, *La femme dans la littérature existentielle* (Paris: P.U.F., 1957).

38 An important example of studies to be made is the question: what epistemological changes did J. S. Mill have to introduce in order to defend women on an apparently Humean model, while sexism may be shown to be deeply rooted in Hume's method itself? Preliminary researches in this line are done in S. Burns, 'The Humean Female' and my comment 'The consistency of Hume's position concerning women', both in *Dialogue* 15 (1976), 415–441 and our joint paper on 'Hume on Women', in L. Clark and L. Lange (eds.), *Sexism of Social and Political Theory* (The University of Toronto Press, 1979), pp. 60–74. Another interesting example would be Ç. Demar, *L'affranchissement des femmes* (1833, reprint in Paris: Payot, 1976) and its relationships to St-Simon and his disciples. On this, see also B. Groulx, *Le féminisme au masculin* (Paris: Denoël-Gonthier, 1977).

39 I refer to "ultimate partition" as the list of strongest consistent potential answers, see I. Levi, 'Acceptance Revisited', *op. cit.*; I refer also to I. Niiniluoto's useful distinction between "answerhood" (what can count as potential answers to a given question) and "questionhood" (what one should ask in the first place on a given occasion) in his 'Inquiries, Problems, and Questions: Remarks on Local Induction', in Bogdan's *Local Induction, op. cit.*, pp. 263–297, esp. p. 264.

40 The expression is from I. Levi in 'Acceptance Revisited', *op. cit.*, see pp. 10–19, 68.

MERRILL B. HINTIKKA AND JAAKKO HINTIKKA

HOW CAN LANGUAGE BE SEXIST?

Prima facie, our title question may seem pointless. Barring bigots, virtually everybody will agree that language is frequently used in a sexist way. Why, then, the question?

We are formulating the title of the paper in this way because it serves to call attention to a general predicament of feminist philosophy as a serious theoretical enterprise. The sexist uses of language which first come to most people's minds are likely to instantiate relatively uninteresting aspects of language. Examples are offered by sexism expressed through purely emotive meaning and by those sexist uses of language which directly reflect sexist customs and institutions, for instance the different ways of addressing a person in Japanese. There is no problem as to how such sexism is possible in language; nor is there any interesting intellectual problem as a how such sexist usages can be diagnosed and cured. Once we have our emotions in line and our institutions and customs freed from sexism, no residual problem remains. Or so it seems.

This discussion illustrates certain criticisms which are often levelled in general at feminist philosophy. While the social problems addressed by feminist philosophy are usually acknowledged to be real and important, it is frequently denied that their diagnosis and solution requires or leads us to any new philosophical, methodological, or other theoretical insights. Hence feminist philosophy comes to seem a misnomer. The problems with which it deals do not appear to have a sufficiently important theoretical component to be labelled philosophical; hence the analyses and solutions it offers are thought not worthy of the designation 'philosophy'.

This is a view we are trying to combat by means of a case study. We suggest that a number of sexist uses of language illustrate interesting general theoretical problems. The diagnosis of such sexist uses hence involves serious problems of theoretical semantics. Even though there is in some cases no question as to how sexist language is possible, in others the very mechanism through which it comes about presents an interesting problem. In this paper, we are less anxious to solve this general theoretical problem we see raising its head here — it is too large for one paper anyway — than to recognize it, and less concerned with the details of instances of sexist language and sexist

139

Sandra Harding and Merrill B. Hintikka (eds.), Discovering Reality, 139–148.

language use than with their connection with the general problem we are posing. Through pointing out this connection, we are trying to give a concrete example of the *theoretical* interest of problems naturally arising from feminist concerns.

The theoretical problem we are posing is the following: In virtually every important current logical or philosophical approach to semantics, a set of representative relations between language and the world it deals with is taken for granted. For instance, in Tarski-type truth definitions, the valuation of nonlogical constants is taken for granted.[1] In Montague semantics, the meaning functions associated with primitive words are likewise taken for granted.[2] And in approaches which rely on translation to some privileged "language of thought", the semantics of the target language is likewise left largely unanalyzed.[3]

What we wish to suggest is, first, that the principles according to which these basic representative relations between language and reality are determined need much more attention than they are now given and that awareness of these principles is vital even for the understanding of and for the applications of contemporary formal semantics. We are tempted to speak of a subsystem of language (a subset of the totality of rules governing language) which is in some sense more fundamental than the subsystem studied in present-day formal semantics. For reasons which emerge somewhat more fully in what follows, we call the latter the *structural* system and the former the *referential* system.[4]

This formulation is somewhat oversimplified, however, in that there is more interplay between the two systems than our schematic first statement leads one to expect. Furthermore, it is not clear that all the phenomena we have in mind are connected closely enough with each other on either side of the fence to justify us in speaking of a real (sub)system. Hence the preliminary formulation of our theme and the term "referential system" must be taken with a grain of salt, and must be considered as being tentative and exploratory in nature. In any case, we shall illustrate the general thesis by means of discussions of a few narrower problems. We shall also indicate how a couple of specific manifestations of sexism of language exemplify our general theoretical problem.

Some aspects of the referential system are sometimes classified as belonging to pragmatics rather than to semantics. Such labels are harmless as long as they do not mislead us into expecting that such "pragmatic" phenomena are somehow intrinsically related to the many other items also relegated to "the pragmatic wastebasket", to use Yehoshua Bar-Hillel's expression.

For instance, we do not see any interesting connection between what we call the referential system and discourse-theoretical (e.g., conversational) phenomena.

As long as the referential system works and does not vary contextually, it remains relatively inconspicuous. (By "working", we mean here sufficing as the sole or main input into the structural system.) This inconspicuousness is one of the reasons why so little attention has been paid to it. For the same reason, the occasions when some aspect of the referential system varies, or proves insufficient for the purpose of understanding the semantics of some natural-language expression, are likely to offer the best quick illustrations of our theses.

We shall first try to give an example where the referential system does not by itself supply enough information to enable the structural system to operate in the way it is in these days usually expected to operate. This example is offered by a word whose force has perhaps been discussed more than that of any other single word: the word "good". Of course we cannot here exhaustively discuss the problems connected with it. We shall simply suggest that the way it operates is to rely on some evaluation principle but to leave it for the context to settle which one this evaluation principle is. On some occasions, the speaker may, e.g., rely on some set of values he or she shares with the audience or at least assumes to be familiar to the audience, whether or not its members actively subscribe to them. But on other occasions, the speaker – who could then be, for instance, a moral reformer – might use the same words to announce a new valuation principle. Preexisting valuations are typically determined by someone's *interests*.[5] But when Socrates claims to be a virtuous man, an *agathos*, while refusing to participate in public life and neglecting his family's welfare, he is not only not relying on an existing valuation principle for one's actions. He is also not relying on any known interest to express his point. He was proclaiming a *new* morality by making judgments which presuppose it (i.e., presuppose the valuation principles which constitute the proclaimed new morality).[6]

The reason why we have classified this context-dependence of "good" as belonging to the referential subsystem of language should be obvious. What is at issue is which cases the predicate "good" can be correctly applied to, i.e., which extension (reference) it has. Since such extensions of our primitive terms are what is assumed to be given prior to the usual (structural) analysis of the semantics of natural language, which is the currently favored type of semantical analysis of any notion, a single evaluation principle would be needed in order for this word to be capable of being handled in the usual approach. However, it is part and parcel of how the referential system operates that in the case of this word no unique scale or principle is forthcoming. This implies that one's actual use of "good" (in the several constructions into which it can enter) may rely on tacit evaluations or interests. These can be present without our noticing their presence. A small but subtle instance is offered by the difference in meaning between the English expression *a good man* and its literal counterparts in other languages, e.g., German, Swedish and Finnish. The difference is strikingly illustrated by comparing a passage from G. E. Moore's autobiography (in the *Library of Living Philosophers* volume devoted to him)[7] with Yrjö Hirn's essay (originally written in Swedish) on 'Voltaire's heart'.[8] Moore tells of one of his schoolteachers that he was not only a *good* man but also a benevolent man. Moore's words indicate clearly that he takes benevolence not to be a component of goodness, which has to do with such things as being conscientious and high-principled. In contrast, Hirn describes at some length Voltaire's noble efforts on behalf of oppressed and persecuted individuals, and goes on to argue that these good works were not only reflections of Voltaire's high humanitarian principles and of his efficiency in putting them into practice. They show, Hirn argues, that Voltaire was a genuinely humane, caring person, in brief, *en god människa* (a good man).

What is going on here is of course that the ambiguity of the English word *man* between a human being and a male of the species has led to the use of *a good man* where the tacitly presupposed interests are not those we presumably have in all our fellow human beings but those which we are likely to have in fellow citizens, business partners and colleagues, who are clearly presumed to be males. The former include primarily at least a minimum of concern with the basic welfare of other human beings. A good man would by this token be a *humane* man, a good representative of mankind, i.e., a kindly or kind man. (Interestingly, these two uses of *kind* are in fact etymologically related.)[9] Indeed, this is precisely what happens in several of those languages which do not exhibit the same ambiguity as English. For instance, for a German *ein guter Mensch* is, well, not unlike a *Mensch* in the colloquial Yiddish sense.

In contrast, the interests of the other kind are what have lent the English words *a good man* their customary force. They signal the virtues a fellow citizen or colleague is expected to exhibit. What of the woman who is citizen and colleague? Can she be a good man? That closely analogous phenomenon has pervaded the psychological concept of a healthy adult: in so far as a human being is a healthy woman, she fails to be a healthy adult, and in so far as she is a healthy adult, she fails to be a healthy woman.[10] Such unwitting sexism cuts much deeper, it seems to us, than e.g., any emotively sexist uses of language.

This diagnosis is supported by the observation that the same ambiguity and the same sexist presupposition is found with vengeance in the ancient Greek.[11] There the relevant interests were predominantly interests in another citizen-soldier, i.e., the military interests, however defensive, that all citizens of a *polis* presumably had.

A much more general part of the referential system are the principles which determine the individuation of the particular entities we talk about in our language. Jaakko Hintikka has argued elsewhere that the best way of conceptualizing these principles is in terms of what is usually (and misleadingly) called possible-worlds semantics, i.e., by considering what the "embodiments" or "roles" of our individuals were in a range of possible situations or possible courses of events.[12] Whatever one can say of this approach in the last analysis, it serves to clarify several aspects of the central conceptual problems in this area. For instance, it leads to the insight that a major role in identifications across the boundaries of possible worlds is played by re-identification, i.e., by the principles which enable us to speak of the same entities as (often) existing at different stages of one and the same course of events. It is characteristic of the state of the art that many of the very best philosophers flatly refuse to consider the details of these principles. W. V. Quine doesn't think that any reasonable, theoretically respectable principles can be discovered,[13] while Saul Kripke claims that we have to postulate temporally persistent individuals as a primitive, unanalyzable presupposition.[14] Notwithstanding such views, we believe that a further analysis of the re-identification and cross-identification principles has a tremendous philosophical and possibly also psychological and automation-theoretical interest. As the basic theoretical situation remains almost completely uncharted, we cannot survey it here. Instead we will discuss some of the related issues.

One pertinent observation here is the following: On the possible-worlds model, the referential system has to include two partly independent components.[15] On the one hand, the references of our primitive non-logical constants such as singular terms, predicates, function symbols, etc. in each possible world have to be specified. On the other hand, the imaginary "world lines" (which connect the roles of the same particulars in different worlds) have to be drawn. Each of these is a part of the objective foundation of ordinary (structural) semantics. The relative independence of these two tasks, the interpretation of nonlogical constants world by world and the drawing of the world lines (which span several worlds), implies that the corresponding two ingredients of the referential system can to some extent be varied independently. This does in fact happen, and such a variation of a part of the referential system is one of the phenomena in our language that can awaken philosophers' and linguists' interest in (or at least attention to) the referential system.

In order to see what such a variation might amount to, we must note a few facts here. In many typical cases, we are dealing with the possible worlds compatible with someone's knowledge, belief, or other propositional attitude. For example, let us consider what Jane knows. This is specified by the set of all possible worlds compatible with what she knows, called Jane's epistemic alternatives or her "knowledge worlds". Whatever is true in all these epistemic alternatives is known by Jane, and *vice versa*. Hence a singular term, say "*b*", picks out the same individual in all of Jane's epistemic alternatives (goes together with a world line) if and only if it is true that

(1) $(\exists x)$ Jane knows that $(b = x)$.

More colloquially, (1) obviously says the same as

(2)   Jane knows who *b* is.

Now the ways in which world lines are drawn can vary without changing the evaluation principles which affect one world at a time. Hence the truth conditions of (1) and (2) can be varied accordingly without affecting the rest of the referential system. More generally, it is (among other things) in the variation of the force of phrases of the form *knows* + *an indirect question* that the variation of world lines can be "seen".

Possible-worlds semantics shows what the cash value of such variation is. It is a question of what the person in question would consider as the same individual in different actual and possible situations, what he or she would "count as" the same individual. Not surprisingly, sexism can rear its head occasionally here, too. The diaries of that inveterate male chauvinist, Evelyn Waugh, offer an example. He quotes there the old saw worthy of Polonius: "Be kind to young ladies. You never know who they will be." The possible-worlds framework instantly reveals the mechanism of Waugh's sexism: Waugh is in effect treating women married to the same gent as being interchangeable, formally speaking, as nodes of one and the same "world line" connecting individuals in future courses of events.

Such variation has been taken by Quine to indicate that something is wrong with the possible-worlds semantics of sentences like (1) and (2). All the variation is in the referential system, however. The structural system, which is the subject matter Quine was in effect commenting on,[16] is of course completely unaffected by this variation.

One of the central problems in this area is how the world lines are drawn, i.e., how we as a matter of fact handle cross-identification in our conceptual system. We are in the process of developing a theory of actual cross-identification.[17] Unfortunately the subject is too large to be expounded here, and we must hence confine ourselves to a promissory note as far as the general problems of identification and individuation are concerned. Instead, let us consider a couple of the many interesting narrower issues involved in the general problem of cross-identification.

David Lewis has in effect claimed that cross-identification takes place according to similarity: those individuals in different possible worlds are as it were declared identical ("counterparts" in Lewis' terminology) which are most closely similar to each other.[18] "Similarity" is not intended to be a primitive notion in this approach. Rather, the relevant comparison may involve several different and differently weighted similarity considerations.

This is not the only a priori possibility, however. Instead of comparing individuals one by one, we may try to compare the structures of the two possible worlds in question at large and try to match them. Individuals corresponding to each other in the closest match we can achieve would be Lewisian counterparts. Such possible cross-world comparisons obviously depend much more on the relational and functional characteristics of the denizens of the different scenarios ("possible worlds") we are envisaging than on the essential properties of the entities involved in the compariosn. For instance, these non-essentialist modes of cross-identification may depend on the continuity properties of the entities in question, which are of course relational rather than essential properties.

What is striking here is that certain psychological studies suggest that there may be sex-linked differences (whether innate or culturally conditioned does not matter for our purposes) in the very matter of such assimilation comparisons. For instance, some studies seem to show that boys tend to bracket together objects (or pictures of objects) whose intrinsic characteristics are similar, whereas girls weight more heavily the functional and relational characteristics of the entities to be compared.[19] For instance, boys frequently bracketed together such entities as a truck, a car, and an ambulance, while girls bracketed such entities as a doctor, a hospital bed, and an ambulance.

More generally, women are generally more sensitive to, and likely to assign more importance to, relational characteristics (e.g., interdependencies) than males, and less likely to think in terms of independent discrete units. Conversely, males generally prefer what is separable and manipulatable.[20] If we put a premium on the former features, we are likely to end up with one kind of cross-identification and one kind of ontology; if we follow the guidance of the latter considerations, we end up with a different one. Moreover, it is not hard to see what the difference between the two will be. All identification which turns on essential properties, weighted similarities, or suchlike, presupposes a predetermined set of discrete individuals, the bearers of those essential properties as similarity relations, and focuses our attention on them. In contrast, an emphasis on relational characteristics of our individuals encourages comparisons of different worlds in terms of their total structure, which leads to entirely different identification methods, which are much more holistic and relational.

The suggestion — and we do not intend it to be more than a suggestion — we make here is now clear: it is not just possible, but quite likely, that there are sex-linked differences in our processes of cross-identification. The differences are such as not to be manifested either very frequently or very blatantly. But in the more refined areas of speculative thought, such differences might very well have their consequences. Indeed, cross-identification methods are in an obvious sense constitutive of our ontology. Hence, what we are suggesting is that language could perhaps be, if not sexist, then at least sexually biased and sensitive to sex differences in the very respects that are most closely related to the structure of our ontology.

Lest this suggestion strike the reader as unrealistic, let us note some of its consequences and ramifications. Quite independently of the perspective from which we are here viewing the problems of ontology and cross-identification, it is arguable that Western philosophical thought has been overemphasizing such ontological models as postulate a given fixed supply of discrete individuals, individuated by their instrinsic or essential (non-relational) properties. These models are unfavorably disposed towards cross-identification by means of functional or other relational considerations. Is it to go too far to suspect a bias here? It seems to us that a bias is unmistakable in recent philosophical semantics and ontology. There we find almost everyone postulating a given domain of discrete individuals whose identity from one model (world) to another is unproblematic. An especially blatant example of this trend is Kripke's notion of a rigid designator,[21] which becomes virtually useless as soon as cross-identification is recognized as a problem. (No wonder Kripke has been led to argue that the re-identification of temporally persisting physical objects must be taken for granted.) Another conspicuous part of the same syndrome is philosophers' surprising slowness in appreciating Jaakko Hintikka's discovery of the duality of cross-identification methods (the descriptive and the perspectival one),[22] which breaks the hegemony of neat prefabricated individuals in philosophical ontology. It is hard not to see in this strong tendency a preference of independent but manipulative units similar to the sex-linked preference several psychologists have noted.

Similar points can be made about earlier history of philosophical ontology. Separability and "thisness" were the characteristic marks of Aristotelian substances,[23] which are historically the most important proposed ontological units of the world. Conversely, we may very well ask whether Leibniz' ontology of monads, whose identity lies in their reflecting the whole universe, has really been given its due.[24] Even though firm documentation is extremely hard in these matters, at the very least we obtain here a challenging perspective on the history of philosophical ontology. At the same time, our questions illustrate the systematic interest within language theory of the referential system we have tentatively postulated. For it is problems of individuation and identification which constitute perhaps the most important ingredient of any serious study of the referential system at large, and hence of philosophical ontology.

NOTES

1 Cf. Alfred Tarski, 'The concept of truth in formalized languages', in Alfred Tarski, *Logic, Semantics, Metamathematics*, Clarendon Press, Oxford, 1956, pp. 152–278.

2 Cf. Richmond H. Thomason (ed.), *Formal Philosophy: Selected Papers of Richard Montague*, Yale U.P., New Haven, 1974; D. R. Dowty, R. E. Wall, and S. Peters, *Introduction to Montague Semantics*, D. Reidel, Dordrecht, 1981.

3 Cf., e.g., Jerry A. Fodor, *The Language of Thought*, Thomas Y. Crowell, New York, 1975.

4 Further observations concerning this distinction are made in Jaakko Hintikka and Merrill B. Hintikka, 'Towards a general theory of individuation and identification', partly forthcoming in the proceedings of the Sixth International Wittgenstein Symposium, Hölder-Pichler-Tempsky, Vienna, 1982.

5 This point is well argued in the seminal last chapter 'The word "good"' of Paul Ziff, *Semantic Analysis*, Cornell, U.P., Ithaca, N.Y., 1960.

6 Cf. here A. Adkins, *Merit and Responsibility*, Clarendon Press, Oxford, 1960.

7 P. A. Schilpp (ed.), *The Philosophy of G. E. Moore* (The Library of Living Philosophers), Tudor, New York, 1952, pp. 3–39, especially p. 9.

8 Yrjö Hirn, 'Voltaires hjärta', in Yrjö Hirn, *De lagerkrönta skoplaggen*, Söderström & Co., Helsinki, 1951.

[9] For an interesting discussion, see chapter 2 of C. S. Lewis, *Studies in Words*, second ed., Cambridge U.P., Cambridge, 1967.

[10] Cf. Inge K. Broverman, Donald M. Broverman, et al., 'Sex-role stereotypes and clinical judgments of mental health', *Journal of Consulting and Clinical Psychology* 34, No. 1 (1970), 1–7.

[11] Cf. Adkins, op. cit. (Note 6 above), especially chapter 3.

[12] See his books *Models for Modalities*, D. Reidel, Dordrecht, 1969, and *The Intentions of Intentionality*, D. Reidel, Dordrecht, 1975.

[13] Cf. W. V. Quine, 'Worlds away', *Journal of Philosophy* 73 (1976), 859–863.

[14] Cf. Saul Kripke, 'Identity through time', paper delivered at the Seventy-Sixth Annual Meeting of APA Eastern Division, New York, December 27–30, 1979.

[15] This point has been implicit in Jaakko Hintikka's work ever since the last chapter of *Knowledge and Belief*, Cornell U.P., Ithaca, N.Y., 1962.

[16] Op. cit. (note 12 above).

[17] 'Towards a general theory of individuation and identification' (note 4 above).

[18] David Lewis, 'Counterpart theory and quantified modal logic', *Journal of Philosophy* 65 (1968), 113–126.

[19] Cf., e.g., J. Kagan, H. A. Moss, and I. E. Sigel, 'The psychological significance of styles of conceptualization', in J. C. Wright and J. Kagan (eds.), *Basic Cognitive Processes in Children* (Society for Research in Child Development Monograph 28, no. 2), 1963.

[20] Cf. e.g. Eleanor E. Maccoby, 'Sex differences in intellectual functioning', in Eleanor E. Maccoby (ed.), *The Development of Sex Differences*, Stanford U.P., Stanford, 1966, pp. 25–55.

[21] Saul Kripke, *Naming and Necessity*, Harvard U.P., Cambridge, Mass., 1980.

[22] Cf. 'On the logic of perception' in *Models for Modalities* (Note 12 above); 'Knowledge by acquaintance – individuation by acquaintance' in Jaakko Hintikka, *Knowledge and the Known*, D. Reidel, Dordrecht, 1974.

[23] Cf. Aristotle, *Categories*, ch. 5.

[24] Cf. Jaakko Hintikka, 'Leibniz on plenitude, relations, and the "Reign of Law"', in Simo Knuuttila (ed.), *Reforging the Great Chain of Being*, D. Reidel, Dordrecht, 1981, pp. 259–286.

JANICE MOULTON

A PARADIGM OF PHILOSOPHY:
THE ADVERSARY METHOD

THE UNHAPPY CONFLATION OF AGGRESSION WITH SUCCESS

It is frequently thought that there are attributes, or kinds of behavior, that it is good for one sex to have and bad for the other sex to have. Aggression is a particularly interesting example of such an attribute. This paper investigates and criticizes a model of philosophic methodology that accepts a positive view of aggressive behavior and uses it as a paradigm of philosophic reasoning. But before I turn to this paradigm, I want to challenge the broader view of aggression that permits it positive connotations.

Defined as "an offensive action or procedure, especially a culpable unprovoked overt hostile attack," aggression normally has well deserved negative connotations. Perhaps a standard image of aggression is that of an animal in the wild trying to take over some other animal's territory or attacking it to eat it. In human contexts, aggression often invokes anger, uncontrolled range, and belligerence.

However, this negative concept, when it is specifically connected to males *qua* males or to workers in certain professions (sales, management, law, philosophy, politics) often takes on positive associations. In a civilized society, physical aggression is likely to land one in a jail or a mental institution. But males and workers in certain professions are not required to physically attack or eat their customers and coworkers to be considered aggressive. In these contexts, aggression is thought to be related to more positive concepts such as power, activity, ambition, authority, competence, and effectiveness — concepts that are related to success in these professions. And exhibition of these positive concepts is considered evidence that one is, or has been, aggressive.

Aggression may have no causal bearing on competence, superiority, power, etc., but if many people believe aggressive behavior is a sign of these properties, then one may have to learn to behave aggressively in order to appear competent, to seem superior, and to gain or maintain power. This poses a dilemma for anyone who wants to have those positive qualities, but does not wish to engage in "culpable unprovoked overt hostile attacks."

Of reluctant aggressors, males have an advantage over females. For as

149

Sandra Harding and Merrill B. Hintikka (eds.), Discovering Reality, 149–164.

members of the masculine gender, their aggression is thought to be "natural."
Even if they do not engage in aggressive behavior, they can still be perceived
as possessing that trait, inherently, as a disposition. And if they do behave
aggressively, their behavior can be excused – after all, it's natural. Since
women are not perceived as being dispositionally aggressive, it looks like they
would have to behave aggressively in order to be thought aggressive. On the
other hand, since women are not expected to be aggressive, we are much
more likely to notice the slightest aggressive behavior on the part of a woman
while ignoring more blatant examples by men just because they are not
thought unusual. But when done by a female, it may be considered all the
more unpleasant because it seems unnatural. Alternatively, it may be that
a woman who exhibits competence, energy, ambition, etc. may be thought
aggressive and therefore unnatural even without behaving aggressively. Since,
as I shall argue, aggressive behavior is unlikely to win friends and influence
people in the way that one would like, this presents a special problem for
women.

Some feminists dismiss the sex distinction that views aggression in a female
as a negative quality and then encourage females to behave aggressively in
order to further their careers. I am going to, instead, question the assumption
that aggression deserves association with more positive qualities. I think it is a
mistake to suppose that an aggressive person is more likely to be energetic,
effective, competent, powerful or successful and also a mistake to suppose
that an energetic, effective, etc. person is therefore aggressive.

Even those who object to sex-roles stereotyping seldom balk specifically
at the assumption that more aggressive people are better suited to "be the
breadwinners and play the active role in the production of commodities
of society", but only at the assumption that aggression is more natural to one
sex than the other.[1] Robin Lakoff assumes that more aggressive speech is
both more effective and typical of males, and objects to the socialization
that forbids direct questions and assertions, devoid of polite phrases, in
women's speech.[2] Lakoff recognizes that the speech she characterizes as
women's speech is frequently used by male academics, but she still assumes
that aggressive speech is more powerful and more effective. She does not
see that polite, nonabrupt speech, full of hesitations and qualifiers can be
a sign of great power and very effective in giving the impression of great
thought and deliberation, or in getting one's listeners on one's side. Although
polite, nonabrupt speech can be more effective and have more power than
aggressive speech, the conceptual conflation of aggression with positive
concepts has made this hard to remember.

Consider some professional occasions where aggression might be thought an asset. Aggression is often equated with energy, but one can be energetic and work hard without being hostile. It may seem that aggression is essential where there is competition, but people who just try to do their best, without deliberately trying to do in the other guy may do equally well or even better. Feelings of hostility may be distracting, and a goal of defeating another may sidetrack one to the advantage of a third party. Even those who think it is a dog-eat-dog world can see that there is a difference between acting to defeat or undermine competition and acting aggressively towards that competition. Especially if one's success depends on other parties, it is likely to be far wiser to *appear* friendly than to engage in aggressive behavior. And in professions where mobility is a sign of success, today's competitors may be tomorrow's colleagues. So if aggression is likely to make enemies, as it seems designed to do, it is a bad strategy in these professions. What about other professional activities? A friendly, warm, nonadversarial manner surely does not interfere with persuading customers to buy, getting employees to carry out directions conscientiously, convincing juries, teaching students, getting help and cooperation from coworkers, and promotions from the boss. An aggressive manner is more likely to be a hindrance in these activities.

If these considerations make us more able to distinguish aggression from professional competence, then they will have served as a useful introduction to the main object of this essay: an inquiry into a paradigm of philosophy that, perhaps tricked by the conflation of aggression and competence, incorporates aggression into its methodology.

SCIENTIFIC REASONING

Once upon a time it was thought that scientific claims were, or ought to be, objective and value-free; that expressions of value were distinguishable from expressions of fact, and that science ought to confine itself to the latter. This view was forsaken, reluctantly by some, when it was recognized that theories incorporate values, because they advocate one way of describing the world over others, and that even observations of facts are made from some viewpoint or theory about the world already presupposed.[3]

Still devoted to a fact-value distinction, Popper recognized that scientific *statements* invoked values, but believed that the *reasoning* in science was objective and value-free.[4] Popper argued that the primary reasoning in science is deductive. Theories in science propose laws of the form "All *A*'s are *B*'s" and the job of scientific research is to find, or set up, instances of

A and see if they fail to produce or correlate with instances of *B*. The test of a theory was that it could withstand attempts to falsify it. A good theory encouraged such attempts by making unexpected and broad claims rather than narrow and expected claims. If instances of *B* failed to occur given instances of *A*, then the theory was falsified. A new theory that could account for the failure of *B* to occur in the same deductive manner would replace the old theory. The reasoning used to discover theories, the way a theory related to physical or mathematical models or other beliefs, was not considered essential to the scientific enterprise. On this view, only the thinking that was exact and certain, objective and value-free was essential to science.

However, Kuhn then argued that even the reasoning used in science is not value free or certain.[5] Science involves more than a set of independent generalizations about the world waiting to be falsified by a single counter-instance. It involves a system, or "paradigm," of not only generalizations and concepts, but beliefs about the methodology and evaluation of research: about what are good questions to ask, what are proper developments of the theory, what are acceptable research methods. One theory replaces another, not because it functions successfully as a major premise in a greater number of deductions, but because it answers some questions that the other theory does not — even though it may not answer some questions the other theory does. Theory changes occur because one theory is more *satsifying* than the other, because the questions it answers are considered more *important*. Research under a paradigm is not done to falsify the theory, but to fill in and develop the knowledge that the paradigm provides a framework for. The reasoning involved in developing or replacing a paradigm is not simply deductive, and there is probably no adequate single characterization of how it proceeds. This does not mean that it is irrational or not worth studying, but that there is no simple universal characterization of good scientific reasoning.

This view of science, or one like it, is widely held by philosophers now. It has been suggested that philosophy too is governed by paradigms.

PHILOSOPHY REASONING – THE ADVERSARY PARADIGM

I am going to criticize a paradigm or part of a paradigm in philosophy.[6] It is the view that applies the now-rejected view of value-free reasoning in science to reasoning in philosophy. On this view all philosophic reasoning is, or ought to be, deductive. General claims are made and the job of philosophic research is to find counterexamples to the claims. And most important, the philosophic enterpriese is seen as an unimpassioned debate between *adversaries* who try to defend their own views against counterexamples and produce counterexamples to opposing views. The reasoning used to discover the claims, and the way the claims relate to other beliefs and systems of ideas are not considered relevant to philosophic reasoning if they are not deductive. I will call this the Adversary Paradigm.

Under the Adversary Paradigm, it is assumed that the only, or at any rate, the best, way of evaluating work in philosophy is to subject it to the strongest or most extreme opposition. And it is assumed that the best way of presenting work in philosophy is to address it to an imagined opponent and muster all the evidence one can to support it. The justification for this method is that a position ought to be defended from, and subjected to, the criticism of the strongest opposition; that this method is the only way to get the best of both sides; that a thesis which survives this method of evaluation is more likely to be correct than one that has not; and that a thesis subjected to the Adversary Method will have passed an "objective" test, the most extreme test possible, whereas any weaker criticism or evaluation will, by comparison, give an advantage to the claim to be evaluated and therefore not be as objective as it could be. Of course, it will be admitted that the Adversary Method does not *guarantee* that all and only sound philosophical claims will survive, but that is only because even an adversary does not always think of all the things which ought to be criticized about a position, and even a proponent does not always think of all the possible responses to criticism. However, since there is no way to determine with certainty what is good and what is bad philosophy, the Adversary Method is the best there is. If one wants philosophy to be objective, one should prefer the Adversary Method to other, more subjective, forms of evaluation which would give preferential treatment to some claims by not submitting them to extreme adversarial tests. Philosophers who accept the AdversaryParadigm in philosophy may recognize that scientific reasoning is different, but think "So much the worse for science. At least philosophy can be objective and value free."

I am going to criticize this paradigm in philosophy. My objection to the Adversary Method is to its role as a paradigm. If it were merely *one* procedure among many for philosophers to employ, there might be nothing worth objecting to except that conditions of hostility are not likely to elicit the best reasoning. But when it dominates the methodology and evaluation of philosophy, it restricts and misrepresents what philosophic reasoning is.

It has been said about science that criticism of a paradigm, however warranted, will not be successful unless there is an alternative paradigm available to replace it.[7] But the situation in philosophy is different. It is not that we have to wait for an alternative form of reasoning to be developed. Nonadversarial reasoning exists both outside and within philosophy but our present paradigm does not recognize it.

DEFECTS OF THE ADVERSARY PARADIGM

The defense of the Adversary Method identified adversary criticism with severe evaluation. If the evaluation is not adversarial it is assumed it must be weaker and less effective. I am going to argue that this picture is mistaken.

As far back as Plato it was recognized that in order for a debate or discussion to take place, assumptions must be shared by the parties involved.[8] A debate is not possible among people who disagree about everything. Not only must they agree about what counts as a good argument, what will be acceptable as relevant data, and how to decide on the winner, but they must share some premises in order for the debate to get started.

The Adversary Method works best if the disagreements are isolated ones, about a particular claim or argument. But claims and arguments about particular things rarely exist in isolation. They are usually part of an interrelated system of ideas. Under the Adversary Paradigm we find ourselves trying to disagree with a system of ideas by taking each claim or argument, one at a time. Premises which might otherwise be rejected must be accepted, if only temporarily, for the sake of the argument. We have to fight our opponents on their terms. And in order to criticize each claim individually, one at a time, we would have to provisionally accept most of the ideas we disagree with most of the time. Such a method can distort the presentation of an opponent's position, and produce an artificially slow development of thought.

Moreover, when a whole system of ideas is involved, as it frequently is, a debate that ends in defeat for one argument, without changing the whole system of ideas of which that argument was a part, will only provoke stronger support for other arguments with the same conclusion, or inspire attempts to amend the argument to avoid the objections. Even if the entire system of ideas is challenged, it is unlikely to be abandoned without an alternative system to take its place. A conclusion that is supported by the argument in question may remain undaunted by the defeat of that argument. In order to alter a *conclusion*, it could be more effective to ignore confrontation on the particular points, not provide counterexamples, however easy they may be to find, and instead show how other premises and other data support an alternative system of ideas. If we are restricted to the Adversary Method we may have to withhold evaluation for a system of ideas in order to find a common ground for debate. And the adversarial criticism of some arguments may merely strengthen support for other ideas in the system, or inspire makeshift revisions and adjustments.

Moreover, the Adversary Paradigm allows exemptions from criticism of claims in philosophy that are not well worked out, that are "programmatic". Now any thesis in philosophy worth its salt will be programmatic in that there will be implications which go beyond the thesis itself. But the claims that have become popular in philosophy are particularly sketchy, and secure their immunity from criticism under the Adversary Paradigm *because* their details are not worked out. A programmatic claim will offer a few examples which fit the claim along with a prediction that, with some modification (of course), a theory can be developed along these lines to cover all cases. Counterexamples cannot refute these claims because objections will be routinely dismissed as merely things to be considered later, when all the details are worked out. Programmatic claims have burgeoned in philosophy, particular in epistemology and philosophy of language. It has become a pattern for many philosophy papers to spend most of the paper explaining and arguing against other claims and then to offer a programmatic claim or conjecture of one's own as an alternative at the end without any support or elaboration. (Perhaps this is the beginning of a new paradigm that is growing out of a shortcoming in the evaluation procedures of the Adversary Paradigm.) Some programmatic claims that were once quite popular are now in disrepute, such as sense-data theories, but not because they were disproved, perhaps more because they failed to succeed -- no one ever worked out the details and/or people gave up hope of ever doing so. The Adversary Method allows programmatic claims to remain viable in philosophy, however sketchy or implausible, as long as they are unrefuted.

MISINTERPRETING THE HISTORY OF PHILOSOPHY

Under any paradigm we are likely to reinterpret history and recast the positions of earlier philosophers. With the Adversary Paradigm we understand earlier philosophers as if they were addressing adversaries instead of trying to build a foundation for scientific reasoning or to explain human nature. Philosophers who cannot be recast into an adversarial mold are likely to be ignored.[9] But our reinterpretations may be misinterpretations and our

choice of great philosophers may be based not so much on what they said
as on how we think they said it.

One victim of the Adversary Paradigm is usually thought to be a model
of adversarial reasoning: The Socratic Method. The Socratic method is
frequently identified with the elenchus, a method of discussion designed
to lead the other person into admitting that her/his views were wrong, to
get them to feel what is sometimes translated as "shame" and sometimes as
"humility". Elenchus is usually translated as "refutation", but this is misleading because its success depends on convincing the other person, not
on showing their views to be wrong to others. Unlike the Adversary Method,
the justification of the elenchus is not that it subjects claims to the most
extreme opposition, but that it shakes people up about their cherished
convictions so they can begin philosophical inquiries with a more open
mind. The aim of the Adversary Method, in contrast, is to show that the
other party is wrong, challenging them on any possible point, regardless of
whether the other person agrees. In fact, many contemporary philosophers
avoid considerations of how of convince, supposing it to be related to trickery
and bad reasoning.

In general the inability to win a public debate is not a good reason for
giving up a belief. One can usually attribute the loss to one's own performance
instead of to inadequacies in one's thesis. A public loss may even make one
feel more strongly toward the position which wasn't done justice by the
opposition. Thus the Adversary Method is not a good way to convince
someone who does not agree with you.

The elenchus, on the other hand, is designed just for that purpose. One
looks for premises that the other person will accept and that will show that
the original belief was false. The discussion requires an acceptance by both
parties of premises and reasoning.

Of course, one could use the elenchus in the service of the Adversary
Paradigm, to win a point rather than convince. And it has been assumed by
many that that is what Socrates was doing, that his style was insincere and
ironic,[10] that his criticisms were harsh and his praise sarcastic. But in fact
Socrates' method is contrasted with that of an antagonist or hostile questioner
in the dialogues.[11] Socrates jokes frequently at the beginning of a dialogue
or when the other party is resisting the discussion, and the jokes encourage
the discussion, which would not be the case if they were made at the expense
of the speaker.[12] Any refusals and angry responses Socrates received occurred
when cherished ideas were shaken and not as a result of any adversary treatment by Socrates.[13] Socrates avoided giving an opinion in opposition to the

one being discussed lest it be accepted too easily without proper examination. His aim is not to rebut, it is to show people how to think for themselves.

We have taken the *elenchus* to be a duel, a debate between adversaries, but this interpretation is not consistent with the evidence in the dialogues. I suspect that the reason we have taken Socrates' method to be the Adversary Method, and consequently misunderstood his tone to be that of an ironic and insincere debater instead of that of a playful and helpful teacher, is that under the influence of the Adversary Paradigm we have not been able to conceive of philosophy being done any other way.

RESTRICTIONS OF PHILOSOPHICAL ISSUES

The Adversary Paradigm affects the kinds of questions asked and determines the answers that are thought to be acceptable. This is evident in nearly every area of philosophy. The only problems recognized are those between opponents, and the only kind of reasoning considered is the certainty of deduction, directed to opposition. The paradigm has a strong and obvious influence on the way problems are addressed.

For example in philosophy of language, the properties investigated are analyzed when possible in terms of properties that can be subjected to deductive reasoning. Semantic theory has detoured questions of meaning into questions of truth. Meaning is discussed in terms of the deductive consequences of sentences. We ask not what a sentence says, but what it guarantees, what we can deduce from it. Relations among ideas that affect the meaning are either assimilated to the deductive model or ignored.[14]

In philosophy of science, the claim that scientific reasoning is not essentially deductive has led to "charges of irrationality, relativism, and the defense of mob rule".[15] Non-deductive reasoning is thought to be no reasoning at all. It is thought that any reasons which are good reasons must be deductive and certain.

In ethics, a consequence of this paradigm is that it has been assumed that there must be a single supreme moral principle. Because moral reasoning may be the result of different moral principles that may make conflicting claims about the right thing to do, a supreme moral principle is needed to "adjudicate rationally [that is, deductively] among different competing moralities".[16] The relation between moral principles and moral decision is thought to be deductive. A supreme moral principle allows one to deduce, by plugging in the relevant factors, what is right or wrong. More than one principle would allow, as is possible if one starts from different premises, conflicting judgments to be deduced. The possibilities that one could adjudicate between conflicting moral percepts without using deduction, that there might be moral problems that are not the result of conflicts in moral principles, and that there might be moral dilemmas for which there are no guaranteed solutions, are not considered.

There is a standard "refutation" of egoism that claims that egoism does not count as an ethical theory and therefore is not worthy of philosophical consideration because an egoist would not advocate egoism to others (would not want others to be egoists too). It is assumed that only systems of ideas that can be openly proclaimed and debated are to count as theories, or as philosophy. Again this is the Adversary Paradigm at work, allowing only systems of ideas that can be advocated and defended, and denying that philosophy might examine a system of ideas for its own sake, or for its connections with other systems.[17]

There are assumptions in metaphysics and epistemology that language is necessary for thinking, for reasoning, for any system of ideas. It is denied that creatures without language might have thoughts, might be able to figure out some things, because the only kind of reasoning that is recognized is adversarial reasoning and for that one must have language.[18]

With the Adversary Paradigm we do not try to assess positions or theories on their plausibility or worthiness or even popularity. Instead we are expected to consider, and therefore honor, positions that are most *unlike* our own in order to show that we can meet their objections. So we find moral theories addressed to egoists,[19] theories of knowledge aimed at skeptics. Since the most extreme opposition may be a denial of the existence of something, much philosophic energy is expended arguing for the existence of some things, and no theory about the nature of those things ever gets formulated. We find an abundance of arguments trying to prove that determinism is false because free will exists, but no positive accounts giving an explanation, in terms of chance and indeterminism, of what free will would be. Philosophers debate and revive old arguments about whether God exists, but leave all current discussions about what the nature of God would be to divinity schools and religious orders.

Philosophy, by attention to extreme positions because they are extreme, presents a distorted picture about what sorts of positions are worthy of attention, giving undo attention and publicity to positions merely because they are those of a hypothetical adversary's and possibly ignoring positions which make more valuable or interesting claims.

THE PARADIGM LEADS TO BAD REASONING

It has mistakenly been assumed that whatever reasoning an adversary would accept would be adequate reasoning for all other circumstances as well.[20] The Adversary Paradigm accepts only the kind of reasoning whose goal is to convince an opponent, and ignores reasoning that might be used in other circumstances: To figure something out for oneself, to discuss something with like-minded thinkers, to convince the indifferent or the uncommitted. The relations of ideas used to arrive at a conclusion might very well be different from the relations of ideas needed to defend it to an adversary. And it is not just less reasoning, or fewer steps in the argument that distinguishes the relations of ideas, but that they must be, in some cases, quite different lines of thought.

In illustration, let us consider the counterexample reasoning that is so effective in defending one's conclusions against an adversary. When an adversary focusses on certain features of a problem, one can use those features to construct a counterexample. To construct a counterexample, one needs to abstract the essential features of the problem and find another example, an analogy, that has those features but which is different enough and clear enough to be considered dispassionately apart from the issue in question. The analogy must be able to show that the alleged effect of the essential features does not follow.

But in order to reach a conclusion about moral issues or scientific theories or aesthetic judgments, one may have to consider *all* the important features and their interactions. And to construct an analogy with all the features and their interactions, which is *not* part of the issue in question, may well be impossible. Any example with all the features that are important may just be another example of the problem at issue. If we construct an analogy using only some of the important features, or ignoring their interactions, a decision based on this could be bad reasoning. It would ignore important aspects of the problem.

Consider a work in the Adversary Paradigm, Judith Thomson's excellent 'A Defense of Abortion'.[21] Thomson says: All right, let's give the "right to lifers" all their premises. Let's suppose, for the sake of argument, that a fetus is a person, and even that it is a talented person. And then she shows by counterexample that it does not follow that the fetus has a right to life. Suppose that you woke up one morning and found that you were connected to a talented violinist (because he had a rare kidney disease and only you had the right blood type) and the Music Lover's Society had plugged you together. When you protested, they said, "Don't worry, it's only for nine months, and then he'll he cured. And you can't unplug him because now that the connection has been made, he will die if you do." Now, Thomson says to the right-to-lifers, surely you have the *right* to unplug yourself. If the time were shorter than nine months, say only nine minutes, you might be an awful person if you did not stay plugged in, but even then you have the *right* to do what you want with your body.

The violinist analogy makes the main point, and Thomson explains it by comparing the right to one's own body to the right to property (a right that the right-to-lifers are unlikely to deny). One's right to property does not stop because some other person needs it, even if they need it to stay alive.

The argument using a counterexample is as effective against adversaries as any argument could be, and therefore a good method for arguing within the adversary tradition. One uses the premises the adversary would accept — property rights, the fetus as a person — and shows that the conclusion — that "unplugging" yourself from the fetus is wrong — does not follow. In general, in order to handle adversaries one may abstract the features they claim to be important, and construct a counterexample which has those same features but in which the conclusion they claim does not hold.

All Thomson tried to show was that abortion would not be wrong just because the fetus were a person.[22] She did not show that abortion would, or would not, be wrong. There are many features beside personhood that are important to the people making a decision about abortion: That it is the result of sexual intercourse so that guilt, atonement or loyalty about the consequences may be appropriate; that the effects only occur to women, helping to keep a power-minority in a powerless position; that the developing embryo may be genetically like others who are loved; that the product would be a helpless infant brought into an unmanageable situation; that such a birth would bring shame or hardship to others. There are many questions connected to whole systems of ideas that need answers when abortion is a personal issue: What responsibility does one have to prevent shame and hardship to others — parents, friends, other children, future friends and future children? When do duties toward friends override duties of other sorts? How is being a decent person related to avoiding morally intolerable situations — dependence, hate, resentment, lying? There is a lot of very serious moral reasoning that goes on when an individual has to make a decision about abortion, and the decisions made are enormously varied. But this moral reasoning has largely been ignored by philosophers because it is different from the reasoning used to address an adversary and it is too complex and interrelated to be evaluated by counterexamples.

A good counterexample is one that illustrates a general problem about some principle or general claim. Counterexample reasoning can be used to rule out certain alternatives, or at least to show that the current arguments supporting them are inadequate, but not to construct alternatives or to figure out what principles *do* apply in certain situations. Counterexamples can show that particular arguments do not support the conclusion, but they do not provide any positive reason for accepting a conclusion, nor can they show how a conclusion is related to other ideas.

If counterexample reasoning is not a good way to reach conclusions about complex issues, and it is a good way to construct arguments to defeat adversaries, then we should be careful when we do philosophy to bear this in mind. Instead, most of the time we present adversary arguments as if they were the only way to reason. The adversary paradigm prevents us from seeing that systems of ideas which are *not* directed to an adversary may be worth studying and developing, and that adversarial reasoning may be incorrect for nonadversarial contexts.

How would discarding the Adversary Paradigm affect philosophy? Any paradigm in philosophy will restrict the way reasoning is evaluated. I have argued that the Adversary Paradigm not only ignores some forms of good reasoning, but fails to evaluate and even encourages some forms of bad reasoning. However, criticism of the Adversary Paradigm is not enough; we need alternatives.

One of the problems with a paradigm that becomes really entrenched is that it is hard to conceive of how the field would operate without it. What other method of evaluating philosophy is there but the Adversary Method?

An alternative way of evaluating reasoning, already used in the history of philosophy and history of science, is to consider how the reasoning relates to a larger system of ideas. The questions to be asked are not just "Must the argument as it stands now be accepted as valid?" but also "What are the most plausible premises that would make this argument a good one?" "Why is this argument important?" "How does its form and its conclusion fit in with other beliefs and patterns of reasoning?" For example, one can consider not only whether Descartes' proofs of the existence of God are valid, but what good reasons there are for proving the existence of God; how Descartes' concept of God is related to his concept of causation and of matter. One can examine the influence of methodology and instrumentation

in one scientific field on the development of a related field.[23] With such
an approach relations of ideas that are not deductive can also be evaluated.
We can look at how world views relate to different philosophical positions
about free will and determinism, about rationality and ethical values, about
distinctions claimed between mind and body, self and other, order and
chaos.

A second way of treating systems of ideas involves a greater shift from
the Adversary Paradigm. It may even require a shift in our concept of reasoning for it to be accepted. It is that experience may be a necessary element
in certain reasoning processes. While many philosophers recognize that
different factual beliefs, and hence basic premises, may arise from different
experiences, it is believed that philosophical discussions ought to proceed
as if experience plays no essential role in the philosophical positions one
holds. Experience may be necessary to resolve factual disputes but aside
from errors about the facts, any differences in experience that might account
for differences in philosophical beliefs are ignored or denied. It is thought
that all genuine philosophical differences can be resolved through language.
This belief supports the Adversary Paradigm, for adversarial arguments could
be pointless if it was experience rather than argument that determined
philosophical beliefs. Yet might it not be possible, for example, that belief
in a supreme deity is correlated with perceived ability to control one's future?
When there is little control, when one is largely powerless to organize one's
environment, then belief in a deity helps one to understand, to be motivated
to go on, to keep in good spirits. When one feels effective in coping with
the world, then belief in a supreme being does not contribute to a satisfactory outlook. Belief in a deity would benefit, would be rational for the very
young, the very old, the poor and the helpless. But for others, with the
experience of being able to control their own lives and surroundings, the
difference in experience would give rise to a different belief.

I am not arguing for this account, but suggesting it as an illustration for
how different experiences could determine different philosophical positions
which are not resolvable by argument. A similar case might be made for
differences in the free will/determinism issue.

These alternatives to the Adversary Paradigm may be objected to by philosophers who are under the delusion that philosophy is different from science,
that unlike science, its evaluation procedures are exact and value-free. But
for those who accept that what philosophers have said about science (that
scientific evaluation is not free from uncertainty and values, because it is

dependent on paradigms) is also true of philosophy, other means of evaluation besides the Adversary Method will not be so objectionable.

I have been criticizing the use of the Adversary Method as a paradigm. And I think one of the best ways to reduce its paradigm status is to point out that it *is* a paradigm, that there are other ways of evaluating, reasoning about and discussing philosophy.

Smith College

NOTES

1 From Ann Ferguson, 'Androgyny as an Ideal for Human Development', in *Feminism and Philosophy*, eds. M. Vetterling-Braggin, F. Elliston and J. English (Totowa, New Jersey: Littlefield, Adams and Co., 1977), p. 47.

2 Robin Lakoff, *Language and Woman's Place* (New York: Harper and Row, 1975).

3 Logical positivism.

4 Sir Karl Popper, *The Logic of Scientific Discovery* (New York: Harper and Row, 1958).

5 Thomas Kuhn, *The Structure of Scientific Revolutions*, 2nd edition (University of Chicago Press, 1962).

6 It may be that the Adversary Method is only part of the larger paradigm that distinguishes reason from emotion, and segregates philosophy from literature, aligning it with science (dichotomies that Martha Nussbaum [*Philosophy and Literature* 1, 1978] attributes to Plato). Believing that emotions ought not to affect reasoning, it may seem to follow that who one addresses and why, ought not to affect the reasoning either. I consciously employ the kinship philosophy claims with science in this paper, arguing that truths we have learned about scientific reasoning ought to hold for philosophic reasoning as well.

7 T. Kuhn, in *Criticism and the Growth of Knowledge*, 'Reflections on My Critics', ed. Imre Lakatos and Alan Musgrave (Cambridge University Press, 1970), 231–278.

8 See the *Meno*, 75d–e.

9 Perhaps this is why Emerson, Carlyle and others are discussed only as part of English literature, and their views are not studied much by philosophers. They are not addressing adversaries, but merely presenting a system of ideas.

10 See Richard Robinson, *Plato's Earlier Dialectic* (Oxford: Clarendon Press, 1953) for this view of Socrates' style. I don't mean to single out Robinson for what seems to be the usual interpretation of Socrates. Robinson, at least thought irony and insincerity objectionable. The term "irony" covers a variety of styles including feigned ignorance to upset an opponent, vicious sarcasm and good natured teasing. It is only the latter that would be justifiably attributed to Socrates from the evidence in the dialogues.

11 See *Euthydemus* 227d, 288d, 295d, where Socrates' method is contrasted with Euthydemus' jeering and belligerent style, and *Meno* 75c–d where Socrates contrasts the present friendly conversation with that of a disputatious and quarrelsome kind. Socrates disapproved of ridicule (*Laches* 1959, *Gorgias* 473d–e, *Euthydemus* 278d, and *Protagorus* 333e).

[12] Socrates teases Polus to get him to change his style (*Gorgias* 461c–462a) and responds to Callicles' insults with praise to get him to agree to a dialogue. Socrates flirts with Meno when he resists questioning (*Meno*, 76b–c) and draws out Lysis by getting him to laugh at his questions (*Lysis*, 207c and ff.).

[13] *Euthydemus* 288b, 259d, 277d.

[14] For example, Donald Davidson, 'Truth and Meaning' *Synthese* 17 (1967), 304–323.

[15] T. Kuhn, 'Reflections on My Critics', op. cit., p. 234. See Feyerabend, Watkins, etc. in that volume and Dudley Shapere's review of *Structure of Scientific Revolutions*, in *Philosophical Review*.

[16] For example, Alan Gewirth, *Reason and Morality* (Chicago University Press, 1978).

[17] See particularly Brian Medlin, 'Ultimate Principles and Ethical Egoism', *Australasian Journal of Philosophy* 39 (1957), 111–18.

[18] See, for example, Ludwig Wittgenstein, *Zettel* 12.9.16 "Now it is becoming clear why I thought that thinking and language were the same. For thinking is a kind of language."

[19] Many people disagree with the universal beneficence and supremacy of moral considerations advocated by current ethical theories and think that they, and many others, by putting their own interests first, are thereby egoists. But their limited beneficence, which Hume thought was the foundation of morality, is very different from the egoism headlined by philosophers. A philosopher's egoist has *no* moral beliefs and not only thinks "me first" but does not care who comes second, third, or last. A philosopher's egoist has no loyalties to ideals or people and is quite indifferent about the survival and well being of any particular individual or thing.

[20] See John Rawls, *A Theory of Justice* (Cambridge, MA: Belknap Press, 1971), p. 191, where he says: "Nothing would have been gained by attributing benevolence to the parties in the original position" rather than egoism because there would be some disagreements even with benevolence. But surely the reasoning needed for people who care about others will be different than for people who do not care about others at all.

[21] Judith Jarvis Thomson, 'A Defense of Abortion,' *Philosophy and Public Affairs* 1, no. 1, 1971.

[22] Thomson, in general, makes it very clear that she is addressing an adversary. Nevertheless, she does claim to reach some conclusion about the morality of abortion, although the central issues for people making the decision are barely discussed – the consequences. See her section 8.

[23] Lindley Darden and Nancy Maull, 'Interfield Theories', *Philosophy of Science* 44 (1977), 43–64.

KATHRYN PYNE ADDELSON

THE MAN OF PROFESSIONAL WISDOM*

1. COGNITIVE AUTHORITY AND THE GROWTH OF KNOWLEDGE

Most of us are introduced to scientific knowledge by our schoolteachers, in classrooms and laboratories, using textbooks and lab manuals as guides. As beginners, we believe that the goal of science is "the growth of knowledge through new scientific discoveries."[1] We believe that the methods of science are the most rational that human kind has devised for investigating the world and that (practiced properly) they yield objective knowledge. It seems to us that because there is only one reality, there can be only one real truth, and that science describes those facts. Our teachers and our texts affirm this authority of scientific specialists.

The authority of specialists in science is not per se an authority to command obedience from some group of people, or to make decisions on either public policy or private investment. Specialists have, rather, an epistemological or cognitive authority: we take their understanding of factual matters and the nature of the world within their sphere of expertise as knowledge, or as the definitive understanding. I don't mean that we suppose scientific specialists to be infallible. Quite the contrary. We believe scientific methods are rational because we believe that they require and get, criticism of a most far-reaching sort. Science is supposed to be distinguished from religion, metaphysics, and superstition *because* its methods require criticism, test, falsifiability.

Our word "science" is ambiguous. Is it a body of knowledge, a method, or an activity? Until recently, many Anglo-American philosophers of science ignored science as an activity and applied themselves to analyzing the structure of the body of knowledge which they conceived narrowly as consisting of theories, laws, and statements of prediction. To a lesser degree, they spoke of "scientific method," conceived narrowly as a set of abstract canons. With such an emphasis, it is easy to assume that it is theory and method which give science its authority. It is *easy* to assume that researchers' cognitive authority derives from their use of an authoritative method, and that they are justified in exercising authority only within the narrow range of understanding contained in the theories and laws within their purview. Everyone knew, of

165

Sandra Harding and Merrill B. Hintikka (eds.), Discovering Reality, 165–186.
Copyright © 1983 by Kathryn Pyne Addelson.

course, that "scientific method" had been developed within an historical situation; but commitment to abstract canons led philosophers to put aside questions of how particular methods were developed, came to dominate, and (perhaps) were later criticized and rejected.[2] Everyone believed that it is an essential characteristic of scientific knowledge that it *grows*, and that new theories are suggested, tested, criticized, and developed. But the narrow focus on knowledge as theories and laws, and the emphasis on analyzing their abstract structure or the logical form of scientific explanation, led philosophers to neglect asking how one theory was historically chosen for development and test rather than another.[3] Most important, philosophers did not ask about the social arrangements through which methods and theories came to dominate or to wither away. Although the "rationality of science" is supposed to lie in the fact that scientific understanding is the most open to criticism of *all* understanding, a crucial area for criticism was ruled out of consideration: the social arrangements through which scientific understanding is developed and through which cognitive authority of the specialist is exercised.

Within the past twenty years, many scientists, historians, and philosophers have begun to move away from the abstract and absolutist conceptions of theory and method. The work which has reached the widest audience is Thomas Kuhn's *The Structure of Scientific Revolutions*. Two major changes in analytic emphasis show in his work. First, Kuhn focuses on science as an activity. Second, given this focus, he construes the content of scientific understanding to include not only theories and laws but also metaphysical commitments, exemplars, puzzles, anomalies, and various other features. Altering the focus to activity does lead one to ask some questions about rise and fall of methods and theories, and so Kuhn could make his famous distinction between the growth of knowledge in normal science and its growth in revolutionary science. However, he makes only the most limited inquiry into social arrangements in the practice of science. And, although he says that under revolutionary science, proponents of the old and the new paradigms may engage in a power struggle he does not explicitly consider cognitive authority.[4] Yet the power struggle in a period of revolutionary change is over which community of scientists will legitimately exercise cognitive authority – whose practices will define the normal science of the specialty and whose understanding will define the nature of the world which falls within their purview. To take cognitive authority seriously, one must ask seriously after its exercise, as embodied in social arrangements inside and outside science.

Within the activity of science in the United States today, researchers exercise cognitive authority authority in various ways.[5] One major way is within the specialties themselves. In accord with the norm of the "autonomy of science," researchers develop hypotheses and theories, discover laws, define problems and solutions, criticize and falsify beliefs, make scientific revolutions. They have the authority to do that on matters in their professional speciality: microbiologists have authority on questions of viruses and demographers, on questions of population changes, though within a specialty some have more power to exercise cognitive authority than others. Researchers have authority to revise the history of ideas in their field, so that each new text portrays the specialty as progressing by developing and preserving kernels of truth and rooting out error and superstition, up to the knowledge of the present.

Researchers also exercise cognitive authority outside their professions, for scientific specialists have an authority to define the true nature of the living and non-living world around us. We are taught their scientific understanding in school. Public and private officials accept it to use in solving political, social, military, and manufacturing problems. The external authority follows the lines of the internal authority. Experts are hired and their texts adopted according to their credentials as specialists in the division of authority-by-specialty within science. But because, within specialties, some people have more power than others, many people never have a text adopted and most never serve as expert advisors.

If we admit Kuhn's claim that metaphysical commitments are an integral part of scientific activity, then we see that scientific authority to define the nature of the world is not limited to the laws and theories printed in boldface sentences in our textbooks. Metaphysical commitments are beliefs about the nature of the living and non-living things of our world and about their relations with us and with each other. In teaching us their scientific specialties, researchers simultaneously teach us these broader understandings. Speaking of Nobel Prize winners in physics from the time of Rontgen to Yukawa, Nicholas Rescher says,

The revolution wrought by these men in our understanding of nature was so massive that their names became household words throughout the scientifically literate world. (Rescher, 1978, p. 27)

The Darwinian revolution, even more thoroughgoing, changed the metaphysics of a world designed by God in which all creatures were ordered in a great chain of being, to a world of natural selection. These scientific breakthroughs weren't simply changes in laws, hypothese and theories. They were changes in scientists' understanding of the categories of reality, changes in the questions they asked, the problems they worked with, the solutions they found acceptable. After the revolution, the changed understanding defined "normal science."

I will use the notion of cognitive authority to argue that making scientific activity more rational requires that criticizing and testing social arrangements in science be as much a part of scientific method as criticizing and testing theories and experiments. In doing this, I will talk a little more about how cognitive authority is exercised within professional specialties (Part 2). To make my case, I assume (with Kuhn) that metaphysical commitments are an essential part of scientific understanding, and that greater rationality in science requires criticizing such commitments. I give a number of examples in Part 3. Part 4 considers whether social arrangements within the sciences limit criticism of scientific understanding. I suggest that prestige hiararchies, power within and without the scientific professions, and the social positions of researchers themselves affect which group can exercise cognitive authority. Thus these features of social arrangements play a major role in determining which metaphysical commitments come to dominate, thus what counts as a legitimate scientific problem and solution. In the end, they affect how we all understand the nature of our world and our selves.

2. COGNITIVE AUTHORITY, AUTONOMY, AND CERTIFIED KNOWLEDGE

Philosophy of science texts, as well as the *New York Times*, talk of "science" and "scientific knowledge" as if there were one unified activity and one stock of information. But, as we all know, there are many scientific specialties, each with its own Ph.D. program. Members of each specialty or subspecialty certify and criticize their own opinions in their own journals and at their own professional meetings. Each specialist shows excellence by climbing the prestige ladder of the specialty.

Only some of the many people who work within a research specialty have epistemological authority within it. Barbara Reskin remarks,

The roles of both student and technician are characterized by lower status and by a technical division of labor that allocates scientific creativity and decision making to scientists and laboratory work to those assigned the role of technician or student. (Reskin, 1978, p. 20)[6]

The role division justifies assigning credit to the chief investigator, but its most important effect is on communication. Technicians and students work on the chief investigator's problems in ways he or she considers appropriate.[7] They rarely communicate with other researchers through conferences or journal articles or by chatting over the WATS line. They are not among the significant communicators of the specialty.

Researchers who *are* significant communicators set categories for classifying their subjects of study, and they define the meaning of what is taking place. With the aid of physicists, chemists define what chemical substance and interaction are. Microbiologists categorize viruses and molecules and explain the significance of electron microscopes. These are different from the understandings we all have of the physical and chemical parts of our world *as we live in it*. We choose honey by taste, smell and color, not by chemical composition, and we meet viruses in interactions we know as flus and colds.[8]

Scientific understandings appear in hypotheses, laws, and theories, but they presuppose metaphysics and methodology. Thomas Kuhn mentions the importance of metaphysics, using physical science in the seventeenth century as an example:

[Among the] still not unchanging characteristics of science are the . . . quasi-metaphysical commitments that historical study so regularly displays. After about 1630, for example, and particularly after the apperance of Descartes' immensely influential scientific writings, most physical scientists assumed that the universe was composed of microscopic corpuscles and that all natural phenomena could be explained in terms of corpuscular shape, size, motion, and interaction. That nest of commitments proved to be both metaphysical and methodological. As methodological, it told them what ultimate laws and fundamental explanations must be like: laws must specify corpuscular motion and interaction, and explanation must reduce any given natural phenomenon to corpuscular action under these laws. More important still, the corpuscular conception of the universe told scientists what many of their research problems should be. (Kuhn, 1970, p. 41)

The seventeenth century scientists also made metaphysical assumptions which most contemporary scientists share. They assumed that because there is *one* reality, there can be only one correct understanding of it. That metaphysical assumption, disguised as a point of logic, took root in western thought more than two millenia ago, when Parmenides said being and thinking are the same; that which exists and that which can be thought are the same. From that maxim, he concluded that the reality which is the object of knowledge, and not mere opinion, is one and unchanging. Differ though they might about the nature of reality, both Plato and Aristotle shared the metaphysical assumption that the object of scientific knowledge is the one, essential, intelligible structure of the one reality. Contemporary scientists share an analogous metaphysical assumption when they presuppose that reality is known through universal laws and predictions, which give the correct description of the world. All admit, of course, that in its present state, scientific knowledge is partial, suffering from inaccuracies, and so on. But, they say, this incompleteness and error is what is to be corrected by the scientific method. In principle the scientific enterprise is based on the metaphysical premises that because there is one reality, there must be one, correctly described truth. This premise is the foundation of the cognitive authority of scientific specialists. The specialist offers the correct understanding of reality while the lay person struggles in the relativity of mere opinion.

In some specialties, researchers deal with living subjects which may have their own understanding of what is going on. But the researcher's understanding is scientifically definitive. Donna Haraway reports on the 1938 field studies of rhesus monkeys done by Clarence Ray Carpenter, an outstanding scientist in his day (Haraway, 1978, p. 30). The studies were designed to answer questions about dominance and social order. Researchers observed an "undisturbed" group for a week as a control, then removed the "alpha male" (the dominant male in terms of priority access to food and sex). Carpenter found that without the apha male, the group's territory was restricted relative to other groups and intra-group conflict and fights increased. As the next two males in order of dominance were removed, social chaos seemed to result. Upon returning the males, researchers observed that social order was restored.

"Alpha male" is obviously a technical term. But understanding primate subjects in terms of dominance and competition is Carpenter's definition of the situation, not the rhesus monkeys' — and not the understanding of their former keepers or the man in the street either, for Carpenter was making scientific discoveries. Underlying Carpenter's definition of the primate interactions is a metaphysics. Haraway puts the understanding this way:[9]

True social order must rest on a balance of dominance, interpreted as the foundation of cooperation. Competitive aggression became the chief form that organized other forms of social integration. Far from competition and cooperation being mutual opposites, the former is the precondition of the latter — on physiological grounds. If the most active (dominant) regions, the organization centers, of an organism are removed, other gradient systems compete to reestablish organic order: a period of fights and fluidity ensues. (Haraway, 1978, p. 33)

The metaphysics clearly enters into questions asked — from "Which male is dominant?" to "What happens to social order when the dominant males are removed?"

If we look at scientific research this way, then theories and explanations can be taken as *the conventional understandings among significant communicators in a scientific specialty of the interactions of researchers with the subjects of the field of study, and of the interactions of those subjects among themselves and with their environments* (whether natural, social, or laboratory).[10]

The conventional understandings are published in journals and presented at conferences, and the significant communicators of the specialty criticize and alter the understandings, correct hypotheses, expand theories and explanations — as part of the process of certifying the knowledge. Some of the understandings are shown to be false in the process. Others emerge as pretty certainly true. Eventually, conventional understandings are ritualized in college texts. To use other language:

The community's paradigms (are) revealed in its textbooks, lectures, and laboratory exercises. By studying them and by practicing with them, the members of the corresponding community learn their trade. (Kuhn, 1970, p. 43)

Through the textbooks and lectures, and through advice to government and industry, the conventional understandings are passed on to the rest of us as part of the exercise of the specialist's external authority.

3. METAPHYSICAL COMMITMENTS AND THE GROWTH OF
SCIENTIFIC KNOWLEDGE

Within a specialty, certified knowledge consists, at any given time of the conventional understandings of researchers not only about the subjects and instruments of their fields of study but about metaphysics, methodology, and the nature of science itself. Scientific discovery and the extension of certified knowledge may therefore sometimes arise from a change in understandings of metaphysics, methodology and the appropriate questions to ask rather than changes in theories, methods or instruments. The example of functionalist metaphysics in some of the life sciences and social sciences indicates how knowledge has grown as a result of such changes as widespread sharing of a metaphysics across disciplines. Clarence Ray Carpenter's work is an example. Concurrent with the sharing of a metaphysics, there may be differences in the understanding of that metaphysics. These differences seem important to the criticism which leads to scientific advance; my example here is Robert K. Merton's theory of deviance.

Finally, although a metaphysics may be shared across disciplines, a single discipline may contain quite different and competing metaphysical commitments. The interactionist theory of deviance in sociology illustrates this and shows how a difference in metaphysics may lead to a difference in scientific questions asked. The interactionist example also indictes that the scientific questions asked may not merely define factual problems, they may also define *social problems*. This raises serious questions about the epistemological authority we grant to research specialists and leads to the analysis of power and cognitive authority in Part 4.

In our own day, advance of knowledge in the life sciences has included a change in metaphysics. The historian of science, Donna Haraway, says,

> Between World War I and the present, biology has been transformed from a science centered on the organism, understood in functionalist terms, to a science studying automated technological devices, understood in terms of cybernetic systems. (Haraway, 1979, p. 207)

Haraway does not talk about metaphysical changes. But the change she mentions involves a move to a new metaphysics, one based not on the organism and a physiological paradigm but on "the analysis of information and energy in statistical assemblages," a "communication revolution."

> A Communication revolution means a retheorizing of natural objects as technological devices properly understood in terms of mechanism of production, transfer, and storage of information ... Nature is structured as a series of interlocking cybernetic systems, which are theorized as communications problems. (Haraway, 1979, p. 222–23)

Individual specialties which share a metaphysics change on a widespread basis when there is a metaphysical change of this sort, and scientific knowledge grows by leaps and bounds. For example, the "communication revolution" made possible the revolutionary discoveries in genetics after the second World War. Those discoveries in turn gave plausibility and prestige to the communications metaphysics. This is one way in which metaphysics spreads.

Clarence Ray Carpenter worked under the earlier metaphysics (based on the organism) in his rhesus monkey studies I mentioned above. He conceived social space to be like the organic space of a developing organism, and he shared functionalist metaphysical presuppositions which were current. Haraway remarks,

Functionalism has been developed on a foundation of organismic metaphors, in which diverse physiological parts of subsystems are coordinated into a harmonious, hierarchical whole. (Haraway, 1978, p. 40)

Carpenter himself was important in the cross-disciplinary spread of metaphysics. Haraway says that, theoretically, Carpenter "tied the interpretations of the laboratory discipline of comparative psychology and sex physiology to evolutionary and ecological field biology centered on the concepts of population and community" (Haraway, 1978, p. 30).

Functionalism (in various forms) was also the metaphysics of some of the most progressive work done in sociology and anthropology during Carpenter's time, and even today it underlies some respected work in those fields. Sociologist Robert K. Merton was a significant communicator in sociology. In an essay written in 1949, he remarks on the widespread use of the "functional approach":

The central orientation of functionalism − expressed in the practice of interpreting data by establishing their consequences for larger structures in which they are implicated − has been found in virtually all the sciences of man − biology and physiology, psychology, economics, and law, anthropology and sociology. (Merton, 1949, p. 47)

In this essay, Merton reviews literature in the social sciences and he clarifies the notion of function − relying on the use of the concept in other fields:

Stemming in part from the native mathematical sense of the term, (the sociological) usage is more often explicitly adopted from the biological sciences, where the term function is understood to refer to the "vital or organic processes considered in the respects in which they contribute to the maintenance of the organism." (Merton, 1949, p. 23)

Merton insists that he is only borrowing a *methodological framework* from the biological sciences. In fact, the framework carries with it a metaphysics − as the change in life sciences reported by Haraway shows. This widespread use of metaphysics confirms its truth, through the internal authority of specialists.[11] It changes dominant world views in a society through the external authority of specialists.

Although a generally functionalist metaphysics was widely shared between the two World Wars, it is more accurate to speak of *varieties* of functionalism. Some of the varieties arose (or were clarified) through metaphysical criticism of earlier varieties.[12] I'll use one of Robert K. Merton's criticisms as an example.

In 'Social Structure and Anomie' (originally published before World

War II), Merton developed a theory of deviance in which he criticized an earlier, widely held metaphysics:

A decade ago, and all the more so before then, one could speak of a marked tendency in psychological and sociological theory to attribute the faulty operation of social structures to failures of social control over man's imperious biological drives. The imagery of the relations between man and society implied by this doctrine is as clear as it is questionable. In the beginning, there are man's biological impulses which see full expression. And then, there is the social order, essentially an apparatus for the management of impulses . . . Nonconformity with the demands of a social structure is thus assumed to be anchored in original nature. It is the biologically rooted impulses which from time to time break through social control. And by implication, conformity is the result of a utilitarian calculus or of unreasoned conditioning. (Merton, 1949, p. 125)

Haraway calls this "management of impulses" perspective the "body politic" view, and she indicates that Clarence Ray Carpenter (and many others) held it.[13] So, for that matter, did Freud and Aristotle.[14] In criticizing this metaphysics, Merton advanced knowledge in a way that was both scientifically enlightened and morally humane.

As a functionalist, Merton does suppose that deviance indicates faulty operation of the social structure. His metaphysics assumes social structures and functions exist in a way that makes them suitable for use in scientific explanation. But social structures are not understood as functioning to restrain biological impulses. Using monetary success goals in American culture as his example, he considers how people in different social positions adapt to those goals. He notes that rates of some kinds of deviance are much higher in the underclasses than in the upper classes. Rather than saying that some members of the underclasses are driven by ungoverned impulses, he suggests they are responding normally to problematic social conditions they face. They accept the goal of monetary success but have little opportunity to achieve it through legitimate, institutionally approved means. So some choose innovative means to reach the goal — perhaps becoming criminally deviant as bank robbers. Others, unable to take the strain, may retreat and become "psychotics, autists, pariahs, outcasts, vagrants, vagabonds, tramps, chronic drunkards, and drug addicts."[15] But it's not due to "ungoverned impulses." Rather, the cause is that "some social structures exert a definite pressure on certain persons in the society to engage in nonconformist rather than conformist conduct." (Merton, 1949, p. 125)

Conformist conduct, in Merton's view, is conduct within an institution which serves a positive function in the society. It's not simply that being a banker is approved and being a bank robber is disapproved. Banking is an institution which contributes to the stability of our society. Bank robbing undermines stability, running a danger of destroying the vital processes necessary to maintenance of our social organism. In this regard, Merton says,

Insofar as one of the most general functions of social structure is to provide a basis for predictability and regularity of social behavior, it becomes increasingly limited in effectiveness as these elements of the social structure become dissociated. At the extreme, predictability is minimized and what may be properly called anomie or cultural chaos supervenes. (Merton, 1949, p. 149)

Scientific knowledge grew by Merton's criticism of one variety of functionalism. This "pluralism" of metaphysics seems as important as the sharing of metaphysics. For example, within sociology, functionalists compete with the tradition called interactionism, whose adherents assume that human society must be explained in terms of acting units which themselves have interpretations of the world. Interactionist Herbert Blumer criticizes functionalist metaphysics in this way:

Sociological thought rarely recognizes or treats human soceites as composed of individuals who have selves. Instead, they (sic) assume human beings to be merely organisms, with some kind of organization, responding to forces that play upon them. Generally, though not exclusively, these forces are lodged in the make-up of the society as in the case of "social system" . . .

. . . Some conceptions, in treating societies or human groups as "social systems" regard group action as an expression of a system, either in a state of balance or seeking to achieve balance. Or group action is conceived as an expression of the "functions" of a society or a group . . . These typical conceptions ignore or blot out a view of group life or of group action as consisting of the collective or concerted actions of individuals seeking to meet their life situations. (Blumer, 1967, p. 143–44)

The interactionist theory of deviance, quite different from Merton's functionalist theory, offers a good case in which to see how a different metaphysics influences the questions asked.

In 1963, Howard Becker's *Outsiders* was published, marking the emergence of what was misleadingly called the "labeling" theory of deviance. In an interview about the development of the theory, Becker said, "The theory, and it really was a pretty rudimentary theory, wasn't designed to explain why people robbed banks but rather how robbing banks came to have the quality of being deviant" (Debro, 1970, p. 167). Merton looked for the *cause* of deviant behavior (as did the "management of impulses" functionalists he criticized) and found the cause in "social structures exerting a definite pressure on some people." Becker asks about deviant behavior as behavior under *ban*, and so he asks about who does the banning, how the ban is maintained, and what effect the ban has on the activity itself.[16] On the basis of interactionist metaphysics, he doesn't assume that deviance is something there for the natural scientific eye to discern. Whether something is deviant or normal in a society is a question of perspective and power within the society. Although bank robbers and marijuana smokers may be considered deviant by "the population at large," that is, in the dominant opinion, the deviants themselves have their own perspective on the matter, and within their perspective, most aren't much interested in looking for the alleged *causes* of their activities so that they can be cured.

The interactionist criticism brings out an important connection between metaphysics, scientific questions, and social problems.[17] Bank robbing, pot smoking, and homosexuality are social problems in the eyes of certain segments of our population, not others. Becker and other sociologists in the interactionist tradition have argued that social problems don't exist for the neutral scientific eye to discern any more than deviance does. Something is a social problem or not depending on one's social position and perspective. It is often a political question – as is the question of what function a social institution serves. In fact, "the function of a group or organization (is sometimes) decided in political conflict, not given in the nature of the organization" (Becker, 1973, p. 7).

The examples I've given in this section indicate that metaphysical commitments have been important to scientific criticism and the growth of knowledge. In some cases at least, they may enter into the definition of a social problem and its solution.

Functionalist metaphysics was widely accepted in the natural sciences before World War II when Merton formulated his anomie. Functionalist sociologists offered a theory of society that was coherent with then current understandings in biology. This isn't simply a case of a "pseudoscience" (sociology) putting on the trappings of a "real science" (biology), for any biological specialty dealing with social organisms requires a theory of society and social behavior.[18] Rather, the sociological case shows that metaphysical understandings in the natural sciences help define the human world in which social problems are categorized and dealt with. In the Merton example, labeling theorist criticism indicates that the cognitive authority of science supported one set of political positions over another by that definition of the human world. This happened not by abuse of authority but by the normal procedures in the normal social arrangements of science. The time has now come to ask more explicitly how those social arrangements influence scientific criticism and the growth of knowledge.

4. COGNITIVE AUTHORITY AND POWER

In Part 2, I suggested that the conventional understandings of significant communicators in science are the definitive understandings of the nature of the world within their spheres of expertise. In Part 3, I suggested that metaphysical commitments are important to the growth of knowledge. In those sections, I spoke as though any researcher with the appropriate certificates of training could serve as a significant communicator, and the reader might think that if one group exercised greater cognitive authority it was on meritocratic or purely rational grounds: their theories and commitments have been shown to withstand test and criticism better than those of their competitors. I believe it is a valuable feature of the scientific enterprise that rational criticism is a factor in determining which group exercises cognitive authority. However, social arrangements are factors as well, and to the degree that we refuse to acknowledge that fact, we limit criticism and cause scientific work to be less rational than it might be. In this section, I shall indicate some social factors which may be relevant, and then I'll close with an example of how scientific understanding has been improved by recent criticisms which did take social arrangements into account.

First, let me take up questions of prestige. The sciences differ in prestige, physics having more than economics, and both having more than educational psychology. Specialties in a science too differ in prestige, experimental having more than clinical psychology, for example. Prestige differences affect researchers' judgments on which metaphysical and methodological commitments are to be preferred. Carolyn Wood Sherif remarks on the "prestige hierarchy" in psychology in the 1950s:

Each of the fields and specialties in psychology sought to improve its status by adopting (as well and as closely as stomachs permitted) the perspectives, theories, and methodologies as high on the hierarchy as possible. The way to "respectability" in this scheme has been the appearance of rigor and scientific inquiry, bolstered by highly restricted notions of what science is about. (Sherif, 1979, p. 98)

Many philosophers of science have not only taken prestige hierarchies to be irrelevant to scientific rationality, they have accepted the hiararchies themselves and in doing so have shared and justified "highly restricted notions of what science is about." This failing was blatant in the work of logical positivists and their followers, for they constructed their analysis of scientific method to accord with an idealization of what goes on in physics, and they discussed the "unity of science" in a way that gave physics star status.[19]

Within specialties, researchers differ in prestige, so that some have access to positions of power while others do not. Some teach in prestigious institutions and train the next generations of successful researchers. Researchers judge excellence in terms of their own understandings of their field, of which problems are important, of which methods are best suited to solving them. Researchers in positions of power can spread their understandings and their metaphysical commitments. Consider the primatology example in Part 3.

Because Robert Yerkes held influential positions, he was able to give an important backing to Clarence Ray Carpenter's career, helping Carpenter to compete successfully for the positions and funding needed to do his research. Haraway says of Carpenter,

From his education, funding, and social environment, there was little reason for Carpenter to reject the basic assumptions that identified reproduction and dominance based on sex with the fundamental organizing principles of a body politic. (Haraway, 1978, p. 30)

Yerkes shared the "body politic" metaphysics. In helping Carpenter, he was helping spread his own metaphysical commitments.

The question is not whether top scientists in most fields produce some very good work but rather the more important question of whether other good work, even work critical of the top scientists, is not taken seriously because its proponents are not members of the same powerful networks and so cannot exercise the same cognitive authority. The question is made particularly difficult because, by disregarding or downgrading competing research, the "top scientists" cut off the resources necessary for their competition to develop really good work. In most fields it is next to impossible to do research without free time, aid from research assistants, secretaries, craftsmen, custodians, and in many cases, access to equipment.

Some very influential philosophers of science have insisted that criticism is an essential part of scientific method, and that criticism requires that there be competing scientific theories.[20] Accepting that as an abstract canon, one might philosophically point out the Yerkes should have encouraged more competition and (if one became particularly moralistic) that he should have been more careful about showing favoritism and bias. But it would be a mistake to describe Yerkes as showing favoritism and bias. As a matter of fact, he did much to set the practice of researchers investigating unpopular subjects and reaching unpopular results in the interests of scientific freedom and research in "pure science." But he made his judgments according to his own understanding of scientific research. Any researcher must do that. Researchers are also the judges of which competing theories it makes sense to pursue or to encourage others in pursuing. If this seems to result in bias, the way to correct it is not by blaming individual researchers for showing favoritisms because they depart from some mythical set of abstract canons. The way to begin to correct it is to broaden rational criticism in science by requiring that both philosophers of science and scientists understand how prestige and power are factors in the way cognitive authority is exercised.

So far, I have talked about influences on the exercise of cognitive authority with the scientific professions. Many people have observed that there are outside influences on scientific research —funding, for example. Given legally dominant understandings of capitalism in the United States, many people consider it proper for private business to fund research on problems that need solving for reasons of economic competition and expansion. Most of us considered it appropriate that public agencies in a democracy should fund research to help solve social problems of the moment (as do private philanthropic foundations for the most part). If we think of science as a stock of knowledge embodied in theories, then the problem of funding does not seem to be a problem having to do with rationality and criticism in science. Instead it may appear to be a question of political or other outside interference with the autonomy of the researchers, at worst preventing them from setting their own problems to investigate.[21] If we use the notion of cognitive authority, however, we may see that the question of funding indeed has to do with scientific rationality and with the content of our scientific understanding of the world.

Metaphysical commitments of a science tell scientists what many of their research problems should be (Kuhn, 1970, p. 41). As we saw from the Merton-Becker example, a difference in metaphysics may bring a difference in *what the problem is taken to be*. In that case, it was a difference between explaining why people rob banks and explaining how robbing banks comes to have the quality of being deviant (Debro, 1970, p. 167). Because problems investigated by a tradition are related to the metaphysical and methodological commitments of its researchers, some understandings of nature will have a better chance for support than others.[22] Those researchers will have a better opportunity to exercise cognitive authority and to help others of their metaphysical persuasion rise in the ranks. They will write the texts and serve as advisors and use their external authority to popularize their metaphysical outlook. So funding influences the content of science at a given historical moment and it influences the way we all come to understand the world.

The influence goes beyond the question of which of a number of competing traditions are to be rewarded. Arlene Daniels traces some of the ramifications in discussing Allan Schnaiberg's remarks on obstacles to environmental research. She says,

Schnaiberg shows us how unpopular socioenvironmental research is within establishment contexts. The science industries won't pay for it, the research foundations won't sanction it. Rewards in the academic market place depend on quick payoffs; accordingly, independent researchers there cannot wait for results that require large expenditures of time in unfunded research. (Daniels, 1979, p. 38–39)

This means not only that some existing metaphysical and methodological traditions will flourish while others are passed over. It means that potentially fruitful metaphysics and methods won't get a chance for development at all because social arrangements in the scientific professions and the influence of funding work against it.

So far, I've suggested that social arrangements within the scientific professions, and between those professions and the larger society, involve factors relevant to the exercise of cognitive authority in the sciences and thus to the content of scientific understanding. Are the professional and social life experiences of researchers also relevant to their metaphysical commitments? This is an extremely interesting question because of the quite general assumption in the United States that there is a privileged definition of reality which scientists capture, a main assumption underlying the authority we give them. Feminists in nearly every scientific field have questioned that assumption.[23] I questioned it from an interactionist perspective in discussing Merton, above. Let me give two suggestive examples here.

The anthropologist E. Ardener suggests that because of their social experience, men and women conceptualize their societies and communities differntly (Ardener, 1972). In most societies, men more frequently engage in political activities and public discourse and have the definitional problem of bounding their own society or community off from others. Models suited to the usual women's experience aren't the object of public discourse and so when circumstances call for it, women will use men's models, not their own.

Ethnographers tend to report the male models for three reasons, according to Ardener. They are more accessible to the researcher. Male models are the officially accepted ones in the ethnographer's home society. And they accord with the metaphysical and theoretical outlook of functionalism which, in the past, many ethnographers have held. Milton reports Ardener's claim this way:

Ethnographers, especially those who have adopted a functionalist approach, tend to be attracted to the bounded models of society, with which they are presented mainly by men and occasionally by women. These models accord well with functionalist theory and so tend to be presented as *the* models of society. (Milton, 1979, p. 48)

We need not accept Ardener's claims as gospel truth to realize that we *cannot* accept without test the empirical assumption that a specialist's social experience has no significant effect on his or her scientific understanding of the world.

Nor can we accept, without test, the empirical hypothesis that the long training and isolating, professional experience of scientific specialists has no significant effect on their scientific understanding. The sociologist Vilhelm Aubert says,

Members of society have, through their own planning and their own subsequent observations, verifications, and falsifications, built up a cognitive structure bearing some resemblances to a scientific theory. ... But ... social man behaves only in some, albeit important, areas in this purposive way. Any attempt, therefore, to stretch the predictability criterion beyond these areas – their limits are largely unknown – may result in a misrepresentation of the nature of human behavior. This danger is greatly increased by origin of most social scientists in cultures which heavily stress a utilitarian outlook, and by their belonging even to the subcultures within these, which are the main bearers of this ethos. A sociology produced by fishermen from northern Norway or by Andalusian peasants might have been fundamentally different. The leading social scientists are people with tenure and right of pension. (Aubert, 1965, p. 135)

The leading physicists, biologists, and philosophers of science are also people with tenure and right of pension. They live in societies marked by dominance of group over group. As specialists, they compete for positions at the top of their professional hierarchies which allow them to exercise cognitive authority more widely. Out of such cultural understandings and social orderings, it is no wonder that we get an emphasis on predictive law and an insistence that the currently popular theories of a specialty represent the one, true, authoritative description of the world. It is no wonder that our specialists continually present us with metaphysical descriptions of the world in terms of hierarchy, dominance, and competition. The wonder is that we get any development of our understanding at all.

But we do. Scientific understanding does seem to grow (in however ungainly a fashion) and our knowledge does seem to "advance" (however crabwise).

In our own century, scientific knowledge has often seemed to grow at the expense of wisdom. However, the corrective isn't to dismiss science as hopelessly biased and wrongheaded and return to some kind of folk wisdom. We can't get along without science any more. The corrective seems rather to ferret out all the irrationalities we can find in scientific activity and to expand our understanding of what science and scientific rationality are. To do this, we should acknowledge metaphysical commitments as part of the content of scientific understanding and thus open them to scrutiny and criticism by specialist and non-specialist alike. Feminist criticism offers a very instructive example here. In the past ten years, political feminists have given lay criticisms of much of our scientific metaphysics. Other feminists have gained specialist training and brought the lay criticisms to bear on technical theories within their fields.[24] This was possible because sexism is a political issue at the moment and funding, journals, etc., are available for this sort of research and criticism. I am suggesting that we should institutionalize this sort of criticism and make it an explicit part of "scientific method". We should also try using the notion of cognitive authority and expanding the range of the criteria of scientific rationality and criticism so that it includes social arrangements within the scientific professions.

If we expand the range of criticism, I believe that philosophers of science and scientists as well will find themselves advocating change in our social system. This would not result in a sudden illegitimate politicization of science or an opening of the floodgates of irrationality. Quite the contrary. Because they have cognitive authority our scientists already *are* politicized. It is the *unexamined* exercise of cognitive authority within our present social arrangements which is most to be feared. Illegitimate politicization and rampant irrationality find their most fruitful soil when our activities are mystified and protected from criticism.

Smith College

NOTES

* Research for this paper was supported in part by a grant from the National Endowment for the Humanities and by the Mellon Foundation grant to the Smith College Project on Women and Social Change. I am very grateful for criticism or advice I received from Howard Becker, Donna Haraway, Arlene Daniels, Sandra Harding, Vicky Spelman, Helen Longino, Kay Warren, Noretta Koertge, Arnold Feldman, and members of two seminars I taught in the Northwestern University sociology department, Fall 1980. I have previously published work under the name Kathryn Pyne Parsons.

1 The quoted remark is from Cole (1979, p. 6n).

2 Instead, philosophers criticized each other's versions of the abstract canons. Positivist Rudolf Carnap was particularly painstaking at criticizing his own and other positivists' analyses of the structure of scientific theories. See, for example, Carnap (1956). Karl Popper also devoted time and energy to criticizing the positivists. See, for example, Popper (1965).

3 At the beginning of the "new wave" in philosophy of science, N. R. Hanson did ask after the choosing of new theories, but he did so by discussing the logic of discovery and the ways in which theories groups of scientists develop are constrained by their patterns of conceptual organization rather than by asking after constraints in the social arrangements within which scientific understanding is developed and criticized. See Hanson (1958).

4 It is there implicitly, however, particularly in his wonderful discussion of science texts.

5 Whether or not a group has authority regarding something depends on social arrangements in the society in which they form a group, thus my restriction to "the United States today." I should make more severe restrictions because there are subgroups in the USA which don't grant "scientists" much authority. We do through our public and major private educational systems, however.

6 One doesn't usually think of artisans as part of science, but one physicist said of his university's craftsmen, "The gadgets they produce for us are just crucial. The reason the work the department does is internationally competitive with major research centers all over the world is in part due to the capabilities of the people in the machine shop. Some of the research simply could not be done without them." *Contact*, August 1980, page 8. (University of Massachusetts, Amherst, publication).

7 Within the hierarchical social relations of the research group, the chief investigator has authority to command obedience from technicians, students, secretaries, and the like. I'm not concerned with that sort of authority in this paper.

8 My remark about honey may still be true, but due to changes in the food industry consequent to the "growth of scientific knowledge," we are learning to choose foods by applying the chemist's categories to lists of ingredients on packages at the supermarket.

9 Haraway doesn't explicitly talk about metaphysics in her paper, and in fact she may use the term in a more limited way than I do. (personal communication)

10 In fields like history or archeology, researchers themselves interact with non-living material not with the (formerly living) subjects of study, but their theories and explanations represent the researchers' conventional understandings of those subjects.

11 Merton himself says, "The prevalence of the functional outlook is in itself no warrant

for its scientific value, but it does suggest that cumulative experience has forced this orientation upon the disciplined observers of man as biological organism, psychological actor, members of society and bearer of culture" (Merton, 1949, p. 47). A whole metaphysic and theory of science underlies that remark, as the reader may see by comparing my remarks and those of other authors in Part 4 of this paper. For example, Merton hints that scientific observation and laboratory and field experience "forces" the outlook, while some authors I report on in Part 4 suggest social experience in the professions and the specialists' society are major influences.

12 Haraway (1978) also indicates that other researchers in the life sciences later criticized the metaphysics, from the standpoint of other varieties of functionalism.

13 See Haraway (1978).

14 See Elizabeth V. Spelman's paper in this volume for an illuminating discussion of the view as Aristotle held it.

15 Merton (1949, p. 142). Merton's theory of deviance is broader and more complex than I am representing it here. I am selecting features for a dual purpose: to show the criticism of the "uncontrolled impulse" view and to contrast with the interactionist theory I give below.

16 Becker himself does not speak of behavior under ban. That conceptualization is David Matza's (1969).

17 See Spector (1977) for a discussion of social problems.

18 Philip Green discusses this issue in criticizing sociobiologists' theories in his (1981). See also the introduction and several of the essays in Addelson (n.d.).

19 Some philosophers of science have insisted that the methods of the physical sciences aren't suitable for historical sciences — see, for example, the whole Verstehen controversy (Collingwood, 1946) as a classic source. For social sciences generally see Winch (1958).

20 See Popper (1965) and Feyerabend (1970).

21 George H. Daniels suggests that the rise of the ideal of pure scientific research in the late nineteenth century led to conflicts with democratic assumptions in 'The Pure-Science Ideal and Democratic Culture,' Science 156 (1967), 1699–1705. My discussion here displays the other side of the conflict.

22 I'm not claiming here that stating a problem in a certain way entails that you'll have a certain metaphysics, or even determines it in some unidirectional way. My point is about the ranges of theories and traditions available at an historical moment and which of them will receive encouragement and support.

23 See, for example, the essays in this volume, Millman (1975), and Sherman (1979).

24 Feminist criticism may seem more obviously politicized than, say, Yerkes's or Merton's or Becker's criticisms I discussed above, but I think that is because feminists themselves insist on the political connections.

REFERENCES

Addelson, Kathryn Pyne and Martha Ackelsberg: (n.d.), An Endless Waterfall: Studies on Women and Social Change, forthcoming.

Ardener, E.: 1972, 'Belief and the Problem of Women,' in J. S. LaFontaine (ed.), The Interpretation of Ritual (London: Tavistock).

Ardener, E.: 1975, 'The Problem Revisited,' in S. Ardener (ed.), *Perceiving Women* (London: Malaby Press).

Aubert, Vilhelm: 1965, *The Hidden Society*, (New Jersey: Bedminster Press).

Becker, Howard S.: 1973, *Outsiders* (New York: The Free Press).

Blumer, H.: 1967, 'Society as Symbolic Interaction,' in J. Manis and B. Mattzer (eds.), *Symbolic Interaction*, (Boston: Allyn and Bacon).

Carnap, Rudolf: 1956, 'The Methodological Character of Theoretical Concepts,' in H. Feigl and N. Scriven (eds.), *Minnesota Studies in the Philosophy of Science*, Vol I, (Minneapolis: University of Minnesota Press).

Cole, Jonathan R.: 1979, *Fair Science, Women in the Scientific Community* (New York: The Free Press).

Cole, J. and S. Cole: 1973, *Social Stratification in Science* (Chicago: University of Chicago Press).

Collingwood, R. G.: 1946, *The Idea of History* (Oxford: Oxford University Press).

Daniels, Arlene: 1979, 'Advocacy Research: Providing New Wares for the Free Marketplace of Ideas,' in *Sociology's Relations with the Community* (Calgary: University of Calgary Colloquium Proceedings).

Debro, Julius: 1970, 'Dialogue with Howard S. Becker,' *Issues in Criminology* 5 (2, Summer).

Feyerabend, P.: 1970, 'Against Method,' in M. Radner and S. Winokur (eds.), *Minnesota Studies in the Philosophy of Science*, Vol. IV. (Minneapolis: University of Minnesota Press).

Green, Philip: 1981, *The Pursuit of Inequality* (New York: Pantheon Books).

Hanson, N. R.: 1958, *Patterns of Discovery* (Cambridge: Cambridge University Press).

Haraway, Donna: 1978, 'Animal Sociology and a Natural Economy of the Body Politic,' Parts I and II, *Signs* 4 21–60.

Haraway, Donna: 1979, 'The Biological Enterprise: Sex, Mind and Profit from Human Engineering to Sociobiology,' *Radical History Review*, Spring-Summer special issue, 206–237.

Kuhn, T.: 1970, *The Structure of Scientific Revolutions*, second edition (Chicago: University of Chicago Press).

Matza, David: 1969, *Becoming Deviant* (Englewood Cliffs, N.J.: Prentice Hall).

Merton, Robert K.: 1949, *Social Theory and Social Structure* (Glencoe, Illinois: The Free Press).

Millman, M. and R. Kanter: 1975, *Another Voice* (New York: Anchor Press/Doubleday).

Milton, Kay: 1979, 'Male Bias in Anthropology,' *Man, The Journal of the Royal Anthropological Institute, London*, N.S. 14, 40–54.

Mullins, N. C.: 1973, *Science: Some Sociological Perspectives* (New York: Bobbs-Merrill Co.).

Popper, Karl: 1965, *Conjectures and Refutations: The Growth of Scientific Knowledge*, second edition (New York: Basic Books).

Rescher, Nicholas: 1978, *Scientific Progress* (Pittsburgh: University of Pittsburgh Press).

Reskin, Barbara F.: 1978, 'Sex Differentiation and the Social Organization of Science,' *Sociological Inquiry* 48 (3–4).

Sherif, Carolyn Wood: 1979, 'Bias in Psychology,' in Sherman and Beck (eds.) (1979), 93–124.

Sherman, Julia A. and Evelyn Torton Beck: 1979, *The Prism of Sex* (Madison: University of Wisconsin Press).

Spector, Malcolm and John Kitsuse (eds.): 1977, *Constructing Social Problems* (Menlo Pk., Cal.: Cummings Press).

Young, R. M.: 1969, 'Malthus and the Evolutionists,' *Past and Present* 43, 109–141.

Winch, Peter: 1958, *The Idea of a Social Science and Its Relation to Philosophy* (London: Routledge and Kegan Paul).

EVELYN FOX KELLER

GENDER AND SCIENCE*

I. INTRODUCTION

The requirements of . . . correctness in practical judgements and objectivity in theoretical knowledge . . . belong as it were in their form and their claims to humanity in general, but in their actual historical configuration they are masculine throughout. Supposing that we describe these things, viewed as absolute ideas, by the single word 'objective', we then find that in the history of our race the equation objective = masculine is a valid one (George Simmel, quoted by Horney, 1926, p. 200).

In articulating the commonplace, Simmel steps outside of the convention of academic discourse. The historically pervasive association between masculine and objective, more specifically between masculine and scientific, is a topic which academic critics resist taking seriously. Why is that? Is it not odd that an association so familiar and so deeply entrenched is a topic only for informal discourse, literary allusion, and popular criticism? How is it that formal criticism in the philosophy and sociology of science has failed to see here a topic requiring analysis? The virtual silence of at least the non-feminist academic community on this subject suggests that the association of masculinity with scientific thought has the status of a myth which either cannot or should not be examined seriously. It has simultaneously the air of being "self-evident" and "nonsensical" – the former by virtue of existing in the realm of common knowledge (i.e., everyone knows it), and the latter by virtue of lying outside the realm of formal knowledge, indeed conflicting with our image of science as emotionally and sexually neutral. Taken seriously, it would suggest that, were more women to engage in science, a different science might emerge. Such an idea, although sometimes expressed by non-scientists, clashes openly with the formal view of science as being uniquely determined by its own logical and empirical methodology.

The survival of mythlike beliefs in our thinking about science, the very archetype of antimyth, ought, it would seem, to invite our curiosity and demand investigation. Unexamined myths, wherever they survive, have a subterranean potency; they affect our thinking in ways we are not aware of, and to the extent that we lack awareness, our capacity to resist their influence is undermined. The presence of the mythical in science seems

187

Sandra Harding and Merrill B. Hintikka (eds.), Discovering Reality, 187–205.
Copyright © 1978 by Psychoanalysis and Contemporary Science, Inc.

particularly inappropriate. What is it doing there? From where does it come? And how does it influence our conceptions of science, of objectivity, or, for that matter, of gender?

These are the questions I wish to address, but before doing so it is necessary to clarify and elaborate the system of beliefs in which science acquires a gender — which amount to a "genderization" of science. Let me make clear at the outset that the issue which requires discussion is *not*, or at least not simply, the relative absence of women in science. While it is true that most scientists have been, and continue to be, men, the make-up of the scientific population hardly accounts, by itself, for the attribution of masculinity to science as an intellectual domain. Most culturally validated intellectual and creative endeavors have, after all, historically been the domain of men. Few of these endeavors, however, bear so unmistakably the connotation of masculine in the very nature of the activity. To both scientists and their public, scientific thought is male thought, in ways that painting and writing — also performed largely by men — have never been. As Simmel observed, objectivity itself is an ideal which has a long history of identification with masculine. The fact that the scientific population is, even now, a population that is overwhelmingly male, is itself a consequence rather than a cause of the attribution of masculinity to scientific thought.[1] What requires discussion is a *belief* rather than a reality, although the ways in which reality is shaped by our beliefs are manifold, and also need articulating.

How does this belief manifest itself? It used to be commonplace to hear scientists, teachers, and parents assert quite baldly that women cannot, should not, be scientists, that they lack the strength, rigor, and clarity of mind for an occupation that properly belongs to men. Now that the women's movement has made offensive such naked assertions, open acknowledgment of the continuing belief in the intrinsic masculinity of scientific thought has become less fashionable. It continues, however, to find daily expression in the language and metaphors we use to describe science. When we dub the objective sciences "hard" as opposed to the softer, i.e., more subjective, branches of knowledge, we implicitly invoke a sexual metaphor, in which "hard" is of course masculine and "soft," feminine. Quite generally, facts are "hard," feelings "soft." "Feminization" has become synonymous with sentimentalization. A woman thinking scientifically or objectively is thinking "like a man"; conversely, a man pursuing a nonrational, nonscientific argument is arguing "like a woman."

The linguistic rooting of this stereotype is not lost among children, who remain perhaps the most outspoken and least self-conscious about its expression. From strikingly early ages, even in the presence of a stereotypic role models, children have learned to identify mathematics and science as male. "Science," my five-year-old son declared, confidently bypassing the fact that his mother was a scientist, "is for men!" The identification between scientific thought and masculinity is so deeply embedded in the culture at large that children have little difficulty internalizing that identification. They grow up not only expecting scientists to be men, but also perceiving scientists as more "masculine" than other male professionals, than, for example, those in the arts. Numerous studies of masculinity and femininity in the professions confirm this observation, with the "harder" sciences as well as the "harder" branches of any profession consistently characterized as more masculine.

In one particularly interesting study of attitudes prevalent among English schoolboys, a somewhat different but critically related dimension of the cultural stereotype emerges. Hudson (1972) observes that scientists are perceived as not only more masculine than are artists, but simultaneously as less sexual. He writes:

The arts are associated with sexual pleasure, the sciences with sexual restraint. The arts man is seen as having a good-looking, well-dressed wife with whom he enjoys a warm sexual relation; the scientists as having a wife who is dowdy and dull, and in whom he has no physical interest. Yet the scientist is seen as masculine, the arts specialist as slightly feminine (p. 83).

In this passage we see the genderization of science linked with another, also widely perceived image of science as antithetical to Eros. These images are not unrelated, and it is important to bear their juxtaposition in mind as we attempt to understand their sources and functions. What is at issue here is the kind of images and metaphor with which science is surrounded. If we can take the use of metaphor seriously, while managing to keep clearly in mind that it is metaphor and language which are being discussed, then we can attempt to understand the influences they might exert − how the use of language and metaphor can become hardened into a kind of reality. One way is through the internalization of these images by scientists them- selves, and I will discuss more explicitly how this can happen later in the paper. As a first step, however, the imagery itself needs to be explored further.

If we agree to pursue the implications of attributing gender to the scientific mind, then we might be led to ask, with what or with whom is the sexual metaphor completed? And, further, what is the nature of the act with which this now desexualized union is consummated? The answer to the first question is immediate. The complement of the scientific mind is, of course, Nature — viewed so ubiquitously as female. "Let us establish a chaste and lawful marriage between Mind and Nature" wrote Bacon (quoted by Leiss, 1972, p. 25), thereby providing the prescription for the birth of the new science. This prescription has endured to the present day — in it are to be found important clues for an understanding of the posture of the virgin groom, of his relation toward his bride, and of the ways in which he defines his mission. The metaphoric marriage of which science is the offspring sets the scientific project squarely in the midst of our unmistakably patriarchal tradition. Small wonder, then, that the goals of science are so persistently described in terms of "conquering" and "mastering" nature. Bacon articulated this more clearly than today's self-consciousness could perhaps permit when he urged: "I am come in very truth leading you to Nature with all her children to bind her to your service and make her your slave" (Farrington, 1951, p. 197).

Much attention has been given recently to the technological abuses of modern science, and in many of these discussions blame is directed toward the distortions of the scientific program intrinsic in its ambition to dominate nature — without, however, offering an adequate explanation of how that ambition comes to be intrinsic to science. Generally such distortions are attributed to technology, or applied science, which is presumed to be clearly distinguishable from pure science. In the latter, the ambition is supposed to be pure knowledge, uncontaminated by fantasies of control. While it is undoubtedly true that the domination of nature is a more central feature of technology, it is impossible to draw a clear line between pure and applied science. History reveals a most complex relation between the two, as complex perhaps as the interrelation between the dual constitutive motives for knowledge — those of transcendence and power. It would be naïve to suppose that the connotations of masculinity and conquest affect only the uses to which science is put, and leave untouched its very structure.

Science bears the imprint of its genderization not only in the ways it is used, but in the very description of reality it offers — even in the relation of the scientist to that description. To see this, it is necessary to examine more fully the implications of attributing masculinity to the very nature of scientific thought.

Having divided the world into two parts — the knower (mind) and the knowable (nature) — scientific ideology goes on to prescribe a very specific relation between the two. It prescribes the interactions which can consummate this union, that is, which can lead to knowledge. Not only are mind and nature assigned gender, but in characterizing scientific and objective thought as masculine, the very activity by which the knower can acquire knowledge is also genderized. The relation specified between knower and known is one of distance and separation. It is that between a subject and object radically divided, which is to say no worldly relation. Simply put, nature is objectified. The "chaste and lawful marriage" is consummated through reason rather than feeling, and "observation" rather than "immediate" sensory experience. The modes of intercourse are defined so as to insure emotional and physical inviolability. Concurrent with the division of the world into subject and object is, accordingly, a division of the forms of knowledge into "objective" and "subjective." The scientific mind is set apart from what is to be known, i.e., from nature, and its autonomy is guaranteed (or so it has been traditionally assumed) by setting apart its modes of knowing from those in which that dichotomy is threatened. In this process, the characterization of both the scientific mind and its modes of access to knowledge as masculine is indeed significant. Masculine here connotes, as it so often does, autonomy, separation, and distance. It connotes a radical rejection of any commingling of subject and object, which are, it now appears, quite consistently identified as male and female.

What is the real significance of this system of beliefs, whose structure now reveals a quite intricate admixture of metaphysics, cognitive style, and sexual metaphor? If we reject the position, as I believe we must, that the associations between scientific and masculine are simply "true" – that they reflect a biological difference between male and female brains – then how are we to account for our adherence to them? Whatever intellectual or personality characteristics may be affected by sexual hormones, it has become abundantly clear that our ideas about the differences between the sexes far exceed what can be traced to mere biology; that once formed these ideas take on a life of their own – a life sustained by powerful cultural and psychological forces. Even the brief discussion offered above makes it evident that, in attributing gender to an intellectual posture, in sexualizing a thought process, we inevitably invoke the large world of affect. The task of explaining the associations between masculine and scientific thus becomes, short of reverting to an untenable biological reductionism, the task of understanding the emotional substructure that links our experience of gender with our cognitive experience.

The nature of the problem suggests that, in seeking an explanation of the origins and endurance of this mythology, we look to the processes by which the capacity for scientific thought develops, and the ways in which those processes are intertwined with emotional and sexual development. By so doing, it becomes possible to acquire deeper insight into the structure and perhaps even the functions of the mythology we seek to elucidate. The route I wish to take proceeds along ground laid by psychoanalysts and cognitive psychologists, along a course shaped by the particular questions I have posed. What emerges is a scenario supported by the insights these workers have attained, and held together, it is to be hoped, by its own logical and intuitive coherence.

II. THE DEVELOPMENT OF OBJECTIVITY

The crucial insight which underlies much of this discussion — an insight for which we are indebted to both Freud and Piaget — is that the capacity for objectivity, for delineating subject from object, is *not* inborn, although the potential for it into doubt is. Rather, the ability to perceive reality "objectively" is acquired as an inextricable part of the long and painful process by which the child's sense of self is formed. In the deepest sense, it is a function of the child's capacity for distinguishing self from not-self, "me" from "not-me." The consolidation of this capacity is perhaps the major achievement of childhood development.

After half a century's clinical observations of children and adults the developmental picture which emerges is as follows. In the early world of the infant, experiences of thoughts, feelings, events, images, and perceptions are continuous. Boundaries have not yet been drawn to distinguish the child's internal from external environment; nor has order or structure been imposed on either. The external environment, consisting primarily of the mother during this early period, is experienced as an extension of the child. It is only through the assimilation of cumulative experiences of pleasure and pain, of gratification and disappointment, that the child slowly learns to distinguish between self and other, between image and percept, between subject and object. The growing ability to distinguish his or her self from the environment allows for the recognition of a world of external objects — a world subject to ever finer discrimination and delineation. It permits the recognition of an external reality to which the child can relate — at first magically, and ultimately objectively. In the course of time, the inanimate becomes released from the animate, objects from their perspective, and events from wishes; the child becomes capable of objective thought and perception. The process by which this development occurs proceeds through sequential and characteristic stages of cognitive growth, stages which have been extensively documented and described by Piaget and his co-workers.

The background of this development is fraught with intense emotional conflict. The primary object which the infant carves out of the matrix of his/her experiences is an emotional object, namely the mother. And along with the emergence of the mother as a separate being comes the child's painful recognition of his/her own separate existence. Anxiety is unleashed, and longing is born. The child (infant) discovers his dependency and need — and a primitive form of love. Out of the demarcation between self and mother arises a longing to undo that differentiation — an urge to re-establish the original unity. At the same time, there is also growing pleasure in autonomy, which itself comes to feel threatened by the lure of an earlier state. The process of emotional delineation proceeds in fits and starts, propelled and inhibited by conflicting impulses, desires, and fears. The parallel process of cognitive delineation must be negotiated against the background of these conflicts. As objects acquire a separate identity, they remain for a long time tied to the self by a network of magical ties. The disentanglement of self from world, and of thoughts from things, requires relinquishing the magical bonds which have kept them connected. It requires giving up the belief in the omnipotence — now of the child, now of the mother — that perpetuates those bonds and learning to tolerate the limits and separateness of both. It requires enduring the loss of a wish-dominated existence in exchange for the rewards of living "in reality." In doing so, the child moves from the egocentricity of a self-dominated contiguous world to the recognition of a world outside and independent of himself — a world in which objects can take on a "life" of their own.

The recognition of the independent reality of both self and other is a necessary precondition both for science and for love. It may not, however, be sufficient — for either. Certainly the capacity for love, for empathy, for artistic creativity requires more than a simple dichotomy between subject and object. Autonomy too sharply defined, reality too rigidly defined, cannot encompass the emotional and creative experiences which give life its fullest and richest depth. Autonomy must be conceived of more dynamically and reality more flexibly if they are to allow for the ebb and flow of love and play. Emotional growth does not end with the mere acceptance of one's own separateness; perhaps it is fair to say that it begins there. Out of a condition of emotional and cognitive union with the mother, the child gradually gains enough confidence in the enduring reality of both him/herself and the environment to tolerate their separateness and mutual independence. A sense

of self becomes delineated — in opposition, as it were, to the mother. Ultimately, however, both sense of self and of other become sufficiently secure
to permit momentary relaxation of the boundary between — without, that
is, threatening the loss of either. One has acquired confidence in the enduring
survival of both self and other as vitally autonomous. Out of the recognition
and acceptance of one's aloneness in the world, it becomes possible to transcend one's isolation, to truly love another.[2]

The final step — of reintroducing ambiguity into one's relation to the
world — is a difficult one. It evokes deep anxieties and fears stemming from
old conflicts and older desires. The ground of one's selfhood was not easily
won, and experiences which appear to threaten the loss of that ground
can be seen as acutely dangerous. Milner (1957), in seeking to understand
the essence of what makes a drawing "alive," and conversely, the inhibitions
which impede artistic expression, has written with rare perspicacity and
eloquence about the dangers and anxieties attendant upon opening ourselves
to the creative perception so critical for a successful drawing. But unless
we can, the world of art is foreclosed to us. Neither love now art can survive
the exclusion of a dialogue between dream and reality, between inside and
outside, between subject and object.

Our understanding of psychic autonomy, and along with it, of emotional maturity, owes a great deal to the work of the English psychoanalyst
Winnicott. Of particular importance here is Winnicott's concept of the
transitional object — an object intermediate between self and other (as,
for example, the baby's blanket). It is called a transitional object insofar
as it facilitates the transition from the state of magical union with the mother
to autonomy, the transition from belief in omnipotence to an acceptance of
the limitations of everyday reality. Gradually, it is given up,

not so much forgotten as relegated to limbo. By this I mean that in health the transitional
object does not "go inside" nor does the feeling about it necessarily undergo repression
. . . It loses meaning, and this is because the transitional phenomena have become dif-
fused, have become spread out over the whole intermediate territory between "inner
psychic reality" and "the external world as perceived by two persons in common,"
that is to say, over the whole cultural field (Winnicott, 1971, p. 5).

To the diffuse survival of the "creative apperception" he attributes what,
"more than anything else, makes the individual feel that life is worth living"
(p. 65). Creativity, love, and play are located by Winnicott in the "potential
space" between the inner psychic space of "me" and outer social space of
"not-me" — "the neutral area of experience which will not be challenged"

(as it was not challenged for the infant) – about which "we will never ask the question: Did you conceive of this or was it presented to you from without" (p. 12).

The inability to tolerate such a potential space leads to psychic distress as surely as the complementary failure to delineate adequately between self and other. "These two groups of people come to us for psychotherapy because in the one case they do not want to spend their lives irrevocably out of touch with the facts of life and in the other because they feel estranged from dream" (p. 67). Both inadequate and excessive delineation between self and other can be seen as defenses, albeit opposite ones, against ongoing anxiety about autonomy.

Emotional maturity, then, implies a sense of reality which is neither cut off from, nor at the mercy of, fantasy; it requires a sufficiently secure sense of autonomy to allow for that vital element of ambiguity at the interface between subject and object. In the words of Loewald (1951), "Perhaps the so-called fully developed, the mature ego is not one that has become fixated at the presumably highest or latest stage of development, having left the others behind it, but is an ego that integrates its reality in such a way that the earlier and deeper levels of ego-reality integration remain alive as dynamic sources of higher organization" (p. 18).

While most of us will recognize the inadequacy of a static conception of autonomy as an emotional ideal, it is easy to fall into the trap of regarding it as an appropriate ideal for cognitive development. That is, cognitive maturity is frequently identified with a posture in which objective reality is perceived and defined as radically divided from the subjective. Our inclination to accept this posture as a model for cognitive maturity is undoubtedly influenced by the definition of objectivity we have inherited from classical science – a definition rooted in the premise that the subject can and should be totally removed from our description of the object. Though that definition has proved unquestionably efficacious in the past, contemporary developments in both philosophy and physics have demonstrated its epistemological inadequacy. They have made it necessary for us to look beyond the classical dichotomy to a more dynamic canception of reality, and a more sophisticated epistemology to support it.

If scientists have exhibited a reluctance to do so, as I think they have, that reluctance should be examined in the light of what we already know about the relation between cognitive and emotional development. Elsewhere (Keller, 1979) I have attempted to show the persistence of demonstrably inappropriate classical ideas even in contemporary physics, where the most dramatic evidence for the failure of classical ideas has come from. There I try to establish some of the consequences of this persistence, and to account for the tenacity of such ideas. In brief, I argue that the adherence to an outmoded, dichotomous conception of objectivity might be viewed as a defense against anxiety about autonomy of exactly the same kind as we find interfering with the capacity for love and creativity. When even physics reveals "transitional phenomena" — phenomena, that is, about which it cannot be determined whether they belong to the observer or the observed — then it becomes essential to question the adequacy of traditional "realist" modes for cognitive maturity as well as for reality. Our very definition of reality requires constant refinement as we continue in the effort to wean our perceptions from our wishes, our fears, and our anxieties; insofar as our conception of cognitive maturity is dictated by our definition of reality, that conception requires corresponding refinement.

III. THE DEVELOPMENT OF GENDER

What, the reader may ask, has all this to do with gender? Though the discussion has led us on a sizable detour, the implicit argument which relates it to the genderization of science should already be clear. Before articulating the argument explicitly, however, we need an account of the development of gender identity and gender identifications in the context of the developmental picture I have presented thus far.

Perhaps the single most important determinant of our conceptions of male and female is provided by our perceptions of and experiences with our parents. While the developmental processes described above are equally relevant for children of both sexes, their implications for the two sexes are bound to differ. The basic and fundamental fact that it is, for most of us, our mothers who provide the emotional context out of which we forget the discrimination between self and other inevitably leads to a skewing of our perceptions of gender. As long as our earliest and most compelling experiences of merging have their origin in the mother-child relation, it appears to be inevitable that that experience will tend to be identified with "mother," while delineation and separation are experienced as a negation of "mother," as "not-mother." In the extrication of self from other, the mother, beginning as the first and most primitive subject, emerges, by a process of effective and affective negation, as the first object.[3] The very processes (both cognitive and emotional) which remind us of that first bond become colored by their association with the woman who is, and forever remains, the archetypal female. Correspondingly, those of delineation and objectification are colored by their origins in the process of separation *from* mother; they become marked, as it were, as "not-mother." The mother becomes an object, and the child a subject, by a process which becomes itself an expression of opposition to and negation of "mother."

While there is an entire world which exists beyond the mother, in the family constellation with which we are most familiar, it is primarily the father (or the father figure) toward whom the child turns for protection from the fear of re-engulfment, from the anxieties and fears of disintegration of a still very fragile ego. It is the father who comes to stand for individuation and differentiation — for objective reality itself; who indeed can represent the "real" world by virtue of being *in* it.

For Freud, reality becomes personified by the father during the oedipal conflict; it is the father who, as the representative of external reality, harshly intrudes on the child's (i.e., boy's) early romance with the mother — offering his protection and future fraternity as the reward for the child's acceptance of the "reality principle." Since Freud, however, it has become increasingly well understood that the rudiments of both gender and reality are established long before the oedipal period, and that reality becomes personified by the father as soon as the early maternal bond comes to be experienced as threatening engulfment, or loss of ego boundaries. A particularly pertinent discussion of this process is presented by Loewald (1951), who writes:

Against the threatening possibility of remaining in or sinking back into the structureless unity from which the ego emerged, stands the powerful paternal force. . . . While the primary narcissistic identity with the mother forever constitutes the deepest unconscious origin and structural layer of ego and reality, and the motive force for the ego's 'remarkable striving toward unification, synthesis,' — this primary identity is also the source of the deepest dread, which promotes, in identification with the father, the ego's progressive differentiation and structuralization of reality (pp. 15, 17).

Thus it is that, for all of us — male and female alike — our earliest experiences incline us to associate the affective and cognitive posture of objectification with masculine, while all processes which involve a blurring of the boundary between subject and object tend to be associated with the feminine.

The crucial question of course is: What happens to these early associations? While the patterns which give rise to them may be quasi-universal (though strongest, no doubt, in our own form of nuclear family), the conditions which sustain them are not. It is perhaps at this point that specific cultural forces intrude most prominently. In a culture which validates subsequent adult experiences that transcend the subject-object divide, as we find for

example in art, love, and religion, these early identifications can be counter-
acted — provided, that is, that such experiences are validated as essentially
human rather than as "feminine" experience. However, in a culture such as
ours, where primary validation is accorded to a science which has been
premised on a radical dichotomy between subject and object, and where
all other experiences are accorded secondary, "feminine" status, the early
identifications can hardly fail to persist. The genderization of science — as
an enterprise, as an intellectual domain, as a world view — simultaneously
reflects and perpetuates associations made in an earlier, prescientific era.
If true, then an adherence to an objectivist epistemology, in which truth
is measured by its distance from the subjective, has to be re-examined when it
emerges that, by this definition, truth itself has become genderized.

It is important to emphasize, even repeat, that what I have been discussing
is a system of beliefs about the meaning of masculine and feminine, rather
than any either intrinsic or actual differences between male and female.
Children of both sexes learn essentially the same set of ideas about the
characteristics of male and female — how they then make use of these ideas
in the development of their gender identity as male or female is another
question. The relation between the sexual stereotypes we believe in and our
actual experience and even observation of gender is a very complex one. It
is crucial, however, to make a vigilant effort to distinguish between belief and
reality, even, or especially, when the reality which emerges is so influenced
by our beliefs. I have not been claiming, for example, that men are by nature
more objective, better suited for scientific work, nor that science, even
when characterized by an extreme objectivist epistemology, is intrinsically
masculine. What I have been discussing are the reasons we might believe
all of the above to be true. These beliefs may in fact lead to observed differences between the sexes, though the question of actual differences between
men and women in a given culture is ultimately an empirical one. The subsequent issue of how those possible differences might be caused by cultural
expectations is yet a separate issue, and requires separate discussion. Without
getting into the empirical question of sex differences, about which there
is a great deal of debate, it seems reasonable to suggest that we ought to
expect that our early beliefs about gender will be subject to some degree of
internalization.

To return, then, to the issue of gender development, it is important
to recognize that, although children of both sexes must learn equally to
distinguish self from other, and have essentially the same need for autonomy,
to the extent that boys rest their very sexual identity on an opposition to

what is both experienced and defined as feminine, the development of their gender identity is likely to accentuate the process of separation. As boys, they must undergo a twofold "disidentification from mother" (Greenson, 1968) – first for the establishment of a self-identity, and second for the consolidation of a male gender identity. Further impetus is added to this process by the external cultural pressure on the young boy to establish a stereotypic masculinity, now culturally as well as privately connoting independence and autonomy. The cultural definitions of masculine as what can never appear feminine, and of autonomy as what can never be relaxed, conspire to reinforce the child's earliest associations of female with the pleasures and dangers of merging, and male with both the comfort and the loneliness of separateness. The boy's internal anxiety about both self and gender is here echoed by the cultural anxiety; together they can lead to postures of exaggerated and rigidified autonomy and masculinity which can — indeed which may be designed to — defend against that anxiety and the longing which generates it. Many psychoanalysts have come to believe that, because of the boy's need to switch his identification from the mother to the father, his sense of gender identity tends always to be more fragile than the girl's. Her sense of self-identity may, however, be comparatively more vulnerable. It has been suggested that the girl's development of a sense of separateness may be to some degree hampered by her ongoing identification with her mother. Although she too must disentangle herself from the early experience of oneness, she continues to look toward her mother as a model for her gender identity. Whatever vicissitudes her relation to her mother may suffer during subsequent development, a strong identification based on common gender is likely to persist — her need for "disidentification" is not so radical. Cultural forces may further complicate her development of autonomy by stressing dependency and subjectivity as feminine characteristics. To the extent that such traits become internalized, they can be passed on through the generations by leading to an accentuation of the symbiotic bond between mother and daughter (see, e.g., Chodorow, 1974).

It would seem, then, appropriate to suggest that one possible outcome of these processes is that boys may be more inclined toward excessive and girls toward inadequate delineation — growing into men who have difficulty loving and women who retreat from science. What I am suggesting, then, and indeed trying to describe, is a network of interactions between gender development, a belief system which equates objectivity with masculinity, and a set of cultural values which simultaneously elevates what is defined as scientific and what is defined as masculine. The structure of this network

is such as to perpetuate and exacerbate distortions in *any* of its parts —
including the acquisition of gender identity.

IV. THE DEVELOPMENT OF SCIENTISTS

Whatever differences between the sexes such a network might, however,
generate — and, as I said earlier, the existence of such differences remains
ultimately an empirical question — they are in any case certain to be over-
shadowed by the inevitably large variations that exist within both the male
and female populations. Not all men become scientists, and we must ask
whether a science which advertises itself as revealing a reality in which subject
and object are unmistakably distinct does not offer special comfort to those
who, as individuals (be they male or female), retain particular anxiety about
the loss of autonomy. In short, if we can take the argument presented thus
far seriously, then we must follow it through yet another step. Would not a
characterization of science which appears to gratify particular emotional
needs give rise to a self-selection of scientists — a self-selection which would,
in turn, lead to a perpetuation of that characterization? Without attempting
a detailed discussion of either the appropriateness of the imagery with which
science is advertised, or of the personality characteristics which such imagery
might select for, it seems reasonable to suggest that such a selection mechanism
ought inevitably to operate. The persistence of the characterization of science
as masculine, as objectivist, as autonomous of psychological as well as of
social and political forces would then be encouraged, through such selection,
by the kinds of emotional satisfaction it provides.

If so, the question which then arises is whether, statistically, scientists
do indeed tend to be more anxious about their affective as well as cognitive
autonomy than nonscientists. Although it is certainly part of the popular
image of scientists that they do, the actual measurement of personality
differences between scientists and nonscientists has proved to be extremely
difficult; it is as difficult, and subject to as much disagreement, as the
measurement of personality differences between the sexes. One obvious
difficulty arises out of the ambiguity of the term scientist, and the enormous
heterogeneity of the scientific population. Apart from the vast differences
among individuals, characteristics vary across time, nationality, discipline,
and, even, with degree of eminence. The Einsteins of history fail, virtually
by definition, to conform to more general patterns either of personality
or of intellect. Nevertheless, certain themes, however difficult they may be
to pin down, continually re-emerge with enough prominence to warrant

consideration. These are the themes, or stereotypes, on which I have concentrated throughout this paper, and though they can neither exhaustively nor even accurately describe science or scientists as a whole — as stereotypes never can — they do acquire some corroboration from the (admittedly problematic) literature on the "scientific personality." It seems worth noting, therefore, several features which seem to emerge from a number of efforts to describe the personality characteristics which tend to distinguish scientists from nonscientists.

I have already referred to the fact that scientists, particularly physical scientists, score unusually high on "masculinity" tests, meaning only that, on the average, their responses differ greatly from those of women. At the same time, studies (e.g., Roe, 1953, 1956) report that they tend overwhelmingly to have been loners as children, to be low in social interests and skills, indeed to avoid interpersonal contact. McClelland's subsequent studies confirm these impressions. He writes, "And it is a fact, as Anne Roe reports, that young scientists are typically not very interested in girls, date for the first time late in college, marry the first girl they date, and thereafter appear to show a rather low level of heterosexual drive" (1962, p. 321) (by which he presumably means sexual, thereby confirming, incidentally, the popular image of scientists as "asexual" which I discussed earlier). One of McClelland's particularly interesting findings was that 90% of a group of eminent scientists see, in the "mother-son" picture routinely given as part of the Thematic Apperception Test, "the mother and son going their separate ways" (p. 323) — a relatively infrequent response to this picture in the general population. It conforms, however, with the more general observation (emerging from biographical material) of a distant relation to the mother,[4] frequently coupled with "open or covert attitudes of derogation" (Roe, 1956, p. 215).

Though these remarks are admittedly sketchy, and by no means constitute a review of the field, they do suggest a personality profile which seems admirably suited to an occupation seen as simultaneously masculine and asexual. Bacon's image of a "chaste and lawful marriage" becomes remarkably apt insofar as it allows the scientist both autonomy and mastery [5] in his marriage to a bride kept at safe, "objectified" remove.

CONCLUSION

It is impossible to conclude a discussion of the genderization of science without making some brief comments on its social implications. The linking of scientific and objective with masculine brings in its wake a host of secondary consequences which, however self-evident, may nevertheless need articulating. Not only does our characterization of science thereby become colored by the biases of patriarchy and sexism, but simultaneously our evaluation of masculine and feminine becomes affected by the prestige of science. A circular process of mutual reinforcement is established in which what is called scientific receives extra validation from the cultural preference for what is called masculine, and, conversely, what is called feminine — be it a branch of knowledge, a way of thinking, or woman herself — becomes further devalued by its exclusion from the special social and intellectual value placed on science and the model science provides for all intellectual endeavors. This circularity not only operates on the level of ideology, but is assisted by the ways in which the developmental processes, both for science and for the child, internalize ideological influences. For each, pressures from the other operate, in the ways I have attempted to describe, to create distortions and perpetuate caricatures.

Neither in emphasizing the self-sustaining nature of these beliefs, nor in relating them to early childhood experience, do I wish to suggest that they are inevitable. On the contrary, by examining their dynamics I mean to emphasize the existence of alternative possibilities. The disengagement of our thinking about science from our notions of what is masculine could lead to a freeing of both from some of the rigidities to which they have been bound, with profound ramifications for both. Not only, for example, might science become more accessible to women, but, far more importantly, our very conception of "objective" could be freed from inappropriate constraints. As we begin to understand the ways in which science itself has been influenced by its unconscious mythology, we can begin to perceive the possibilities for a science not bound by such mythology.

How might such a disengagement come about? To the extent that my analysis rests on the crucial importance of the gender of the primary parent, changing patterns of parenting could be of special importance.[6] But other developments might be of equal importance. Changes in the ethos that sustains our beliefs about science and gender could also come about from the current pressure, largely politically inspired, to re-examine the traditionally assumed neutrality of science, from philosophical exploration of the boundaries or limitations of scientific inquiry, and even, perhaps especially, from events within science itself. Both within and without science, the need to question old dogma has been pressing. Of particular interest among recent developments *within* science is the growing interest among physicists in a process description of reality — a move inspired by, perhaps even necessitated by, quantum mechanics. In these descriptions object reality acquires a dynamic character, akin to the more fluid concept of autonomy emerging from psychoanalysis. Bohr himself perspicaciously provided us with a considerably happier image than Bacon's — one more apt even for the future of physics — when he chose for his coat of arms the yin-yang symbol, over which reads the inscription: *Contraria Sunt Complementa*.

Where, finally, has this analysis taken us? In attempting to explore the significance of the sexual metaphor in our thinking about science, I have offered an explanation of its origins, its functions, and some of its consequences. Necessarily, many questions remain, and it is perhaps appropriate, by way of concluding, to articulate some of them. I have not, for example, more than touched on the social and political dynamics of the genderization of science. This is a crucial dimension which remains in need of further exploration. It has seemed to me, however, that central aspects of this problem belong in the psychological domain, and further, that this is the domain which tends to be least accounted for in most discussions of scientific thought.

Within the particular model of affective and cognitive development I have invoked, much remains to be understood about the interconnections between cognition and affect. Though I have, throughout, assumed an intimate relation between the two, it is evident that a fuller and more detailed conception is necessary.

Finally, the speculations I offer raise numerous questions of historical and psychological fact. I have already indicated some of the relevant empirical questions in the psychology of personality which bear on my analysis. Other questions of a more historical nature ought also to be mentioned. How, e.g., have conceptions of objectivity changed with time, and to what extent have these conceptions been linked with similar sexual metaphors in other, prescientific eras, or, for that matter, in other, less technological cultures? Clearly, much remains to be investigated; perhaps the present article can serve to provoke others to help pursue these questions.

Northeastern University,
Boston, MA.

NOTES

* First appeared in *Psychoanalysis and Contemporary Thought* 1, 3 (1978), New York: International Universities Press, Inc. Reprinted by permission of *Psychoanalysis and Contemporary Science*, Inc.

[1] For a further elaboration of this theme, see 'Women in Science: A Social Analysis' (Keller, 1974).

[2] See, e.g., Kernberg (1977) for a psychoanalytic discussion of the prerequisites for mature love.

[3] To the extent that she personifies nature, she remains, for the scientific mind, the final object as well.

[4] These studies are, as is evident, of male scientists. It is noteworthy, however, that studies of the relatively small number of female scientists reveal in women scientists a similar, perhaps even more marked, pattern of distance in relation to the mother. For most, the father proved to be the parent of major emotional and intellectual importance (see, e.g., Plank and Plank, 1954).

[5] Earlier I pointed out how Bacon's marital imagery constitutes an invitation to the "dominance of nature." A fuller discussion of this posture would also require consideration of the role of aggression in the development of object relations and symbolic thought processes – an aspect which has been omitted from the present discussion. Briefly, it can be said that the act of severing subject from object is experienced by the child as an act of violence, and it carries with it forever, on some level, the feeling tone of aggression. For insight into this process we can turn once again to Winnicott, who observes that "it is the destructive drive that creates the quality of externality" (p. 93), that, in the creation and recognition of the object there is always, and inevitably, an implicit act of destruction. Indeed, he says, "it is the destruction of the object that places the object outside the area of the subject's omnipotent control" (p. 90). Its ultimate survival is, of course, crucial for the child's development. "In other words, because of the survival of the object, the subject may now have started to live a life in the world of objects, and so the subject stands to gain immeasurably; but the price has to be pain in acceptance of the ongoing destruction in unconscious fantasy relative to object-relating" (p. 90). It would seem likely that the aggressive force implicit in this act of objectification must make its subsequent appearance in the relation between the scientist and his object, i.e., between science and nature.

[6] In this I am joined by Dinnerstein (1976), who has recently written an extraordinarily provocative analysis of the consequences of the fact that it is, and has always been, the mother's "hand that rocks the cradle." Her analysis, though it goes much further and much deeper than the sketch provided here. happily corroborates my own in the places where they overlap. She concludes that the human malaise resulting from the present sexual arrangements can be cured only by dividing the nurturance and care of the infant equally between the mother and the father. Perhaps that is true. I would, however, argue that, at least for the particular consequences I have discussed here, other changes might be of comparable importance.

REFERENCES

Chodorow, N. (1974), 'Family Structure and Feminine Personality,' in *Woman, Culture and Society*, ed. M. Z. Rosaldo & L. Lamphere (Stanford: Stanford University Press), pp. 43–46.

Dinnerstein, D. (1976), *The Mermaid and the Minotaur* (New York: Harper & Row).

Farrington, B. (1951), *Temporus Partus Masculus*, an Untranslated Writing of Francis Bacon. *Centaurus* 1 193–205.

Greenson, R. (1968), 'Disidentifying from Mother: Its Special Importance for the Boy,' *Explorations in Psychoanalysis* (New York: International Universities Press, 2978), pp. 305–312.

Horney, K. (1926), 'The Flight from Womanhood,' in *Women and Analysis*, ed. J. Strouse (New York: Dell, 1975), pp. 199–215.

Hudosn, L. (1972), *The Cult of the Fact* (New York: Harper & Row).

Keller, E. F. (1974), 'Women in Science: A Social Analysis,' *Harvard Magazine*, October, pp. 14–19.

Keller, E. F. (1979), 'Cognitive Repression in Contemporary Physics,' *Amer. J. Physics* 47, 718–721.

Keller, E. F. (1980), 'Lewis Carroll: A Study of Mathematical Inhibition,' *J. Amer. Psychoanalytic Assn* 28, 133–160.

Keller, E. F. (1980), 'Baconian Science: A Hermaphroditic Birth,' *Philosophical Forum* 11, 000–000.

Keller, E. F. (1979), 'Nature as "Her",' *Proceedings of the Second Sex Conference*, New York University.

Keller, E. F. (1980), 'Feminist Critique of Science: A Forward or Backward Move?' *Fundamenta Scientiae*, Summer, 1980.

Keller, E. F. (1983), 'The Mind's Eye' (with C. R. Grontkowski), this volume.

Kernberg, O. (1977), 'Boundaries and Structure in Love Relations,' *J. Amer. Psychanal. Assn.* 25, 81–114.

Leiss, W. (1972), *The Domination of Nature* (Boston: Beacon Press, 1974).

Loewald, H. (1951), 'Ego and Reality,' *Internat. J. Psycho-Anal.* 32, 10–18.

McClelland, D. C. (1962), 'On the Dynamics of Creative Physical Scientists,' in *The Ecology of Human Intelligence*, ed. L. Hudson (London: Penguin Books), pp. 309–341.

Milner, M. (1957), *On Not Being Able to Paint* (New York: International Universities Press).

Plank, E. N., & Plank, R. (1954), 'Emotional Components in Arithmetic Leaning as Seen Through Autobiographies,' *The Psychoanalytic Study of the Child* 9 274–293.

Roe, A. (1953), *The Making of a Scientist* (New York: Dodd, Mead).

Roe, A. (1956), *The Psychology of Occupations* (New York: Wiley).

Winnicott, D. W. (1971), *Playing and Reality* (New York: Basic Books).

EVELYN FOX KELLER AND CHRISTINE R. GRONTKOWSKI

THE MIND'S EYE

Feminist thought in the 1970's and 80's echoes a number of themes familiar from radical thought of the 60's. One such theme appears in the revolt against the traditional Western hierarchy of the senses. In this view, the emphasis accorded the visual in Western thought is not only symptomatic of the alienation of modern man, but is itself a major factor in the disruption of man's "natural" relation to the world. The logic [1] of Western thought is too rooted in the visual; its failure, it is implied, derives from an unwholesome division of the senses.

Today, these themes appear in a new context and with a new specificity. There is a movement among a number of feminists to sharpen what, until now, had only been a vague sentiment weaving in and out of the major theme. The gist of this sentiment is that the logic of the visual is a male logic. According to one critic, what is absent from the logic which has dominated the West since the Greeks, and has been covered over by that logic, is woman's desire. "Woman's desire", writes Luce Irigaray, "does not speak the same language as man's desire. In this logic, the prevalence of the gaze ... is particularly foreign to female eroticism. Woman find pleasure more in touch than in sight ... "[2] In the same vein, Hélène Cixous dismisses Freudian and Lacanian theory of sexual difference for its "strange emphasis on exteriority and the specular. A Voyeur's theory, of course."[3]

The notion that vision is a peculiarly phallic sense, and touch a woman's sense, is, of course, not new. Indeed, it accords all too well with the belief in vision as a "higher" and touch a "lower" sense. As such, it has a long tradition, although not necessarily one that should be accepted. But the suspicion that the pervasive reliance on a visual metaphor marks Western philosophy as patriarchal is a more general one. As such it needs to be explored. At the same time, however, such exploration needs to avoid the facile identifications of women with the "lower" senses which are part of that same tradition.

This paper is written out of the conviction that, before these suspicions − feminist or otherwise − can be addressed explicitly, a thorough reexamination of the role which vision has played in Western thought needs to be undertaken. Vision is itself a complex phenomenon, with multiple subjective

Sandra Harding and Merrill B. Hintikka (eds.), Discovering Reality, 207–224.
Copyright © 1983 by D. Reidel Publishing Company.

meanings. If it has had an historically phallic association, surely this association is rooted, not in biology, but in the cultural and psychological meanings
attached to the visual. The following is an inquiry into the history of these
meanings, entwined as they are in the role which the visual metaphor has
played in Western epistemology. Only at the end do we consider, although
briefly, the possible bearing that history has on the phallocentricity of our
philosophical tradition. What we present is, therefore, primarily a philosophical and historical analysis of an issue which has important implications for
feminist theory – implications which will need further elaboration elsewhere.

In one sense, philosophy, feminist theory – and the natural sciences too –
are all engaged in a common endeavor, an endeavor which might be described
as archeological. Each is concerned with revealing those basic assumptions
which have been hidden; with making explicit that which has been merely
implicit, even often barely conscious. Once unearthed, such assumptions
can then, and only then, be challenged.

Some underlying assumptions escape our attention by virtue of being
too familiar. Unnoticed, they can form both our concepts of knowledge and
the language in which those concepts are formulated. To take one obvious
example, an example which in fact constitutes the subject of this paper,
we speak of knowledge as illumination, knowing as seeing, truth as light.
How is it, we might ask, that vision came to seem so apt a model for know-
edge? And, having accepted it as such, how has that metaphor "colored"
our conceptions of knowledge. This paper is written in what might also be
called an archeological spirit; it is devoted to an attempt to make explicit
those assumptions which might have inadvertently crept into our conceptions
of knowledge as a consequence of our reliance on the visual metaphor.

The tradition of grounding our epistemological premises in visual analogies
dates back to the Greeks and is most vividly evident in Plato's theory of
knowledge. If we can cease to accept the visual metaphor as necessarily
natural or intrinsic to the meaning of knowledge, then it is essential to inquire
into the ways in which our reliance on it has informed and shaped this
meaning – to ask what particular relation between us as knowers and the
nature to be known is implied by such a metaphor, and to ask how that
relation affects our conception of reality.

We will proceed by examining the moves by which vision and knowledge
have become intertwined in Western thought; how that process has served,
as it were, to ennoble them both, and yet, though paradoxically, how knowledge, while fashioned after the visual, has come to transcend all the senses.

This is a story which began with Plato and continues to this day. In the course of these 2500 years, our conception of vision has changed radically. How, we further need to ask, have these changes affected our conceptions of knowledge, modelled as they are on the visual?

Beginning with an examination of Plato's treatment of the senses, we will show that, from the start, two different, even paradoxical functions of the visual can be discerned in its metaphoric uses — a connective and a dissociative. Vision connects us to truth as it distances us from the corporeal. As we trace the use of the visual metaphor through history, we find that these functions, originally intertwined, become quite distinct — splitting, finally, into functions of two different eyes, the body's eye and the mind's eye. This split is paralleled by the division of the functions of science into the objectifiability and knowability of nature. For Plato such a division was not necessary. This, however, anticipates our account, and we should perhaps begin by asking how vision came to assume so central a position in our thinking about knowledge.

Hans Jonas has observed that, while Greek philosophy assumes the preeminence of vision among the senses from Plato on, "neither he (Plato) nor any other of the Greek thinkers, in the brief treatments of sight which we have, seems to have really explained by what properties sight qualifies for these supreme philosophical honors."[4] This may be because those properties are self-evident — certainly it has come to seem so to us. Yet it may not always have been so. In an elegant analysis of the transition from an oral tradition to a literate culture in ancient Greece, occuring between Homer and Plato, Eric Havelock has argued that not only has "the eye supplanted the ear as the chief organ"[5] but that in the process a host of other changes was induced — changes from identification and engagement to individualization and disengagement, from mimesis to analysis, from the concrete to the abstract, from mythos to logos.[6] With the growing emphasis on the visual eye comes the growing development, even birth, Havelock argues, of the personal "I." It is perhaps no accident, then, that we first find a clear articulation of the preeminence of vision in Plato. But articulation does not mean argument, and it may be as important to note the general absence of Plato's direct argument on this subject as his indirect comments, which, by contrast, are amply present.

By Plato's time, much could apparently be taken for granted about the privileged status of the visual organ. So evident was it to him that vision enjoys an elevated status over the other senses that he was able merely to assert its preeminence, at least implicitly, without feeling it necessary to argue the point. In his only extensive analysis of the senses *per se*, he entirely separates the discussion of vision from the discussion of the other four senses.[7] He describes the creation of the sense of sight in the same context as the creation of soul and intelligence in human beings; all of the other senses are described in the context of the creation of man's material nature. He also says, in *Timaeus*, that "the first organs they (the gods) fashioned were those that gave us light."[8] He comments elsewhere on the intrinsic nobility of this construction, as for example, in *Phaedrus*, when referring to "the keenest of all the senses"[9] and in the *Republic* when he observes: "Have you noticed how extremely lavish the designer of our senses was when he gave us the faculty of sight?"[10]

The particular preeminnce which the visual enjoys is related, for Plato, to the preeminence light enjoys as a medium of perception, as well as to the preeminence which the sun enjoys among the divinities (or heavenly bodies). Of the special status of light he says: "if light is a thing of value, the sense of sight and the power of being visible are linked together by a very precious bond, such as unites no other sense and its object."[11] whereas of the special status and power of the sun he speaks constantly, albeit obliquely.

It is important here to consider how Plato's particular conception of vision affects its metaphoric uses. Vision is accomplished by a matching of like with like, first through a correspondence by likeness between the eyes and the sun ("the eye is the sense organ most similar to the sun" and "The eye's power of sight is a kind of effusion dispensed to it by the sun."[12]), and then through a matching of the various lights emanating from the eyes, the object, and the sun.

So when there is daylight around the visual stream, it falls on its like and coalesces with it, forming a single uniform body in the line of sight, along which the stream from within strikes the external object. Because the stream and daylight are similar, the whole so formed is homogeneous, and the motions caused by the stream coming into contact with an object or an object coming into contact with the stream penetrate right through the body and produce in the soul the sensation which we call sight.[13]

The mediation of perception (recognition) through likeness becomes a model for intellection as much as the eye itself becomes a model for the intellect. That is, the correspondence between the visual and the mental operates on several levels simultaneously. The eye is likened to the intellect, the "eye of the mind," the sun to the Good, objects of sight to truth, and knowledge itself to the meeting of like with like which, according to Plato's theory of vision, accounts for perception. All three components of the visual system

– eye, the sun, and light – are used by Plato, both metaphorically and directly, to establish the characteristics of intelligibility.

The sun furnishes to visibles the power of visibility. . . . In like manner, . . . the objects of knowledge receive from the presence of the good their being known, but their very existence and essence is derived from it . . . [14]

As the good is in the intelligible region to reason and the objects of the reason, so is this (the sun) in the visible world to vision and the objects of vision.[15]

The correspondence between the visual and the mental interlocking of organ and object persists in Plato's conception of the origins of knowledge. This is perhaps most evident in his theory of *anamnesis*, recollection. In an attempt to explain the process of learning, Plato offers the following suggestion. The soul, before entering the body, once dwelt with the gods. There it enjoyed the same pure understanding of the cohesiveness of all things which was understood by the gods themselves. This knowledge was untrammeled by the senses; i.e., the senses were neither limitations to the grasp of real and true being, nor were they required as avenues of approach to it. Nevertheless, visual imagery is used for the description of the state of pure knowledge:

Every human soul has, by reason of her nature, had contemplation of true being . . . Beauty it was ours to see in those days when, amidst that happy company, we beheld with our eyes that blessed vision, our selves in the train of Zeus, others following some other god . . . ; pure was the light that shone around us, and pure were we, without taint of that prison house which now we are encompassed, and call a body . . . [16]

Thus knowledge in its purest form is, for Plato, a state of being which is essentially divine. The soul as we experience it in this life is no longer unfettered but has to deal with the body which it inhabits. For living human beings the process of learning is recollection. The contact of our senses with some object or set of objects reminds us of the essential reality which we knew before. Plato uses the illustration of the slave boy who does not know, in the colloquial sense; that is, he has never been taught the rudiments of geometry. Nevertheless, he can construct the diagram which illustrates the Pythagorean theorem when properly questioned.[17] Plato maintains, in effect, that since the untutored boy had not learned in this life, he could not "know" unless his soul recalled the structures from a former existence in which he understood them.

Further, in the same breath with his affirmation of learning as recollection, Plato touches again on two of the other themes we have been stressing in this context: the affinity between seeing and knowing and the principle of kinship, or the meeting of like with like. This allows him to postulate the in-principle intelligibility of all things:

Thus the soul . . . since it has seen all things both here and in the other world, has learned everything that is. So we need not be surprised if it can recall the knowledge of . . . anything . . . it once possessed. All nature is akin, and the soul has learned everything, so that when a man has recalled a single piece of knowledge . . . there is no reason why he should not find out all the rest . . . [18]

The union, or reunion, of the soul with the Forms then constitutes knowing, just as the uniting of the light from the eye with the light from the sun constitutes seeing. Though that which mediates the meeting of the soul with the Forms is not specified, its analogy to light is often implicit. The terms which Plato uses for the Forms are *eidos* and *idea*, i.e., things which are seen.

The knower and that which is known, in this metaphor, are essentially kindred. They are both parts of the whole of being itself. Thus kinship with the universe and its structures constitutes Plato's metaphysical presupposition. His epistemological assumption is that we, who were originally part of the lawful divine structure, are thereby in principle able to see into (intuit) it fully again.[19]

Modern science's confidence that nature, (properly objectified), is indeed knowable is surely derived from these Platonic concepts. Its confidence in the objectifiability of nature is, however, only partly derived from Plato. Two features of the scientific conception of objectifiability need to be distinguished. The first is the separation of subject from object, i.e., the distinction between the individual who perceives and the object which is perceived. The second is the move away from the conditions of perception, i.e., the separation of knowledge from the unreliability of the senses or, so to speak, the dematerialization of knowledge. The first is a move begun, as Havelock argues,[20] by Plato, but not completed until Descartes. The second is more thoroughly Platonic, and the greater part of two dialogues, the *Protagoras* and the *Theaetetus*, is devoted to the explication of the impossibility of basing knowledge on perception. In sum, he makes Theaetetus say: "Taking it all together then, you call this perception . . . a thing which has no part in apprehending truth . . . nor, consequently, in knowledge either."[21] It is precisely this aspect of the Platonic endeavor which makes the reliance on the visual metaphor for true knowledge most curious. We must ask whether there are not characteristics of vision, at least as conceived by Plato, which simultaneously invite the retreat from the body sought in Plato's epistemology and the maintenance of the moral-mystical character of his thought, in short, which constitute a paradox which pervades his work.

Indeed, there are several paradoxes implicit here. One is intrinsic to the conceptualization of knowledge as simultaneously objective and transcendent. The "cool light of reason" establishes, in a single move, worldly distance and divine communion. Further, in allowing for the dissociation of truth from process, the visual metaphor ironically allows for the dissociation of the mental from the sensory. Vision is that sense which places the world at greatest remove; it is also that sense which is uniquely capable of functioning outside of time. It lends itself to a static conception of "eternal truths." Although itself one of the senses, by virtue of its apparent incorporeality, it is that sense which most readily promotes the illusion of disengagement and objectification. At the same time, it provides a compelling model for intangible communication offering the most profound and primitive satisfactions.

What appears to us so paradoxical in the different metaphoric functions served by vision and light must have been considerably less so to Plato, or for that matter, to all thinkers of the next two thousand years who accepted Plato's theory of vision. In Plato's understanding incorporeality and communion were not in conflict; the distinction between two kinds of looking or seeing which we have introduced had no place there — in fact, the very nature of the visual process he postulated incorporated both.[22] The light emitted by the eye was itself transcendent, and provided the very vehicle needed for the meeting of soul with soul. It is the "stream from within" which, through its sympathetic coupling with the "stream from without" can "produce in the soul the sensation which we call sight." As theories of vision underwent change, however, the different functions which the visual metaphor performed, and continued to perform, became considerably more paradoxical. With modern theories of optics, the eye becomes a passive lens, no longer thought to be emitting its own stream, and the transcendent coupling between inside and outside which Plato had imagined to occur was gone. With perception regarded as a passive recording, vision becomes a more suitable model for objectifiability and, at least ostensibly, a less adequate one for knowability. Nonetheless, it continues as a model for both. How it does so constitutes something of a puzzle — a puzzle which, as we shall see, is somewhat resolved by the Cartesian split, but only by invoking a kind of "seeing" other than that which the physicists had described.

By the seventeenth century these connections between light and knowledge were so much a part of the intellectual fabric that Descartes scarcely had to acknowledge his debt to Platonism in any of its forms. The analogy with the sun is no longer explicit — perhaps because the concept of light has become so closely affiliated with the process of intellection that no reference to a physical source is required. Vision and light, however, are frankly recognized by Descartes as analogous to the process of intellection. For example, in speaking of mental intuition, Descartes suggests that "we shall learn how to employ our mental intuition from comparing it with the way in which we employ our eyes."[28] And later, "understanding apprehends by means of an inborn light" and inner perception must be perfected by the "natural light" of reason. Ideas are reliable to the extent that they are "clear," where

I term that 'clear' which is present and apparent to an attentive mind, in the same way that we see objects clearly when, being present to the regarding eye, they operate upon it with sufficient strength.[29]

Mind, for Descartes, is not only ontologically primary but a priori reliable; only the validity of aspects of our experience which are *not* purely mental requires explanation; the senses are repudiated *ab initio*. Descartes speaks of mental vision which is the means by which we know everything from the simplest to the most complex objects of knowledge.

those things which relatively to our understanding are called simple, are either purely intellectual or purely material . . . Those are purely intellectual which our understanding apprehends by means of a certain inborn light, and without the aid of any corporeal image.[30]

The inborn light receives its metaphysical dignity and stability by being totally and in a markedly Augustinian manner derived from divinity:

And so I very clearly recognize that the certainty and truth of all knowledge depends alone on the knowledge of the true God, in much that, before I knew Him, I could not have a perfect knowledge of any other thing. And now that I know Him I have the means of acquiring a perfect knowledge of an infinitude of things, not only of those which relate to God Himself, but also to those which pertain to corporeal nature . . . [31]

It is, in fact, with this kind of confidence that Descartes undertook the writing of his three treatises, the *Geometry*, the *Dioptrics*, and the *Meteors*, which he prefaced with the *Discourse on Method*. His purpose in the second of these texts was to examine in detail the existing science of optics, the theory of vision, and the actual workings of the human eye. The very first line of the *Dioptrics* reveals his attitude toward the importance of the faculty of sight: "the whole of the conduct of our lives depends on our senses, among which that of sight is the most universal and the most noble . . . "[32]

His work on vision, perhaps even motivated by his commitment to both its literal and metaphoric importance, in fact led to an undermining of the suitability of sight as a metaphor for knowledge. Descartes' inquiries into the nature of vision and optics were of paramount importance in the Western acceptance of the copy theory of perception.[33] He, perhaps more than any other Western thinker, was responsible for laying the emission theory to rest,[34] with the result that the eye was henceforth regarded as a purely passive lens which simply receives the images projected upon it from without.

The consequences of this shift for our theories of knowledge were critical. It would seem, at this juncture, that we either accept the conclusion that knowledge itself is passive, or we abandon the visual metaphor. Not so, Descartes provided us with another alternative. He enabled us to retain *both* the conception of knowledge as active and the use of the visual metaphor by severing the connection between the "seeing" of the intellect and physical seeing — by severing, finally, the mind from the body. He says, "We know for certain that it is to the soul that that sense belongs, not to the body."[35] and later, "It is the soul that sees, not the eye"[36] although *not*, he tells us emphatically, with "another pair of eyes."[37]

To repeat then, we are arguing that our continuing reliance on the visual metaphor for knowledge inevitably implied that a change in our theories of one would induce changes in our theories of the other. Insofar as it does not seem possible to conceive of knowledge as a passive recording of data — human pride alone would seem to preclude such an epistemological posture — then a sharper division between visual and mental sight was necessitated. It is this necessity which Descartes' dualistic philosophy provides a response to, which, we would argue, becomes a wedge for the mind-body dichotomy itself. As light and vision become more explicitly technical, physical phenomena, the eye itself a more mechanical device, the active knower is forced ever more sharply out of the bodily realm. The subject becomes finally severed from the objects of perception. With this move, the knowing agent has lost its last links to the percipient organism whose sense organs can now be relegated safely to the "purely material." Having made the *eye* purely passive, all intellectual activity is reserved to the "I," which, however, is radically separate from the body which houses it.

The division of the world into mind and body has served to protect the active nature of an understanding which "apprehends by means of an inborn light," of the "natural light" of reason, from the passivity of a lens which merely records — in short, it has salvaged the possibility of knowledge. Nature may be visible by a mechanical process which leaves both the knower and the known disengaged, but it only becomes "knowable" by virtue of an "inborn" or "natural" light which connects the mind's eye to truth. It is that light which reestablishes the subject's relation — a relation now totally and finally dematerialized — to the objects of perception.

The implications of this division for the concept of objectivity are traceable throughout the *Meditations. Meditation I* begins with the methodological doubt engendered by Descartes' suspicion of the subjectivity of the senses: their unreliability over the short term, their relationship to dreams and hallucinations, in short, of their epistemological uselessness. However, it is the conclusion of *Meditation VI*, that the material world really exists and can be known, which gives Descartes the scientific advantage. Having reached God through pure intellection, he can return, assured, to the material world which is now accessible through scientific, i.e., objective reasoning. Specifically, experience is reliable to the extent to which it yields, in the final analysis, to the clear and distinct idea, i.e., to mathematics; thus objectified and measured, it is separated from the subject and can be evaluated accordingly.

The incorporation of these ideas into the scientific world view is evident throughout the subsequent history of science, but perhaps most notably in Newton's work. By the time of Newton, "modern" theories of vision and optics were well ensconced. No longer can the visual mode provide the means of intermingling inside and outside which had been possible for Plato. More securely than ever, however, it can and does provide the means of establishing a total and radical severance of subject from object. The metaphor seems now to be cleansed of its "coniunctio fantasies." Communion with the Gods, or with Truth, must be established elsewhere. For Newton as for Descartes, pure thought, now emancipated from its visual dependence, is proposed to mediate this communion. The reliance on the senses with which Plato was ultimately and inextricably saddled, can finally be superceded, or so it would seem. Rational inquiry requires no sensory, physical intermediate to establish its one to one correspondence with the truths of nature. For example, Newton writes:

I do not define time, space, place and motion as being well known to all. Only I must observe that the vulgar conceive those quantities under no other notions but from the relation they bear to sensible objects. And thence arise certain prejudices, for the removing of which, it will be convenient to distinguish them into absolute and relative, true and apparent, mathematical and common.[38]

Nevertheless, both the eye and light remain of central importance to Newton, literally as well as metaphorically. In his methodology, psychology, metaphysics, even in his theology, the visual assumes throughout a position of prominence. Neither the thoroughgoing nor consistent rationalist that Descartes was, his strong commitment to empirical data is well known. Manuel has observed that "Newton's eye was his first scientific instrument,"[39] and so, in many ways, it remained. His first researches were indeed in optics, where his quasi-religious devotion to the measurement of optical diffraction patterns, performed to superhuman exhaustion and completion has been well documented.[40] These phenomena, as well as the many other optical phenomena he investigated, were studied with the naked eye. His early curiosity about colors led him to long bouts of direct gazing into the sun:

In a few hours I had brought my eyes to such a pass that I could look upon no bright object with either eye but I saw the sun before me, so that I durst neither write nor read to recover the use of my eyes shut myself up in my chamber made dark for three days together and used all means to divert my imagination from *the Sun. For if I thought upon him I presently saw his picture* though I was in the dark. But by keeping in the dark and employing my mind . . . [41]

His outrage, should any of his measurements be called into question, is also well known. Mannel comments: "Scientific error was assimilated with sin, for it could only be the consequences of sloth on his part and a failure in his divine service."[42]

But Newton's profound preoccupation with "looking" seems to have had more to do with the inner eye than with the outer eye which he, as a physicist, knew well to be a mere lens — even though much of his "looking" was done with the outer eye. Visual imagery is prominent in his remarks about theoretical inquiry as it is in his theory itself. "When asked how he came to make his discoveries: 'I keep the subject constantly before me, and wait till the first dawnings open slowly by little and little into the full and clear light'."[43] Indeed, as Heelan and others have noted, for Newton, "The ideal of science was to 'see' what God 'saw',"[44] a belief illustrated so vividly in his many discussions of the Sensorium of God. In one passage, now famous, from the *Opticks*, he wrote:

there is a Being incorporeal, living, intelligent, omnipresent who in infinite space, as it were in his sensory, sees things themselves intimately . . . of which things the images only (i.e., on the retina) carried through the organs of sense into our little sensoriums are there seen and beheld by that which in us perceives and thinks.[45]

Thus it seems evident that fantasies of union — union mediated by vision

– retained a powerful hold on even so late a thinker as Newton. In spite of Newton's success in objectifying science, and in spite of the rejection of an interaction model of vision, we can see, in this quote, the residual influence of the Platonic metaphor. Communion through knowledge remains, even now, a central goal of science, and for both Newton and his successors, vision continues to provide an acceptable metaphoric model. To the extent that, for Newton, this communion could be achieved through both the physical and mental eye, vestiges of an earlier conception of vision are still evident. The subsequent history of science reveals a more thorough incorporation of the implications of modern theories of vision as the communicative functions of science are relegated more and more completely to thought.

Indeed, the history of science appears to have taken us on a long road of emancipation from the physical – somewhat paralleling Aristotle's hierarchy of the senses. Where Kepler experienced science as an opportunity to "grasp (God), as it were, with my hand,"[46] it was Newton's ambition to "see" as God "saw." Einstein perhaps came closest to the ancient ideal when he concluded: "I hold it true that pure thought can grasp reality."[47] Throughout the history of scientific thought, then, the impact of the visual tradition continues to make itself felt, however residually. It is a tradition so deeply internalized by scientists that it no longer requires direct expression; indeed direct expression is perhaps no longer possible. Its indirect expression does, however, persist. The dual paradigm behind the promise of the visual – clarity and communion – survives as the root aspiration behind the dual tenets of modern science. In *objectifiability* the world is severed from the observer, illuminated as it were, by that sense which could operate, it was thought, without contaminating. In *knowability*, communion is re-established, mediated by a now-submerged but still evident dimension of the same sense. The persistence of theses tenets,[48] no longer quite appropriate, needs to be understood. One way of doing so is to seek to identify the philosophical moves which gave rise to them as we have attempted to do here. It would seem, from the present analysis, that our continuing commitment to the visual metaphor for knowledge is at least implicated in that persistence, and perhaps even in part responsible.

We have yet, however, to answer the basic question of why vision came to seem so apt a model for knowledge in the first place. A partial answer has been provided by Hans Jonas[49]. In an attempt to understand the characteristics of vision which are responsible for its particular appeal to classical philosophy, Jonas has conducted a phenomenological analysis of the senses. He finds three basic aspects of vision which provide grounds for its philosophical centrality. Under what he calls "simultaneity of presence" he notes the distinctively spatial rather than temporal character of vision — a property uniquely responsible for our capacity to grasp the "extended now." He says:

Indeed only the simultaneity of sight, with its extended 'present' of enduring objects, allows the distinction between change and the unchanging and therefore between being and becoming. All the other senses operate by registering change and cannot make that distinction. Only sight therefore provides the sensual basis on which the mind may conceive the idea of the eternal, that which never changes and is always present.[50]

Under the heading of "dynamic neutrality" he notes the peculiar lack of engagement entailed by seeing, the absence of intercourse. "I see without the object's doing anything."[51] "I have nothing to do but to look, and the object is not affected by that: . . . and I am not affected." To the "neutralization of dynamic content" he attributes the expurgation of "all traces of causal activity," . . . "The gain," he says, "is the concept of objectivity"[52] but also, he notes earlier, the distinction between theory and practice. Indeed, he argues that it is through this very freedom effect by which the "separation of contained appearance from intrusive reality gives rise to the separableness of *essentia* from *existentia* underlying the higher freedoms of theory."[53] At the same time, however, it is precisely by virtue of its causal detachment that sight is the "least 'realistic' of the senses," and Jonas departs radically from Plato in concluding that when the "underlying strata of experience, notably motility and touch" are rejected, "sight becomes barren to truth."[54]

Finally he notes a third dimension of vision which contributes critically to "objectivity" and that is its uniquely advantageous dependence on distance. "To get the proper view we take the proper distance,"[55] . . . "and if this is great enough it can put the observed object outside the sphere of possible intercourse," and at the same time he observes that "the facing across a distance discloses the distance itself as something I am free to traverse . . . The dynamics of perspective depth connects me with the projected terminus,"[56] (even with infinity).

Jonas' analysis makes a number of important points. As we have said above, there is little doubt that vision, by virtue of its apparent atemporality, both invites and lends itself to an atemporal description of truth and reality. Similarly, there is little question that its equally apparent disengagement from action, experience, and dynamic interaction invites and lends itself to a model of truth which transcends the more body bound, materially contingent senses. And, of course, the possibilities of perspective it grants us, and the gain the visual sense derives from distance further contribute to a model of truth based on distance between subject and object, knower and known.

However, this analysis neglects the ways in which vision as a model for knowledge can promote the sense of communion, of meeting of like with like, so central in Plato's understanding, which continues to survive in contemporary scientific belief. Though Jonas touches on an aspect of seeing which is connective, he fails to take account of an entire dimesnion of the visual experience not centrally contained in the experience of looking at, or surveying. That dimension is most dramatically captured in the experience of looking into, or "locking eyes" – a form of communication and communion which is primitive and universally formative. In direct eye contact, we have a visual experience quite different from and in many ways even opposed to the sense of distance and objectivity evoked by merely looking at an object. The often highly charged experience of "locking eyes" seems to do away with distance. As such, it may remain for all of us as a kind of paradigm for communion, for the connective aspects of vision.

In view of the preceeding discussion, the question of a possible male bias to an epistemology modelled on vision has become considerably more complex than originally might have been thought. In particular, two facets of the metaphoric functions of vision need to be separated. The emphasis on the "objectifying" function of vision, and the corresponding relegation of its communicative – one might even say erotic – function, needs to be separated from the relaince on vision as distinct from other sensory modalities. We suggest that if sexual bias has crept into this system, it is more likely to be found in the former than in the latter. Whatever germ of truth lies in Cixous' allegation of voyeurism may be more readily traced to the de-eroticization of the visual than to the traditional preoccupation with vision as such – a de-eroticization in fact promoted by classical theories of vision. Furthermore, in relegating the now submerged communicative aspects of the visual metaphor to the realm of thought, the latent eroticism in such experience is protected against by total disembodiment. The net result of such disembodiment is the same (as one of us has argued elsewhere [57]) as that implied in the radical division between subject and object assumed to be necessary for scientific knowledge: Once again, knowledge is safeguarded from desire. That the desire from which knowledge is so safeguarded is so intimately associated with the female (for social as well as psychological reasons) suggests an important impetus which our patriarchal culture provides for such disembodiment.[58] It is in this sense that Cixous is right.

Of course, the implications of ennobling vision above and beyond all the other senses also need to be examined, but we see no evidence to suspect in this an explicitly patriarchal move. Rather, it seems to express more diffuse cultural biases — biases which may, however, prove consonant with other more explicitly patriarchal biases. But even the last claim would be meaningless in the absence of alternative models. Our effort to articulate some of the influences the visual metaphor has had on our views of knowledge in general and scientific knowledge in particular would be futile if there were no other ways to describe knowledge. For us, the visual model seems almost inescapable. Yet, the question this paper must end with is whether or not that is so. Many authors have suggested that it is not a universal model, that it is not so prominent in other intellectual traditions. These suggestions lead us to ask: How might a conception of knowledge based on another metaphor differ? Some implications are immediately evident. Knowledge likened to the sense of hearing, for example, could not have made the same claims to atemporality, and might well lend itself more readily to a process view of reality. It is interesting to note in this regard that Heraclitus, our earliest temporal ontologist, evidently had a different metaphor in mind. In fact the verbal form of "know" used by him, *ksuniemi*, originally meant "to know by hearing".[59] Similarly, a theory of knowledge which invokes the experience of touch as its base cannot aspire to either the incorporeality of the Platonic Forms, or the "objectivity" of the modern scientific venture; at the very least it would have necessitated a more mediate ontology. We might agree with Vesey, who writes:

We can imagine a disembodied mind having visual experiences but not tactile ones. Sight does not require our being part of the material world in the way in which feeling by touching does. ... The directness of seeing when contrasted with hearing, its non-involvement with the object when contrasted with feeling by touching, and its apparent temporal immediacy when contrasted with both feeling and hearing are features that may partly explain the belief that sight is the most excellent of the senses.[60]

But the crucial question which remains is whether it is possible to reconsider the criteria which lead to that conclusion. In a time when physics has once again altered our conception of vision and light, when we know that neither the apparent atemporality nor the "dynamic neutrality" of vision are features of reality, but only of our relatively coarse daily observations it seems appropriate to reassess our commitment to the ideals which these features imply.

Northeastern University, Boston, Mass. (EFK)
State University of New York, College at Purchase (CRG)

NOTES

[1] The use of the term "logic" here is colloquial; it refers to the general structure of thought rather than to a system of formal language.

[2] Luce Irigaray, 'This Sex Which Is Not One', in *The New French Feminisms* (Amherst: University of Massachusetts Press, 1980), p. 101.

[3] Hélène Cixous, 'Sorties', *The New French Feminisms*, p. 92.

[4] Hans Jonas, 'The Nobility of Sight' in *Philosophy and Phenomenological Research* 14 (1954), 507.

[5] Eric A. Havelock, *Preface to Plato*, (Universal Library, 1967).

[6] For development of the same theme in different terms, see Bruno Snell, *The Discovery of the Mind*, T. G. Rosenmeyer, trans. (Cambridge, Mass.: Harvard University Press, 1953), especially chapter 9.

[7] *Timaeus*, 61d–68e.

[8] *Ibid.*, 45b.

[9] *Phaedrus*, 250d.

[10] *Republic*, 507c.

[11] *Ibid.*, 508a.

[12] *Ibid.*, 508b.

[13] *Timaeus*, 45c–d.

[14] *Republic*, 509b.

[15] *Ibid.*, 508c.

[16] *Phaedrus*, 250a–c.

[17] *Meno*, 82c–86b.

[18] *Ibid.*, 81c–d.

[19] Actually, "intuit" is etymologically more closely related to contemplation (*theoria*) than to insight. Both, however, are communion-concepts. *Theoria* or contemplation suggests that we can only gaze at that which is infinitely superior to us, as in the Aristotelian contemplation of the Prime Mover or the Christian contemplation of God in the Beatific Vision. Is there a subtle sense in which intuition as "having insight" is for science essentially what contemplation is for theology, with this one crucial difference: that looking *at* God fulfills our being while looking *into* nature's secrets increases our power?

[20] He attributes to Plato in particular the "self imposed task" of establishing "two main postulates: that of the personality which thinks and knows, and that of a body of knowledge which is thought about and known."

[21] *Theaetetus*, 186e.

[22] For an analysis of the Greek vocabulary of seeing, cf. Snell, ch. 1.

[23] *On Free Will*, II, p. 43.

[24] *Soliloquies*, From Whitney J. Oates, *Basic Writings of St. Augustine* (New York, 1948), Vol. L, pp. 165–66.

[25] *Ibid.*, p. 266.

[26] *Ibid.*, p. 267.

[27] Existence, of course, is not a property but St. Augustine treats it as though it were. Strictly speaking, God also has no properties.

[28] René Descartes, *Regulae* IX from Haldane and Ross (eds.), *Philosophical Works of Descartes* (Dover, 1934), Vol. 1, 000–000.

[29] *Principles* I, 45–46, Haldane and Ross (eds.), p. 237.

[30] *Regulae XII*, Haldane and Ross (eds.), Vol. I, p. 41.

[31] *Meditation V*, Haldane and Ross (eds.(, Vol. I, p. 183.

[32] Our translation from *Descartes, Ouevres et Lettres* (Paris: Bibliorèque de la Pleiade, 1953), p. 181.

[33] For a discussion of a much earlier articulation of the copy theory of perception, see Vasco Ronchi, *Optics: The Science of Vision*, translated by Edward Rosen, (New York, 1957).

[34] Some recent critics have argued for a residual presence of the emission theory in Descartes work see, for example, Stephen H. Daniel 'The Nature of Light in Descartes' Physics,' *The Philosophic Forum* 7 (1976), 341, but this point is debatable and does not in any case mitigate against his essential and explicit rejection of optical emanations.

[35] *Dioptrics in Descartes: Philosophical Writings* translated and edited by Elizabeth Anscombe and Peter T. Geach (London, 1954), p. 242.

[36] *Ibid.*, p. 253.

[37] *Ibid.*, p. 246.

[38] Sir Isaac Newton, *Principia*, quoted by Alexandre Koyré in *From the Closed World to the Infinite Universe* (New York, 1958), p. 161.

[39] Frank Manuel, *A Portrait of Newton* (Cambridge: Harvard University Press, 1968), p. 78.

[40] *Ibid.*, pp. 78–79.

[41] *Ibid.*, p. 85. "Newton," Manuel says, "had confronted a god face to face and had been preserved." The psychological dimensions of this experience are provocatively explored by Manuel, and need to be explored further. The personalization of the sun, evident in this quote, makes its presence felt again in Newton's commitment to absolute space. Here Newton conjoins the Greek tradition with another, the Hebraic, in which space itself is identified with God. Jammer, *Concept of Space* (Cambridge, Mass.: Harvard University Press, 1954) provides an excellent account of the confluence of these traditions in Newton's thought. The relation between them needs to be further explored in connection with Newton's acknowledged problematic commitment to a fixed point in space, originally though not finally identified with the sun. See also the discussion of the "symbolic identification of the sun and God" in Thomas Kuhn's *The Copernican Revolution* (Harvard, 1957).

[42] *Ibid.*, p. 141.

[43] *Ibid.*, p. 86.

[44] Patrick A. Heelan, 'Horizon, Objectivity and Reality in the Physical Sciences,' *International Philosophical Quarterly* 7 (1967), 407). In this context of cf. also E. A. Burtt, *The Metaphysical Foundations of Modern Science*, Ch. VII (New York, 1932).

[45] *Opticks*, 3rd edition, London, 1721, p. 344.

[46] Gerald Holton, *Thematic Origins of Scientific Thought* (Harvard, 1973), p. 86.

[47] *Ibid.*, p. 234.

[48] See, e.g., E. F. Keller, 'Cognitive Repression in Contemporary Physics', where ongoing dispute over the interpretation of quantum mechanics is traced to the retention of one or the other of these dual tenets. Some psychological sources of the appeal to belief in the "objectifiability" and the "knowability" of nature are suggested.

[49] Jonas, 'The Nobility of Sight'

[50] *Ibid.*, p. 513.

[51] *Ibid.*, p. 514.

[52] *Ibid.*, p. 515.

[53] *Ibid.*, p. 514.

[54] *Ibid.*, p. 517.

[55] *Ibid.*, p. 518.

[56] *Ibid., loc. cit.*

[57] Evelyn Fox Keller, 'Gender and Science', reprinted in this volume.

[58] See ref. 48 for further discussion of this point, as well as E. F. Keller, 'Nature as "Her" ', *Proceedings of the Second Sex Conference* (New York University, 1979).

[59] Cf., Kurt Von Fritz, '*Noûs, Noein* and Their Derivation in Pre-Socratic Philosophy' in *The Pre-Socratics*, edited by Alexander P. D. Mourelatos, (New York, 1974).

[60] G. N. A. Vesey, 'Vision' *The Encyclopedia of Philosophy* (New York, 1967), Vol. 8, p. 252.

NAOMI SCHEMAN

INDIVIDUALISM AND THE OBJECTS OF PSYCHOLOGY

> ... it was men mostly who did the talking and what
> they were talking about was themselves although they
> used such generic terms as people or mankind these
> terms were really a euphemism for men but we didn't
> know that since the men didn't think it was necessary
> to say so and the women permitted the men to do
> most all the talking it was easy to conclude that we
> were all humans and when one human spoke that
> human spoke for all of us all of which means that until
> recently very few of us realized we were women.
>
> Jill Johnston, *Lesbian Nation*

Much philosophical discussion has been devoted to questions about what sort of existence to attribute to the objects[1] of psychology. Recent focus on scientific realism as a way of answering ontological questions[2] has subtly shifted the center of these questions. Thus, Descartes claimed to have demonstrated that psychological states were of (or in) a mind, a substance wholly different from the body. The question of causal interaction between the two arose, but he took his ultimate inability to answer it to indicate not the inadequacy of his dualism but the limits of metaphysical investigation. In contrast, for modern scientific realists what exists is whatever has to exist for our best theories to be true, and causality plays a central role in these accounts. Psychological states are whatever they have to be to have the (physical and psychological) causes and effects that they do.[3]

This focus has generally led to some sort of physicalism, construed very broadly: types of psychological states (like being angry or in pain) actually are types of physical states (like certain patterns of neurons firing), or, more weakly, each particular psychological state (an occasion of anger or pain) is a particular (though perhaps each time a different type of) physical state. There are a plethora of arguments for positions that are variations of one or the other of these two, and some arguments for why one or the other must be true, given the causal roles of psychological states.[4]

What there are no arguments for, to the best of my knowledge, is the underlying assumption that, whatever they may be, psychological states can be assigned and theorized about on an individualistic basis.[5] Here, as in

225

Sandra Harding and Merrill B. Hintikka (eds.), Discovering Reality, 225–244.
Copyright © 1983 by D. Reidel Publishing Company.

discussions in political theory and social science methodology, it is difficult to make clear what is meant by 'individualism' or 'individualistic'. What I have in mind is the assumption that my pain, anger, beliefs, intentions, and so on are particular, (in theory) identifiable states that I am in, which enter as particulars into causal relationships. Some examples of individualistic states are being five feet tall, having pneumonia, missing three teeth, and having some immediate subjective experience (though how to describe the last is by no means clear). Being the most popular girl in the class or a major general or divorced are not individualistic states: nor, I want to argue, are being in love or angry or generous, believing that all eels hail from the Sargasso Sea, knowing how to read, intending to be more honest, or expecting an explosion any minute now.

This largely unquestioned assumption, that the objects of psychology — emotions, beliefs, intentions, virtues and vices — attach to us singly (no matter how socially we may acquire them) is, I want to argue, a piece of ideology. It is not a natural fact, and the ways in which it permeates our social institutions, our lives, and our senses of ourselves are not unalterable. It is deeply useful in the maintenance of capitalist and patriarchal society and deeply embedded in our notions of liberation, freedom, and equality. It is connected with particular features of the psychosexual development of males mothered by women in a patriarchal society, with the development of the ego and of ego-boundaries. It is fundamentally undercut by an examination of female experience, if that experience is seen in its own terms and not as truncated male experience.

My aim in this paper is to make the claims of the last paragraph reasonably clear and at least somewhat plausible. I want to argue that (1) what I will call the individualist assumption does underlie contemporary philosophical accounts of the nature of the objects of psychology, as much as these differ from each other, (2) this assumption is substantive, not merely formal, and the underlying reasons that might be advanced for it are inadequate, (3) we can illuminate the grip the assumption has on us by seeing it as forming part of the ideology of liberal individualism, and (4) part of the functioning of that ideology is the structure of the bourgeois family, producing men who see themselves as conforming to the assumption and men and women who see such conforming as natural.

I

Why is this assumption so nearly universal? I want to suggest four different sorts of answers to this question, not as alternatives, but as complementary attempts to grasp the depth and power this assumption has for us. The first two are reasons that might be offered for thinking the assumption true, one a straightforwardly theoretical demand and the other an appeal to (philosophically colored) common sense.

Nearest to the argumentative surface, the assumption that psychological objects are particular states of individuals is required by the claim that they are physiological objects, since, presumably, physiology needs to ascribe states to us singly. This claim is in turn connected with the demands of physicalism, that the world be one causally closed system, containing one kind of stuff, governed at bottom by one coherent system of laws, and that those laws be the ones of the physical sciences. Reduction on the level of explanation has fared ill enough that it has, even as a possibility, been widely rejected in favor of the autonomy of levels of explanation, along with the irreducibility of the natural kinds of one explanatory level to those of a lower. But there is hope for another sort of reduction that is much more tenacious – the view that the objects of one theory are complex objects of a more basic, lower level theory, and that as such each particular one will be explicable in terms of that lower level theory.

This view is plausible whenever we have some idea of how to individuate the objects in question in terms of the lower level theory whose elements are their parts. We need to be able to do this in a way that shows the objects as appropriate objects of explanation, that is, as particulars rather than as motley conglomerations. Typically (always?) this will be done in terms of causal connections among the parts and ways in which the complex objects enter as wholes into causal relationships. Thus, for example, even if chemistry is as a theory nonreducible to physics, it could still be that all the objects of chemistry – molecules, chemical bonds, and such – are each particular physical objects or states, subject as coherent wholes to physical law. I suspect that this sort of reduction will always be possible except when the higher level theory is ontologically holistic, except, that is, when in order to individuate a particular object of the theory one needs to refer to structural features of the theory as a whole. Thus, chemistry is not generally thought to be ontologically holistic: it is assumed that its objects can be specified singly as complex objects of physics. Economics and sociology are – debatably – holistic; that is to say that, e.g., classes cannot be picked out independently of the framework of those theories, not, say, as the collection of all the people in them.[6]

In general, the objects of the social sciences, including psychology, are, I want to suggest, objects only with respect to socially embodied norms, and thus any reduction would have to proceed via the whole social system, explaining a particular object as an object-with-respect-to-that-system. Since such a project is, to say the least, unrealistic, we must acknowledge that our explanations in such fields are of objects whose existence as particulars is relative to a social framework.

This claim is likely to seem obviously false; in fact, the projects of philosophers of psychology in the empiricist tradition can often be seen as attempts to explain how we can best theorize about such things as emotions, beliefs, occasions of understanding, thoughts, and pains in a way that is simultaneously responsible to the social complexity of our ordinary attributions and explanations and to what they take to be the obvious, particular existence of such things in each of us. And typically our ordinary talk is seen as more likely to be wrong or misleading or theoretically unhelpful than is the clear conviction that we are in particular states of belief or emotion, do experience particular episodes of coming to understand, having a thought, or feeling a pain.

Similarly, in attempting to understand Quine's thesis about the indeterminacy in the actual reference of words many people are most disturbed by his claim that "there is no fact of the matter" about what our own words refer to. This also, for much the same reason, seems obviously false: surely there is some real difference to be marked in me corresponding to my referring to rabbits or to their undetached parts or temporal stages. Many people relatively comfortable with indeterminacy as an epistemological thesis about the limits of our knowledge of others balk at it as an ontological thesis denying the real, particular existence in us of determinate states of belief.

Worries of this sort are connected with the second reason for the individualist assumption, namely, that it seems so obviously, and so importantly, true. Arguments against Quinean philosophy or behavioral psychology often center on the conviction that there is something important we are not being enabled to account for or allowed to take adequately seriously: our inner lives, subjective experiences, the richness of what seems to each of us to mediate between stimulus and response. There is a great deal that is going on in us that is expressed in our behavior only inadequately, if at all, states that we seem clearly and definitely to know ourselves to be in, as well as some we may be ignorant of or wrong or confused about — it seems clear to us that there is much about our inner lives to *discover*, much that is true of us though no one may be able to observe it.

Wittgenstein is also usually taken as denying that we can make any sense of these claims, talk coherently about inner lives or private experiences. Although he is deeply critical of what, particularly as philosophers, we are inclined to say about such things, including our calling them "inner" or "private", I think it is a mistake to take him this way. (He was aware that people would; he asks throughout the *Investigations*, "what gives the impression we mean to deny anything?" It is not a rhetorical question.) I want to offer in response to these worries what I believe to be a reading of Wittgenstein, although I cannot here defend it as such.[7]

There are things, sensations, for example, that are definite particulars, events, states, or processes that may be introspectively accessible, immediate objects of awareness, whose identity or relation to physical events, states, or processes is (perhaps) an open philosophical or even scientific question. But *most* of what we care about – emotions, beliefs, understanding, motives, desires – are not such particulars. We make one sort of mistake (that it is the task of Wittgenstein's private language argument to address)[8] when we are tempted to think that objects of introspective awareness, like sensations or color patches in our visual fields, are the objects our words most directly hook on to, independently of interpersonal criteria, but we make quite another (the one I am here concerned to address) when we assimilate all psychological objects to objects of introspection or to nonintrospectible bodily states.

The problem with this assimilation is that it ignores the nature of the complexity of our identification of our own (let alone others') complex psychological objects. What we take to be our emotion, our belief, our desire is a bundle of introspectible states and behavior, unless we are simply assuming that some one thing underlies them all. What it is that we know, what it is that is so definitely and particularly there in us, is not the thing itself (our *feeling* of anger is not the anger itself, surely not all of it) but, we usually think, some sign of it. We can, I think, maintain that our twinges, pangs, and so on are particular events no matter what our social situation, but it does not follow that the same is true for more complex psychological objects, such as emotions, beliefs, motives, and capacities. What we need to know in order to identify *them* is how to group together introspectible states and behavior and how to interpret it all. The question is one of meaning, not just at the level of what to call it, but at the level of there being an "it" at all. And questions of meaning and interpretation cannot be answered in abstraction from a social setting.[9]

For all I've said so far, it might of course be that what connects introspectible states and behavior to make them manifestations of a psychological

object could be an underlying causal mechanism, or even some one underlying
state. I don't think that that could be the case, but I don't intend to argue
here that it couldn't. Rather, I want to suggest that the nearly universal
failure even to see the need to argue that emotions, beliefs, etc. *are* particular
states, that is, the tendency to make the individualist assumption, is in need
of explanation.

Our attachment to the reality of the contents of our own minds and to our
often unshared (and in an important way always unshareable) experience of
them is, I think, part of such an explanation. But it is not enough: we can
acknowledge the existence, even the importance, of such things, without its
following that emotions and so on are themselves among them — as, in fact,
they do not obviously seem to be. It may well be misleading in many or most
cases (surely not in all) to say that I infer on the basis of evidence that I am
angry or jealous or that I understand calculus or believe that primates can
learn language, but nothing I immediately encounter just *is* my emotion, my
understanding, or my belief, the way something I encounter *is* my itch or my
sudden awareness of music in the room.

What I want to suggest in the remainder of this essay is that there is an
underlying ideology of individualism and a set of child-rearing practices
connected to it that can help us to understand the seeming obviousness of
the individualist assumption and to account for the difficulty we feel about
relinquishing it.

II

My first two answers to the question of the universality of the individualist
assumption took the form of reasons for believing the assumption to be
true: its apparently being demanded by a commitment to physicalism and its
seeming to follow from our access to and commitment to the reality of the
constituents of our inner lives. The other two answers I would like to suggest
are of another sort: rather than purported justifications of the assumption,
they are attempts at explaining how it has come to have the hold on us that it
does. Thus the third answer articulates the connections between the individualist assumption in the philosophy of mind and the notion of the self
embodied in the ideology of liberal individualism, and the fourth answer
ties that notion of the self to the psychosexual development of males in a
patriarchal culture where childcare is primarily in the hands of women.

To see the individualist assumption as stemming from the ideology of
liberal individualism is to see that what purports to be a statement about how

things naturally are is instead an expression of a historically specific way of structuring some set of social interactions. The supposed naturalness and the various theories that support it are essential components of an ideology. Thus, it is supposed to be a natural fact about human beings, and hence a constraint on any possible social theory, that, no matter how social our development may be, we exist essentially as separate individuals – with wants, preferences, needs, abilities, pleasures, and pains – and any social order has to begin by respecting these as attaching to us determinately and singly, as a way of respecting *us*.

Classical liberal social theory gets off the ground with the observation that individuals so defined are in need of being enticed – or threatened – into enduring and stable association with one another. The societies thus envisioned aim at maximally respecting the separateness of their members by providing mechanisms for adjudicating the claims that one member may make against another, while leaving as intact as possible the rights of each to be self-defining.

Central to this liberal vision, as Ronald Dworkin persuasively argues,[10] is the conviction that the state· ought not to discriminate among conceptions of the good life. The state ought not to embody or even favor any among the alternative pictures of human nature or human flourishing that may be favored by individuals or groups within the society. This evenhandedness is seen as constitutive of the equal respect that all citizens can expect from the liberal state: no vision of how human beings ought to live is to be favored over any other. Such visions, and the people who hold them, cannot always, of course, be *treated* equally: I may respect your desire to eat the whole cake as much as the desires of the rest of us to share it, without giving it all to you, and I can continue to respect your desire for cake even if I decide that the children should get it all.

Making these distinctions is no easy matter, but they are central to liberal political theory because, among other reasons, liberals need to affirm an individualism of *method*, a counting of each as one and only one, a respect for separateness, without being committed to an individualism of *substance*, an anti-communitarian view of how society ought to work. This for two reasons: we may be required by facts about developmental psychology to see human beings as springing from, perhaps even necessarily continuing to see themselves as members of, particular social groups, and communitarian ideals may well be among the possible visions the state is committed to respecting equally.

The feasibility of this project (as, for example, John Rawls pursues it in

A Theory of Justice) has been called into question: whether, that is, methodological individualism in political and social theory does not discriminate against communitarian ideals.[11] An alternative way of framing this concern is to ask whether in claiming neutrality among views of human nature and human flourishing the state is in fact expressing some particular set of views to the exclusion of others. A view of human beings as socially constituted, as having emotions, beliefs, abilities, and so on only insofar as they are embedded in a social web of interpretation that serves to give meaning to the bare data of inner experience and behavior, would in fact seem to be incompatible with a social and political theory that sees social groups as built on the independently existing characteristics of individuals.

If this incompatibility is real (as I think it is) the liberal has good reason to resist the view of the social construction of the objects of psychology: only if psychological states can be seen as attaching to individuals in abstraction from their social setting can we expect to appeal to them to justify forms of social organization; otherwise, we are in the position of attempting to evaluate a system that is constructing the data on which the evaluation is based. I would argue that we are in fact in this position, and the attendant circularity is unavoidable, but one of the hopes of liberalism is precisely to avoid this situation. As Larry Scidentop argues,[12] English liberalism (the sort we have come to think of as liberalism) is based on empiricist epistemology, using the individual as a foundation on which to justify both scientific and political theories.

Even if liberals give up this project of constructing social theory on an asocial foundation, the individual as self-defining and the state as neutral among these definitions remain important: the problem becomes one of characterizing these self-definitions in individualistic terms. Although psychological individualism as a realist thesis (emotions, beliefs, and so on are really there as particular states) is a natural way of doing this, one could attempt to salvage liberal individualism without it. I suspect, however, that the difficulties inherent in such an attempt contribute to the attraction of the realist thesis.

Non-realist individualism would acknowledge that psychological objects are constructions, ways of making sense of the observed regularities of individual and social life (and even that social considerations, such as the meanings of the words we use, must enter in) but reserve to the individual the ultimate authority over these constructions: my psychological life may not be simply as I *find* it, but respect for me entails that it is as I construct it. On such a view privileged access (the epistemological thesis that I and only I have direct knowledge of my psychological states) becomes a social thesis affirming a sort of property right.

But consider what this entails. The realist will see psychological states as definite particulars analogously to the way the liberal social theorist sees individuals, with identities and entitlements attached to them. Internal respect, like political respect, would demand that evenhanded consideration be given to all these states. But (as Charles Taylor argues)[13] this position is untenable: we don't value all our feelings, inclinations, and so on equally, or even equally identify with them, and we would be abandoning an important aspect of our intelligence and humanity were we to do this. We need morally to evaluate our psychological states even if we are realists about them, and this need for moral evaluation is of course increased if the states themselves are not simply facts about us. We cannot treat our own feelings, inclinations, etc. as though we were liberal states and they the citizens.

How are we to make these evaluations? For the liberal they must be made in terms of one's own true nature, deepest desires, or self-definition. But it is difficult to give these notions sufficient content to do the work we need them for. We may start by thinking of our true nature as an objective fact about us: we can be wrong in how we evaluate our feelings; we can be self-deceived, victims of false-consciousness, or simply muddled. But if we can appeal neither to psychological realism nor to social constraints, it is hard to see how to make sense of these claims; there seems to be nothing in terms of which to settle their truth.

Alternatively, we can focus on the idea of self-definition, on the freedom to constitute ourselves in any way we choose. But this sort of voluntarism (exemplified most starkly by Sartre) makes this ultimate choice irrational and inaccessible to criticism. What we choose today as central to our self-definition we may repudiate tomorrow. And we are unable to account for our deep, shared conviction that some activities of self-constitution are misguided, silly, futile, or immoral. The liberal doesn't want us (certainly not in the name of the state) to make these judgments about each other, but without psychological realism to fall back on it's unclear how we can give any content to such judgments about *ourselves*, or why it should matter to us, as it clearly does, that we make them.

The problem is this: if individuals, their identities and life-plans are to be identifiable independently of forms of social organization (as the liberal needs them to be),[14] then we are hard pressed to come up with anything that could make this identification non-arbitrary — unless we accept psychological realism. The idea that psychological states are definite particulars is the

natural mirror to the liberal conception of individuals, since on this conception we are deprived of anything to guide the choosing and the weighing that
would need to be involved were these states to be seen as constructions.

This situation is reflected in disputes in psychological theory. Freudian
meta-psychology needs to postulate real forces within each individual in
terms of which the organization of the self occurs. In contrast, object-relations
theory (which I will discuss below), sees the self as developing essentially
in relation to particular others — and the principles of self-organization
as arising out of those relationships. The more individualist the theory the
greater the need for psychological realism, particularly since the competing
notion of free choice makes even less sense for infants than it does for adults.
I want to suggest that these connections account for some of the strength of
psychological individualism: our liberal view of persons as separable individuals would seem to require, or at least to fit most naturally with, a view of
psychological objects as existing brutely in us.

III

My fourth reason for the widespread acceptance of the individualist assumption ties this view of the self to patriarchal child-rearing practices. I want to
suggest that if certain recent accounts of gender differences in psychosexual
development are correct, they would lead us to expect that precisely such an
individualistic view of the self would come to be both exemplified by men
and taken by men and women alike as essentially human. Although most of
the work I know in this area follows Freud in giving what the authors take to
be causal accounts, I think there are grave difficulties with this approach. In
what follows I mean the account to function as a sort of narrative framework,
in terms of which we can make sense of the functioning of institutions such
as the bourgeois family. We can think of the family as serving, *inter alia*, to
produce heterosexual adults who will go on to form families like the ones
they grew up in: in any particular case the causal story, whether or not it has
this outcome, is very much more complicated.

Two recent books — Dorothy Dinnerstein's *The Mermaid and the Minotaur*
and Nancy Chodorow's *The Reproduction of Mothering*[15] — explore the
consequences of early, intimate child-rearing's being nearly exclusively in the
hands of women. My discussion will rely heavily on both books, in particular
on Chodorow's discussion there and elsewhere[16] on gender differences in
the development of the self. (Chodorow in turn draws heavily from object-
relations theorists).[17] I will argue that the view of a separate, autonomous,

sharply individuated self embedded in liberal political and economic ideology and in individualist philosophies of mind can be seen as a defensive reification of the process of ego development in males raised by women in a patriarchal society. Patriarchal family structure tends to produce men of whom these political and philosophical views seem factually descriptive and who are, moreover, deeply motivated to accept the truth of those views as the truth about themselves. In turn, the acceptance by all of us of those views as views about *persons* sustains these child-rearing practices by leading us to devalue, to see as truncated, as less than fully, healthily adult, the very different psychical structures of *women* raised by women. Since men (tend to) exemplify the psychical structures declared by political and philosophical theories to be universal (or if they don't, to see that as a personal failing), we are kept from criticizing those structures and from considering alternatives based on female experience. This interrelationship is characteristic of an ideology: a set of views purports to tell us the facts, what is "naturally" true, in the nature of things, and through doing this helps to structure social institutions in such a way as to produce people who tend to exemplify those views, thereby providing evidence for their own truth.

Chodorow and Dinnerstein offer theories of psychosexual development based on those of Freud, but differing sharply from his in emphasizing particular, historically and socially contingent features of the family and its relationship to male-dominated society. The principal feature is the nearly exclusive role of women in early, intimate child-rearing. It is obvious that this fact plays an important role in Freud's account, but since he doesn't treat it as an alterable social arrangement, it remains insufficiently examined, and its effects remain undifferentiated: we seem to be presented with an account of what it's like to grow up, period. And although this account is different for girls and for boys, Freud never considers the development of girls sufficiently in its own terms, but rather in relation to, or as a truncated and altered version of, the development of boys.

The usual reason given for this failure on Freud's part (one he recognized) is that, although a large proportion of his patients were women, they remained for him, as for most men, a mystery. A deeper reason, I want to suggest, is that he was concerned to chart the path from infancy to civilized adulthood, and his model of civilized adulthood was that of bourgeois patriarchy, a model even ideally applicable only to men.

Related to his concern with this particular path is Freud's emphasis on the Oedipal period, on the vicissitudes of desire and sexual orientation, at the expense of the pre-Oedipal period and questions of individuation and core gender identity (that is, a sense of oneself as fitting in one of two utterly distinct social categories — the masculine or the feminine). One casualty of this relative lack of attention has been the question of gender differences in the infant's achievement of a sense of self and the connections between that achievement and the achievement of core gender identity. What we learn from the recent work that addresses these questions supports my hypothesis that philosophical and political individualism are connected with the psychosexual development of males raised by women. The lack of attention by psychologists, philosophers, and social theorists to the specific effects of this social practice and to gender differences in the pre-Oedipal period have led us to see individualism as somehow natural, as reflecting brute facts about the nature of persons.

When we focus on the pre-Oedipal period, two striking achievements in the way of psychosexual development stand out: before (possibly well before) the age of three children become aware of themselves as social persons who are distinct from all others, in particular from their mothers, and they become aware of themselves as feminine or masculine, an identity it is next to impossible to change, even when it contradicts biological sex (difficult as that is to provide clear criteria for). Chodorow argues that these two feats are connected and are accomplished differently for girls and for boys.

Given our child-rearing arrangements, all infants start out intimately connected with a woman, and their development of a sense of self flows from their relationship with her. For a boy that relationship is colored by the mother's sense of difference from her son: since in our society the genders are utterly and profoundly distinct, a son is experience and treated as other, as different. Achieving masculinity thus becomes a central part of what the male infant must do to establish his independence. In becoming himself he defines himself as both separate and different from his mother, and his achievement of masculinity is emblematic of this separation and difference.

As Freud argues that the girl has a harder time in establishing a heterosexual orientation because of the need to switch the gender of her object choice, so the boy has a harder time establishing core gender identity, since this identity is different from that of the person with whom he has identified. His initial experience of gender is an experience of difference. Thus a boy's sense of self is and remains reactive and defensive, something to be protected by an emphasis on his differences from others, his separateness and distinctness from them. It is through this emphasis that he comes to fit the picture underlying individualism in the philosophy of mind and in political theory:

he is defined as a person by those properties that he senses as uniquely his and that seem somehow internal to him.

This process is profoundly affected by the relative absence of fathers in early childcare. Boys are urged to identify with their only distantly experienced fathers, and later, in the Oedipal stage, the power promised by this identification becomes a bribe to the boy to give up his desire for his mother. But this identification remains relatively abstract: in industrial society the father seems to have the independence the boy craves; he moves out of the family into the wider world, in ways that the boy knows little about. Being a man is exciting and attractive, but extremely vague and scary. About the only clear thing he knows about it is that it is *not* being a woman, and the importance of this fact is underscored by the social devaluation of women.

A boy's father serves, of course, not only as an attractive figure to identify with, but also as the embodiment of punishment and threat: during the Oedipal phase he is seen as the agent of social authority, and it is his voice that becomes internalized as the boy's superego. The intense fear generated by the power of this threat, which Freud sees as the threat of castration, accounts, on his view, for the strength of the male superego necessary for the maintenance of morality and civilization. And the relative abstractness of the father gives the superego an air of impersonality that I believe is connected with the centrality of objectivity and universalizability in moral theory. Thus, unsurprisingly, our moral notions reflect the socially dominant view of the nature of persons: if individuals are distinct and not essentially connected with one another, then morality can be expected to concern itself not with the particularity of relationships among people, but with abstractly characterizable features of interactions among individuals whose natures are taken as given.

The emphasis on externally conceived, rule-governed morality and the development of liberal individualist politics are connected with the increasing absence of fathers from the home with the development of industrial capitalism. Before this development, what it was to be a man was more closely connected with becoming like one's father, in relatively concrete and easily understood ways. [The speculative nature of these remarks can be summed up by saying that this view commits one to holding that serfs had weaker ego boundaries than workers in modern industrial society.] It may be significant in this regard that moral worth was often, in European societies, defined in religious ways, to be punished or rewarded not in this world, but in the next, and to be determined by how well one's life conformed to the model of the life of Christ, an absent, distant, relatively abstract "role model". Only when the demands of industrialization moved fathers far enough away to play this role did individual merit become attached to earthly achievement.

In turn, as Chodorow argues, sons growing up with absent fathers were better suited for industrial life — more responsive to abstract and impersonal demands, more likely to internalize a need to conform to authority.[18] Putting the matter this way emphasizes the extent to which, for the vast majority of men, the ideal of autonomous self-creation has been a myth. Very few men have a significant amount of control over their lives, but the myth encourages the illusion, generated by the promise held out by identification with their fathers, that the obedience to authority required by industrial capitalism is their free choice and will be rewarded.

Turning to the psychosexual development of girls, we can discover the psychological meaning of the claim made by Marxists in economic terms that women's position in capitalist society is essentially feudal. A girl achieves a sense of herself much less sharply differentiated from her mother. In part she is responding to the fact that in a society in which gender categories structure all social relations, her mother is likely to identify with her. There is furthermore no particular sharp difference she can seize upon to break the primitive identification all infants must develop with their primary caretaker. Nor does her father typically provide this for her as he does for her brother: he urges her not to identify with him, but to relate to him, and to do so as a woman, i.e., as someone like her mother. And the person she is to grow up to be like is someone she knows intimately, concretely, and who is defined for her not as a separate, mysteriously self-actualizing individual, but only through family relations.

Thus, for a girl neither a sense of herself nor her gender identity consists, as it does for a boy, in a free leap in the dimly perceived direction of self-definition, defined mainly by difference from the mother. Her gender identity is characteristically less tenuous (compare the relative scarcity of female to male transsexuals) and less connected with heterosexual orientation, which, if it occurs, is a later, more difficult achievement. (Consider here the fact that, whereas the "macho" element in much of gay male life is connected with an understandable need to assert to a sexist world that they are "real men", the butch and femme roles lesbians have adopted at certain times in certain societies have more to do with survival than with a need to be seen as "real women".) But although core gender identity tends to be easier to achieve, a girl's sense of self is typically weaker than a boy's; her ego-boundaries are less strong. Who she is is much more closely bound up with intimate relationships and with how she is perceived by others; it is less natural for her to separate how others react to and treat her from how she perceives herself to be.

Given that the masculine model of a sharply distinguished self is our cultural norm (since men have had both the need and the power to define as fully human this sort of self, to turn their experience of infancy on its head and define 'feminine' as 'not masculine'), women have been perceived as less than fully human, or at least as less than fully adult. We are less likely to consider ourselves, or to be considered by others, as having an identity, a character, talents, and virtues independently of our particular intimate relationships and of how we are perceived by others.

Thus, if we accept the picture of a person embedded in political and philosophical individualism, the picture of healthy autonomous adulthood embedded in psychological theory, and the picture of conscience and morality embedded in moral philosophy, we come to the conclusion that there is something deficient in the natural character of girls or in their upbringing. Traditionally, of course, this deficiency has been seen as natural: women have quite consistently been excluded from the centrally important metaphysical, epistemological, moral, and political conceptions of personhood.[19] Freud is often read as being in this tradition, but he is at least from time to time explicit about the unnaturalness and painful deformation involved in the making of women and men from female and male children, and even to some extent of how this process is more destructive of females than of males.[20]

Liberal feminism, in line with liberal social theory in general, sees most women's failure to exemplify fully the culture's ideals of personhood as due not to our nature but to constraints imposed upon us by that culture. As a rule, the more radical the thinking the further back in our lives these constraints are located and the deeper and more difficult to eradicate their effects are taken to be. Thus, the use made by Dinnerstein and Chodorow of the Freudian stress on the family, infancy, and early childhood. As Juliet Mitchell has argued,[21] we can read Freud as providing a theory of how children develop under patriarchy, a theory we have reason to learn both better to explain ourselves and how we came to be the way we are and to help discover the constraints on changes we might want to make in the mechanisms of enculturation in order to allow women better to conform to the norms of adult personhood. (For Freud the constraints were absolute: he writes as though our only alternatives were the Hobbesian state of nature or the turn-of-the-century bourgeois patriarchal family.)

Chodorow and Dinnerstein are far more revisionist than Mitchell in their neo-Freudianism, and, in particular, Chodorow's use of object-relations theory can lead us to question not only the mechanisms of enculturation and their crippling effects on females, but the norms of adult personhood those mechanisms are designed to enable some men to achieve (and others to strive for).

In the final section I would like briefly to question those norms: there are, of course, many changes necessary in the upbringing of girls, but we ought not to accept the masculinist pictures of persons, of healthy adulthood, or of morality as the goals of those changes.

IV

These norms have been at least tacitly questioned, in the forms of women's lives, work, and conversations, for as long as they have existed. The recent flourishing of feminism is having two effects: this questioning is more and more going on in (a deliberate subversion of) the "father tongue", the allegedly universal but socially masculine language of philosophy and science.[22] Women who have been allowed and trained to "think like men" are using that training to think more clearly — which means more radically — like women, that is, like people who are living real, embodied lives, shared in deep and important respects with others of our culturally devalued gender, at a particular time in history involving, crucially, a re-evaluation of gender.

The other effect is a backlash of fear that liberals are ill-equipped to deal with. The fear arises from a recognition of the fact that men have been free to imagine themselves as self-defining only because women have held the intimate social world together, in part by seeing ourselves as inseparable from it. The norms of personhood, which liberals would strive to make as genuinely universal as they now only pretend to be, depend in fact on their not being so — just what we should expect from an ideology. Thus, the fear aroused by liberal feminism's ideal of opening to women the sort of autonomy previously reserved for men is, I think, a real one.

There is every reason to react with alarm to the prospect of a world filled with self-actualizing persons pulling their own strings, capable of guiltlessly saying 'no' to anyone about anything, and freely choosing when to begin and end all their relationships. It is hard to see how, in such a world, children could be raised, the sick or disturbed could be cared for, or people could know each other through their lives and grow old together.

Liberal feminism does have much in common with this sort of "human potential" individualistic talk, but it is my suspicion that it was in reaction to the deeper, and more deeply threatening, insights and demands of feminism that the current vogue for self-actualization developed – urging us all back inside the apolitical confines of our own heads and hearts and guts. Because what I hope I have begun to suggest here is that the psychological individualists might be wrong, and we are responsible for the meaning of each other's inner lives, that our emotions, beliefs, motives, and so on are what they are because of how they – and we – are related to the others in our world – not only those we share a language with, but those we more intimately share our lives with.

We cannot and do not want to see this because men, who have traditionally had the power to define what it is to be human, to be adult, to be moral, have done so in response to their own experience of and need for separateness and distinctness. And, as women, we have accepted this view of the self as truly, fully human – despite our own inevitable sense of failure in the face of it – because (as Dinnerstein powerfully argues) we share with men deep ambivalence about birth, death, dependence, the body, its needs and demands. We too have our earliest and deepest associations of these things with a woman and we too evade the difficult resolution of this ambivalence by splitting off these aspects of experience. We allow – and require – men to express a vision, which even at the cost of self-denigration we need as they do, of pure, clean, free, uncontaminated humanness.

Chodorow's and Dinnerstein's books are not particularly optimistic. Unlike Freud, they allow for the possibility of changes in the deepest structures of enculturation, but the path to these changes is not well marked, and all that is clear about it is that it is difficult both to find and to keep to. Part, though by no means all, of finding this path is saying what we can about its destination. We may not be able to say very much about it since, as Marx has argued, the very ideas of new forms of human existence are created out of forms of practice. But we can learn something about where we are or ought to be going by looking at the practices that form our lives as women, by taking them seriously, listening to what we do, and finding the voices with which to speak what we hear.

My contention in this paper has been that one of the things we will learn is a radically different conception of the nature of persons and a deep suspicion of some of the underpinnings of philosophical psychology, metaphysics, epistemology, ethics, and political theory: the essential distinctness of persons and their psychological states, the importance of autonomy, the value of universal principles in morality, and the demand that a social theory be founded on an independent theory of persons, their natures, needs, and desires.

These issues have traditionally been discussed in Western culture by upper or middle class white males who have taken themselves to be speaking in a universal human voice. Our very varied experiences as women have been crucially different from theirs, in part because of the often limited and limiting social roles we have been constrained to fill (defined by our bodies' sexual appeal and reproductive capacity and by our immersion in the intimate social world) as well as by what we have chosen to do with our lives (in, for example, art and in the interconnectedness of experience and perception in consciousness-raising groups). We are less likely to speak naturally in voices at once abstractly disembodied and autonomously self-defining.

Rather than claim our right to speak in such voices, to transcend our experience as women, I would urge us to speak out of that experience, in part as a way of changing it, but also out of a recognition of what there is to learn from the perspectives on human life that have been distinctively ours.[23]

University of Minnesota

NOTES

[1] I'm using the term 'objects' very broadly, in the sense of objects of attention or study. Often it will serve as short for 'events, states, and processes'. I am not attending particularly to the distinctions among these, in part because I'm not convinced that they are genuine or genuinely helpful and in part because, be that as it may, my discussion will, I think, apply equally to all three. I will most often speak of states, as they have figured most prominently in recent discussions.

[2] See especially Hilary Putnam, *Mind, Language and Reality: Philosophical Papers*, Vol. 2, Cambridge Univ. Press, 1975.

[3] In addition to Putnam, this has been argued (in different ways and to rather different ends) by David K. Lewis, 'An Argument for the Identity Theory,' in *Materialism and the Mind-Body Problem*, ed. David M. Rosenthal (Prentice Hall, 1971), pp. 162–71; and Donald Davidson, 'Mental Events,' in *Experience and Theory*, ed. L. Swanson and J. W. Foster, (Univ. Massachusetts Press, 1970), pp. 79–101.

[4] Lewis and Davidson in particular argue for the claim that some sort of identity theory must be true. These various positions are well represented in Rosenthal's collection and laid out in the introduction.

[5] For arguments very different from mine against this claim, see Putnam, 'The Meaning of Meaning,' in *Mind, Language and Reality*, and Tyler Burge, 'Individualism and the Mental,' in *Midwest Studies in Philosophy* 4 (1979), pp. 73–121.

[6] It is this sort of holism that Quine has in mind when he marks, with the thesis of the indeterminacy of translation, a sharp break between all the natural sciences and psychology, linguistics, and related fields. His theory of analytical hypotheses is meant to make the relativization precise for the case of translation. See *Word and Object* (Massachusetts Institute of Technology Press, 1960), esp. Ch. 2.

[7] For various reasons I would not expect this response to be taken as support by a Quinean or a behavioral psychologist, but it is no part of my intent to offer such support: I want to argue that the realm of the psychological is essentially social, not that it is in any way unreal or unworthy of serious investigation.

[8] Ludwig Wittgenstein, *Philosophical Investigations*, trans. G. E. M. Anscombe (Oxford: Basil Blackwell, 1967), esp. §§ 269–315.

[9] Hilary Putnam's work on the division of linguistic labor, the irreducibly social dimension of meaning, has similar consequences for individualistic functionalist accounts. In Putnam's writing, as in Davidson's, an acute perception of this social dimension undercuts his programmatic work in philosophical psychology. (Putnam himself has acknowledged this effect, though perhaps not as extremely as I argue.) See esp. 'The Meaning of Meaning'.

[10] Ronald Dworkin, 'Liberalism,' in *Public and Private Morality*, ed. Stuart Hampshire (Cambridge: Cambridge Univ. Press, 1978), pp. 113–43.

[11] This point was argued by Mary Gibson in a session of the Radical Caucus of the American Philosophical Association on 'Rawls and the Left,' New York, December 1979.

[12] Larry Scidentop, 'Two Liberal Traditions,' in *The Idea of Freedom: Essays in Honour of Isaiah Berlin*, ed. Alan Ryan (London: Oxford Univ. Press, 1979), pp. 153–75.

[13] Charles Taylor, 'What's Wrong with Negative Liberty,' in *The Idea of Freedom*, pp. 175–95.

[14] Strictly speaking, the liberal wants individuals to be prior to forms of *political* organization. This split allows free reign to forces of social and economic coercion. The liberal needs to argue that these forces are less powerful or less reprehensible than political forces, or, quite implausibly as soon as one considers the socialization of children, extend liberalism to social forces as well. Problems of this sort are raised by Marx in 'On the Jewish Question' and discussed in relation to pluralism by Robert Paul Wolff in 'Beyond Tolerance' (in *Critique of Pure Tolerance*, with Herbert Marcuse and Barrington Moore, Jr., Boston: Beacon Press, 1965), and by Lorenne Clark in 'Sexual Equality and the Problem of an Adequate Moral Theory: The Poverty of Liberalism' (unpub. ms.).

[15] Dorothy Dinnerstein, *The Mermaid and the Minotaur: Sexual Arrangements and Human Malaise* (New York: Harper and Row, 1976); Nancy Chodorow, *The Reproduction of Mothering: Psychoanalysis and the Sociology of Gender* (Berkeley: Univ. of Calif. Press, 1978).

[16] See especially 'Being and Doing: A Cross-Cultural Examination of the Socialization of Males and Females,' in *Woman in Sexist Society: Studies in Power and Powerlessness*, ed. Vivian Gornick and Barbara K. Moran (Basic Books, 1971); 'Family Structure and Feminine Personality,' in *Woman, Culture and Society*, ed. Michelle Z. Rosaldo and Louise Lamphere (Stanford: Stanford Univ. Press, 1974); and 'Mothering, Male Dominance, and Capitalism,' in *Capitalist Patriarchy and the Case for Socialist Feminism*, ed. Zillah R. Eisenstein (New York: Monthly Review Press, 1979).

[17] Chodorow cites principally Alice Balint, Michael Balint, W. R. D. Fairbairn, Harry Guntrip, Hans Loewald, Margaret Mahler, Roy Shafer, and D. W. Winnicott.

[18] For the discussion on the need for men's personality structures to change to accommodate the demands of industrial capitalism, see Heidi Hartmann, 'Capitalism, Patriarchy, and Job Segregation by Sex,' in Eisenstein.

[19] This point is argued and historically illustrated by Susan Moller Okin, *Women in Western Political Thought* (Princeton Univ. Press, 1979).

[20] See especially Sigmund Freud, ' "Civilized" Sexual Morality and Modern Nervous Illness,' 1908, Standard Edition, vol. 9, ed. James Strachey (London: Hogarth Press, 1959), esp. pp. 194–5.

[21] Juliet Mitchell, *Psychoanalysis and Feminism* (New York: Pantheon Books, 1974).

[22] The notion of a father tongue and its difference from a mother tongue is Thoreau's. See Stanley Cavell, *The Senses of Walden* (New York: The Viking Press, 1972), pp. 15ff.

[23] It has been pointed out to me (by Jane Gallop) that this paper seems to have been written by at least two different people. I have made no attempt to change that: it reflects something important about feminist scholarship today — the complexities of our relationships to traditional disciplines, their problems and methodologies, and to politics, to our histories and to our audiences. I am grateful to my colleagues in the Philosophy Department and in the Women's Studies Program at the University of Minnesota for their criticism and suggestions and, especially, for their making it possible for me comfortably to be all the authors of this paper. I read earlier versions at the University of Wisconsin at Milwaukee and the University of Minnesota and profited from those discussions, as well as from extensive ones with Peter Shea, Burton Dreben, and Sandra Harding, and from Kathryn Morgan's work on autonomy. Adam Morton brought to my attention various flaws in the penultimate draft.

JANE FLAX

POLITICAL PHILOSOPHY AND THE PATRIARCHAL UNCONSCIOUS: A PSYCHOANALYTIC PERSPECTIVE ON EPISTEMOLOGY AND METAPHYSICS*

> The windiest militant trash
> Important Persons shout
> Is not so crude as our wish:
> What mad Nijinsky wrote
> About Diaghilev
> Is true of the normal heart;
> For the error bred in the bone
> Of each woman and each man
> Craves what it cannot have,
> Not universal love
> But to be loved alone.
>
> W. H. Auden, from 'September 1, 1939'**

The denial and repression of early infantile experience has had a deep and largely unexplored impact on philosophy. This repressed material shapes by its very absence in consciousness the way we look at and reflect upon the world. The repression of early infantile experience and the oppression of women are linked by the fact that *women* (and only women) "mother," that is, assume primary responsibility for the physical care and psychological nurturance of young children. While there is considerable variation in the extent to which men participate in child care, to my knowledge there is no known society in which men assume the primary responsibility for the care of children under six.[1] However, the negative consequences of this fact, boh for the status of women and for psychological development, may be mitigated by a number of factors including the extent of male participation in child care, the degree to which the household is isolated from other social functions (such as production), the rigidity of the distinction between public and private life, and the degree to which women are permitted to participate in socially valued tasks other than child rearing. The philosophers discussed below represent "pure case" examples, since they each lived in an historical period in which these mitigating factors were largely absent.

Both individual male development and patriarchy are partially rooted in a need to deny the power and autonomy of women. This need arises in part

245

Sandra Harding and Merrill B. Hintikka (eds.), Discovering Reality, 245–281.
Copyright © 1983 *by D. Reidel Publishing Company.*

out of early infantile experience. The experience of maturing in a family in which only women mother insures that patriarchy will be reproduced.[2] Males under patriarchy must repress early infantile experience for several reasons: patriarchy by definition imputes political, moral and social *meanings* to sexual differentiation. Women are considered inferior in all these dimensions. The social world is thus both gender differentiated and stratified. (Men and women are different types of humans; men are superior to women.) Differentiation need not lead to stratification, but under patriarchy it does and must. Men want very much to attain membership in the community of men, in order to attain both individual identity and social privilege, even at great psychic pain. Patriarchal society depends upon the proper engendering of persons, since gender is one of the bases of social organization.[3]

Male identity under these social conditions requires the rejection of the mother by the son and a shift of libido and identification from her to the father. This is not easily accomplished, however, since the self is formed in part in and through relations with others. These persons and feelings about them are internalized; they become an "internal object" and the self is formed out of internal objects, the relations between them, and one's innate constitution (see section on object relations psychoanalysis, below). The mother and the son's relationship with her are literally part of the son's self. He is psychically originally female (since in the traditional patriarchal family the father does not care for infants). He must *become* male. In order to do so he must become *not female*, since under patriarchy gender is an exclusionary category. He must repress part of himself – his identification with the mother and memories of his relation with her (which are now internalized) and identify with the father, who as an adult has repressed his female self. The son can also devalue his mother and his relationship with her in order to make these aspects of himself less powerful within his own psyche. This devaluation is reinforced by the outside world. Females are less likely to repress infantile experience because of their gender identity with the mother.

Infancy (no matter how the social world is organized) is characterized by a state of dependence and powerlessness as well as intense wishes. The infant becomes aware of its dependence on others and its inability to satisfy its own needs. Frustration is projected onto whomever is present for the child, and when the person becomes an internal object, these frustrations, fears, desires, etc., are introjected along with the person. Therefore the mother is internalized along with the child's own powerful feelings about her. The boy, as he represses the mother, must repress all these feelings, too, since they are part of his experience of her. Given how powerful she is in infancy, the son must carefully guard against her power (since it is part of his self). Thus, repression must be as complete as possible so that this internal object can be kept separate from the conscious self. Otherwise, it would threaten masculine identity and ego boundaries.

As long as the infantile drama (resolving the desire for fusion with the mother as well as the fear of it and the desire for separation and the fear of it) is played out only with one gender (the mother), the child, including the child within the adult, will not resolve its ambivalence about growing up and taking responsibility for its self and its actions in a world in which complete knowledge and control is not possible.[4] Rather than confronting this ambivalence and resolving it, the child can turn the mother (including the internalized mother) into a bad object or he can split her into a good and bad object. The bad object is then responsible for personal dilemmas and painful feelings. Since the bad object is split off from the central self, the ambivalence is not resolved.[5] If both genders were present in infancy, this splitting off would be a less available defensive maneuver. If patriarchy did not exist, it would not be reinforced by social reality and both genders would be more likely to grow up — that is, to become persons who develop individually within reciprocal relationships and who can accept responsibility for their acts without the need to create an illusion of complete control over other persons and the environment.

The repression of early infantile experience is reflected in philosophy since its subject matter is primarily the experience and actions of male human beings who were created in and through patriarchal social relations. (I assume here that while there are many forms of male dominance, inequality between men and women has been a persistent feature of human history.[6]) This experience is seen not as typically male but constituitive of human experience itself. The problem is thus much deeper than the existence of consciously held misogynous ideas within philosophy but extends to fundamental issues of epistemology and ontology, to the very essence of philosophy as such.

Adopting a feminist viewpoint that seeks to include infantile experience and women's activity within the realm of the social and knowable, permits and requires a critique of philosophy in which previously unacknowledged assumptions are revealed. This critique also reveals fundamental limitations in the ability of philosophy to comprehend women's and children's experiences and thus forces us to go beyond existing theories and methods.

The development of the argument to be presented here requires an unusual amount of preparatory work. The paper begins with a brief note on method and then moves to an introduction of the fundamental tenets of object relations (psychoanalytic) theory, from which many of my own assumptions are derived. Following this preparation is a feminist interpretation of Plato, Descartes, Hobbes and Rousseau. This analysis is meant to be exemplary of the possibilities of a feminist-psychoanalytic approach and exhausts neither the richness of the works discussed nor the possibilities of the approach suggested here. I regard the analysis of each philosopher as a preliminary sketch which would require considerable development (far beyond the constraints of a single essay) to be fully convincing.

A NOTE ON METHOD

... Just as the psychiatrist must proceed from the fragmentary and deceptive verbalizations of his patient's conscious mind to the more complex levels of subconscious experience, so must his political confrere use the potentially misleading but indispensible statements of political theorists, whose awareness of political matters is uncommonly acute, as a clue to the less fully articulate experiences and reactions of ordinary men.[7]

What forms of social relations exist such that certain questions and ways of answering them become constitutive of philosophy? This question is important both to understanding the current state of philosophy and the interpretation of previous philosophic work. In philosophy, being (ontology) has been divorced from knowing (epistemology) and both have been separated from either ethics or politics. These divisions were blessed by Kant and transformed by him into a fundamental principle derived from the structure of mind itself.[8] A consequence of this principle has been the enshrining within mainstream Anglo-American philosophy of a rigid distinction between fact and value which has had the effect of consigning the philosopher to silence on issues of utmost importance to human life.[9] Furthermore, it has blinded philosophers and their interpreters to the possibility that apparently insoluble dilemmas within philosophy are not the product of the immanent structure of the human mind and/or nature but rather reflect distorted or frozen social relations.[10]

I assume here that knowledge is the product of human beings. Thinking is a form of human activity which cannot be treated in isolation from other forms of human activity including the forms of human activity which in turn shape the humans who think. Consequently, philosophies will inevitably bear the imprint of the social relations out of which they and their creators arose. Philosophy can thus be read (among many other ways) as a stream of social consciousness. The very persistence and continuing importance of certain philosophies and philosophic issues can be treated as evidence of their congruence with fundamental social experiences and problems. Philosophy must at least resonate with central social and individual wishes and offer some solution to deeply felt problematics.[11] In philosophy however, as in psychotherapy, what is not said, or what is avoided, is often as significant as the manifest content of thought.

A focus on social relations is especially important for feminist philosophy for it enables us to analyze the influence of patriarchy on both the content and process of thought. The social relations of childrearing are especially important to feminist analysis because such arrangements are both among the roots of patriarchy and continue to sustain its existence.[12] Feminist philosophy thus represents the return of the repressed, of the exposure of the particular social roots of all apparently abstract and universal knowledge. This work could prepare the ground for a more adequate social theory in which philosophy and empirical knowledge are reunited and mutually enriched.

Psychoanalysis, especially object relations theory, is a crucial tool for feminist philosophy. Its content represents a systematic attempt to understand human nature as the product of social relations in interaction with biology.[13] Object relations theorists differ somewhat among themselves, but more important for my purposes are their differences with Freud and "orthodox" psychoanalysis. These differences include: a radical questioning of Freud's "instinct theory" (especially the reduction of libido to sexuality and the notion that instincts are a- or anti-social); the construction of a more dynamic model of psychological development which stresses the interaction of child and parents, rather than the immanent unfolding of instinctive zones — a model actually suggested by Freud's concept of the super-ego, especially as articulated in *The Ego and the Id*; and a reconceptualization of the first three years of life which emerges as the crucial period of psychodynamic development, rather than the Oedipal period. Consequently (given patriarchal child-rearing patterns), the focus is on the mother-child relationship rather than the father-child one.

Also important for philosophy is the breakdown within object relations theory of the rigid distinction between primary process (id) and secondary process (ego-reason), so that reason is seen as an innate potential capacity rather than a faculty painfully acquired through the internalization of the authority of the father and as a defense against frustration and threats from the external world. Reason no longer appears as a fragile, tentative acquisition, dependent upon the existence of patriarchal authority.

Furthermore, the logic of object relations theory suggests that there may be more than one form of "human nature"; as social relations change, so too (over very long periods of time) would "human nature." This possibility forces us to avoid the determinism present in certain contemporary theorizing which claims an invariant form of human nature rooted in biology without falling into the equal and opposite fallacy of vulgar Marxism (e.g., unmediated responsiveness to changes in the political-economy).[14]

The method of psychoanalysis is equally valuable.[15] From this perspective, certain questions immediately arise: What aspects of social experience are repressed and why? What distortions are introduced into the structure of mind and our accounts of it by acts of repression and the defensive mechanisms against the return of the repressed? How do social prohibitions and power relations enter into the construction of individual personality and thus partially determine the individual's acts and thought? The contents of the unconscious, its influences and consequences, can be revealed by an analysis of conscious thought processes, their form, contradictions, implicit assumptions and significant avoidances and by how (and by whom) human experience has been reconstructed through acts of interpretation.

The texts discussed here will be interpreted from this perspective, with absolutely no claims made as to the particular psychodynamics of individual philosophers. Nor do I mean to claim that philosophy can or should be treated as the mere rationalization of unconscious impulses and conflicts. Rationalization is only one of the many roots and purposes of philosophy. To discuss this particular root is not to deny the others, or to somehow "reduce" the importance of philosophy. This could only be the case if the existence of the unconscious is considered shameful.

OBJECT RELATIONS THEORY AND ITS POTENTIAL
CONTRIBUTION TO PHILOSOPHY

The most basic tenet of object relations theory is that human beings are created in and through relations with other human beings. The theory claims:

(1) The "psychological birth of the human infant"[16] does not occur simultaneously with physical birth. While physical birth is a distinct event, occurring within a finite and easily determined period of time, psychological birth is a complex process which stretches over roughly the first three years of life. Psychological birth emerges out of the interaction of physical and mental processes. The character of the process calls into question the simple separation of mind and body and any form of determinism built

upon these distinctions such as mechanism (Hobbes and modern variants), idealism (Plato or Husserl, for example) or instinct theory (early Freud or utilitarianism).

(2) "Psychological birth" can only occur in and through social (object) relations.[17] While there appear to be certain innate potentials and character traits within human beings (for example, the ability to walk and talk and differing levels of stress toleration), these potentials are most adequately achieved within good object relations. Sufficiently bad object relations can retard or distort the developmental process, including such "physical" achievements as walking. The necessarily social and interactive character of early human development calls into question certain philosophies of mind and being, especially radical individualism and the "monads" of Spinoza, early Sartre and others.

(3) The most important tasks of the first three years of life are first, establishing a close relationship with the caretaker – usually the mother – (symbiosis) and then moving from that relationship through the process of separation and individuation.[18] Separation means establishing a firm sense of differentiation from the mother, of possessing one's own physical and mental boundaries. Individuation means establishing a range of characteristics, skills and personality traits which are uniquely one's own. Separation and individuation are the two "tracts" of development; they are not identical, but they can reinforce or impede each other.

By the end of the third year a "core identity," or a distorted one, will have been established. Gender is a central element of this core identity. The child's sense of gender is firmly established by one and one-half to two years of age and has little to do with an understanding of sexuality or reproduction.[19] Under patriarchy, this sense of gender is not neutral. Becoming aware of gender means recognizing that men and women are not valued equally, that in fact, men are socially more esteemed than women. Being engendered, therefore, entails a coming to awareness of and to some extent internalizing asymmetries of power and esteem.

(4) Children's psychological development is a dialectical process played out in and through a changing relationship between mother and child. Both members of the dyad must learn to be sensitive to the needs and feelings of the other while also attempting to have their own needs met. Early development occurs between two poles: symbiosis and separation-individuation. In symbiosis (one to six or seven months), "the infant behaves and functions as though he and his mother were an omnipotent system – a dual unity within one common boundary."[20] The infant has no sense of its own body

boundaries and is extremely sensitive to its mother's moods and feelings. In this state of fusion with the mother, I and not-I are not yet differentiated, and inside and outside the self are only gradually distinguished. This phase is "the primal soil from which all subsequent human relationships form."[21]

In order for this phase to be adequate, the mother must be emotionally available to the child in a consistent, reasonably conflict-free way. She should be able to enjoy the sensual and emotional closeness of the relationship without losing her own sense of separateness. She should be concerned for the child's well being without developing a narcissistic overinvestment in the child as a mere extension of her own self. Her infantile wishes for a symbiotic relationship should have been adequately gratified in childhood. If this was not the case, resentment and hostility may be aroused in her by the infant's needs. The mother requires adequate support, both emotional and material, during this period from adults who are able both to nurture her and reinforce her own sense of autonomy.

Separation-individuation begins at about six months and continues to about the end of the second year. The child gradually develops an autonomous ego, practices and takes pleasure in its locomotor skills (which allow it to physically distance itself from the mother), explores the possibilities of being its own separate person. The initial euphoria present in the discovery of the child's own powers and skills diminishes as it discovers the limitations as well as the possibilities of its developing skills. The child painfully learns that not only is it not omnipotent, but that the mother, too, is not all powerful.

The child explores and continually develops its separateness, then returns to the mother for "emotional refueling." The potential presence of the relationship between child and mother allows the child to leave it. Gradually the relationship is internalized and becomes part of the child's internal psychic reality. Both members of the dyad must learn to let go of the early bond without rejecting the other. The ambivalence present throughout this process gradually intensifies. The child both wants to return to the symbiotic state and fears being engulfed by it. In "good enough" social relations a resolution is achieved in which both members of the dyad come to accept their bond (mutuality) and their separateness. This is the basis of a truly reciprocal relationship between the pair, which creates the possibility for the child to then establish reciprocal relations with others.

However, under patriarchy it is not possible to fully achieve this satisfactory synthesis. The girl child never resolves her ambivalent tie to the mother.[22] The boy child must identify with the father to consolidate his differentiation. Mahler notes that by the age of 21 months, there were significant development differences between boys and girls:

The boys, if given a reasonable chance, showed a tendency to disengage themselves from mother and to enjoy their functioning in the widening world. The girls, on the other hand, seemed to become more engrossed with mother in her presence; they demanded greater closeness and were more persistently enmeshed in the ambivalent aspects of the relationship.[23]

The boy by age five will have repressed the "female" parts of himself, his memories of his earliest experience and many relational capacities. He will have developed the "normal contempt" for women that is a fundamental part of male identify under patriarchy.[24] The girl, precisely because of her continuing ambivalent tie to the mother (which remains, in part, because of their gender identity) cannot so thoroughly repress her experience and relational capacities. The boy deals with the ambivalence inherent in the separation-individuation process by denial (of having been related), by projection (women are bad; they cause these problems) and by domination (mastering fears and wishes for regression by controlling, depowering and/or devaluing the object).

These defenses become part of ordinary male behavior toward adult women and to anything which seems similar to them or under their (potential) control — the body, feelings, nature. The ability to control (and be in control) becomes both a need and a symbol of masculinity. Relations are turned into contest for power. Aggression is mobilized to distance oneself from the object and then to overpower it. The girl, on the other hand, seeks relationships, even at the expense of her own autonomy. The two genders thus come to complement each other in a rather grotesque symmetry.[25]

(5) The social context of development includes not only the immediate child-caretaker(s) relation but also more general social relations which affect the child through its interaction with the caretakers. The caretaker(s) brings to the relationship a complex series of experiences which include not only personal history but also the whole range of social experience — work, friends, interaction with political and economic institutions and so on. Thus seemingly abstract and supra-personal relations such as class, race and patriarchy enter into the construction of "individual" human development.

The relation between these more general social relations and individual development is never simple and direct. The relationship is mediated not only by the particular qualities of each child-caretaker relationship but also by what the child brings to the world (its own innate constitution), by the inevitable permutations and distortions which occur in the incorporation of experience in the preverbal and prerational state of infancy and later in the ongoing unconscious process, and by the particular characteristics of each child's family (e.g., the number of family members present, and cultural, religious, class and ethnic norms as they affect childrearing patterns).[26]

(6) This long period of development is unique to the human species. No other species is so physically helpless at birth and remains so for such a long time. Physical helplessness makes us dependent on the good will of others. This dependence is made more significant and deeply felt by the fact that we rapidly develop a consciousness of it (beginning around the age of three months). This discovery is inextricably bound up with our growing consciousness of other human beings and the relationship between us. Our wishes and will to act far exceed in complexity and strength our ability to act upon them. The tension between desire and capacity is played originally in relation to only one gender – women – since only women take care of young children. Women are experienced as both the physical memory of this struggle and the cause of it (since very young children think their mother is omnipotent and the source of all their feelings). Women also embody the residual infantile ambivalence – the wish to be cared for and totally fused with another person and the dangers which this wish poses to the distinctness of self.

(7) This period is marked not only by physical and emotional dependency but by an intensity of experience which will never be repeated except in psychosis and perhaps in altered states of consciousness such as religious or drug experiences. Precisely because it is prerational and preverbal, it is difficult for the infant to screen, sort and modify its experience. Every new experience and life itself is a stream of feelings, stimuli and impressions which cannot be preshaped or categorized and thus easily organized and made coherent. This sort of organization of experience is a function of the ego which is itself developing during this period.

However, our early experiences are not lost as we develop; they never disappear. They are retained in the unconsciousness and continue to reverberate throughout adult life. We are often unaware of these reverberations since they are expressed in feeling or bodily forms (such as psychosomatic illness), not thought. Their roots are not immediately accessible to consciousness and their very existence may be so threatening and/or disorganizing that they must remain repressed. Many pleasures, too, however, are deepened by the unconsciousness awakening of infantile memories.

Object relations theory and psychology as a whole takes for granted that the mother (and/or other women) is the primary caretaker.[27] They do not usually point out the negative consequences this arrangement entails, nor the fact that it derives not from biological necessity but from a series of social relations and structures, the replication of which is essential for the existence and maintenance of patriarchy.[28]

THE RETURN OF THE REPRESSED: UNCONSCIOUS REVERBERATIONS IN PHILOSOPHY, OR, THE METAPHYSICS OF MALE IDENTITY

The repression of early infantile experience is reflected in and provides part of the grounding for our relationship with nature [29] and our political life, especially the separation of public and private, the obsession with power and domination and the consequent impoverishment of political life and theories of it. The repression of our passions and their transformation into something dangerous and shameful, the inability to achieve true reciprocity and cooperative relations with others, and the translation of difference into inferiority and superiority can also be traced in part to this individual and collective act of repression and denial. The following analysis of Plato, Descartes, Hobbes and Rousseau will show that philosophy is not immune from these problems and the influence of the unconscious.

1. Plato: Regression and Light

A crucial element in Plato's philosophy is the distinction between mind and body, knowledge and sense, reason and appetite. The model for the Republic [30] is the well-ordered soul which provides the principles for the organization of the city. The primary basis of social class in the Republic is the ability to reason. The purpose of the educational system is not only to prepare potential philosophers to see the forms, but to determine *who* has the capacity to do so. The capacity to reason depends upon the control and sublimation of the passions. In turn, "it is for control of the passions which threaten to erupt into public life that society exists. The state itself thus resembles a vast dehydrating plant for 'drying up' the passions of men through education and restraint."[31] Women are clearly identified by Plato with the most dangerous and disruptive forms of passion, especially sexuality. Thus, there is a deep, covert link between the very *purpose* of Plato's ideal state and his fear and dislike of women which is so evident in other dialogues (e.g., the *Symposium* and the *Timaeus*).

Plato stresses the importance of instilling a lawful spirit at an early age; "if a sound system of nurture and education is maintained, it produces men of a good disposition; and these in turn, taking advantage of such education, develop into better men than their forebears."[32] Such a spirit is a defense against the outbreak of disorder and chaos which constantly threatens even the best ordered personality or state. Even children's games must be carefully controlled or else:

little by little, this lawless spirit gains a lodgement and spreads imperceptibly to manners and pursuits; and from thence with gathering force invades men's dealings with one another, and next goes on to attack the laws and the constitution with wanton recklessness, until it ends by overthrowing the whole structure of public and private life.[33]

Reason, laboriously won, must dominate the "lower" aspects of the mind just as the philosopher must rule the state. These lower aspects, although repressed, retain their power and threaten to return. Even the philosopher cannot be fully trusted; the social arrangements of the Republic reinforce internal control of the passions (communal property, wives, controlled mating, parents who do not know which children are theirs). These restrictions apply only to the guardians. The other groups, less rational and therefore not rulers, can indulge their appetites more fully.

The acquisition of knowledge also depends upon the purification of the mind. The sensual present leads to confusion and belief, not to knowledge:

Your lovers of sights and sounds delight in beautiful tones and colours and shapes and in all the works of art into which these enter; but they have not the power of thought to behold and to take delight in the nature of Beauty itself.[34]

The forms are pure, abstract, eternal, universal, all that the flesh and the passions are not. "The forms are objects of thought, but invisible."[35] The ultimate Form (the Good), although invisible, is the light by which all the other forms become intelligible. To be able to see the light and stand its glory, one must be in control of the body, its claims and demands. The Ruler must be able to sublimate his "high spirits" into a passion for knowledge and be willing to "live a quiet life of sober constancy."[36]

The images of light and dark and the fear of losing reality recur in the *Republic*. Mere belief is compared to living in a dream, in which semblance is mistaken for the reality it resembles.[37] The "dark" is within the individual as well as in the threat of chaos and decay from outside. The very qualities that suit a person to be a philosopher king, under adverse conditions, may lead to ruin. The more noble the soul, the greater the danger and the potential harm to the self and others.[38]

The imagery of the cave in the *Republic*; the world of shadows, of the unconscious and of the womb, which the light of reason cannot penetrate or dispel, reveals the fear of regression to that preverbal state where feelings, the needs of the body and women (mothers) rule.[39] One must ascend from the cave to be free from it, and the philosopher must be forced to return to it, for of course, having escaped, he would rather live in the light.

Ideal love is also pure, uncorrupted by materiality and the body. In the *Symposium*,[40] for example, true love is distinguished from sexuality, and the love of women from the love of boys, and ultimately love of persons from love of knowledge (the highest form of love). Socrates' teacher in these matters was an old (and thus presumably asexual) woman. Women seem to be most dangerous when young and capable of stirring up the passions of both men and women.[41] In the *Laws* women are not eligible for even the limited number of offices open to them until age 40, although men may assume a much more extended range of offices at age 30. Once women are old enough to be married and bear children, they are largely excluded from public life, including military training. Pregnancy is, of course, an unavoidable reminder of the existence of sexuality and it is not surprising that Plato would want to banish pregnant (or potentially pregnant) women from public life.

In the *Symposium*, Socrates is presented as a hero, in part because he does not succumb to lust for one of the most beautiful (male) youths, despite Alcibiades' attempts at seduction. Socrates' ability to withstand the physical hardships of war and to continue to dwell in the world of thought despite them is also praised.

The transcendence of the body, especially sexuality, enables Socrates to attain the knowledge of the beautiful itself and be "quickened" with the true, and not the seeming virtue:

And once you have seen it, you will never be seduced again by the charm of gold, of dress, of comely boys, or lads just ripening to manhood; you will care nothing for the beauties that used to take your breath away and kindle such a longing in you . . . and when he has brought forth and reared this perfect virtue, he shall be called the friend of god and if ever it is given to man to put on immortality, it shall be given to him.[42]

Thus reason is in part a defense against regression against those "longings" which threaten to ensnare us forever in the body and the material world. A well ordered state will reinforce this internal defense.

Experientially the first body we escape (physically, then emotionally) is that of our mother. Our relation with our own body is mediated through our continuing ambivalence about separating and differentiating from her. Part of the power and terror possessed by the "longings" Diotima describes is derived from the rekindling of unconscious infantile memories of the first erotic love relationship – with our mother. These longings contain many elements, including a desire to fuse with the lover,[43] to lose one's own ego boundaries and the fear of doing so, since fusion evokes powerlessness and vulnerability.

It is far safer to reside in the world of light, beyond embodied physicality, for ultimately the body, its memories and desires, brings not only entanglements with other persons and the material world, but the final physical fusion – death. The pure soul, however, will not fear death, for from Plato's perspective it is merely deliverance from the flesh; it is the radical (and desired) decoupling of mind and body.

Thus in Plato's philosophy, the purposes of the state and the unconscious wishes of the individual are inextricably linked. Both seek to restrain and channel passion and to defend against chaos so that earthly materiality, which is subject to decay, may be transformed. The eternal, unchanging forms assure freedom from the cave, the womb, the unending cycle of birth and death, the realm of necessity and of women (mothers). The ideal state represents the closest possible approximation of this eternal realm on earth. The ideal soul is a miniature state and the ideal state the well ordered soul writ large.[44] Unless both are present, each will not be able to achieve its full potential excellence, and neither justice nor happiness will be possible. Yet, what sort of justice and happiness does Plato offer? Does justice require the sublimation and repression of sexuality and the denial of any significant differences between men and women? These questions, central to feminist theory, remain unresolved.

2. Descartes and the Narcissistic Position

Descartes' philosophy can also be read as a desperate attempt to escape from the body, sexuality, and the wiles of the unconscious. His philosophy is important not only in itself but also because it defined the problematics for much of modern Western philosophy.

In the *Discourse on Method*,[45] the problem of the cogito emerges in relation to the problem of distinguishing reality from a dream. For Descartes, the solution to the problem of certainty and the confusion generated by the senses is a radical reduction of consciousness to pure ego, to that which thinks. The ego is emptied of all content, since in principle there is nothing it can know a priori about its life situation or history, all of that having been cast into doubt.

Consider the assumptions behind and implications of this statement:

The very fact that I thought of doubting the truth of other things, it followed very evidently and very certainly that I existed while on the other hand, if I had only ceased to think, although all the rest of what I had ever imagined had been true, I would have had no reason to believe that I existed; I thereby concluded that I was a substance, of which the whole essence or nature consists in thinking, and which in order to exist, needs no place and depends on no material thing; so that this "I" that is to say, the mind by which I am what I am, is entirely distinct from the body, even that it is easier to know than the body, and moreover, that even if the body were not, it would not cease to be all that it is.[46]

My essence and the only thing of which I can be certain, is thought. This self needs "no place and depends on no material thing" including, one presumes, other human beings. It is thus completely self-constituting and self-sustaining. The self is created and maintained by thought. This view of the self entails a denial of the body and any interaction between body and self (except somehow through the pineal gland). Social relations are not necessary for the development of the self. The self is a static substance. Although it may think new thoughts, it is not transformed by them. It appears to come into the world whole and complete, clicking into operation like a perpetual motion machine.

Descartes' ego contemplates the material world, a material world emptied of particularity and subjective content. Thought contemplates not nature as lived through, how this particular orange tastes or smells, for example, but nature as mathematics. Only when nature is reduced to extension and motion can it be known with certainty. Nature cannot be known in its full concreteness, but only as the abstract object of an abstract cogito. Any knowledge not built on the foundations of mathematics is like the "moral writings of the ancient pagans, the most proud and magnificent palaces, built on nothing but sand and mud."[47]

Underlying the concern for certainty is a desire for control, control both of nature and of the body. Descartes was convinced that:

it is possible to arrive at knowledge which is most useful in life and that instead of the speculative philosophy taught in the schools, a practical philosophy can be found by which, knowing the power and the effects of fire, water, air, the stars, the heavens, and all the other bodies which surround us as distinctly as we know the various trades of our craftsmen, we might put them in the same way to all the uses for which they are appropriate and thereby make ourselves, as it were, masters and possessors of nature.[48]

The purpose of science is to capture the power of nature and hence to make it one's own, thus compensating for the weakness of mortal flesh.

Such a science might even overcome death, that reminder of the materiality of life, of the independence of the body:

We could free ourselves of an infinity of illnesses, both of the body and of the mind, and even perhaps also of the decline of age, if we know enough about their causes and about all the remedies which nature has provided us.[49]

There is a deep irony in Descartes' philosophy. The self which is created and constituted by an act of thought is driven to master nature, because ultimately the self cannot deny its material qualities. Despite Descartes' claim, the body reasserts itself, at least at the moment of death. In order to fully become the substance it is, the cogito must master nature and possess its secrets, "the remedies nature has provided us," so that the self will never "cease to be all that it is," that is, die.

The desire to know is inextricably intermeshed with the desire to dominate. Nature is posited as pure otherness which must be conquered to be possessed and transformed into useful objects. The posture of Descartes' cogito replicates that of a child under two in its relation to a caretaker. The child originally believes that it and its mother are one person, a symbiotic unity.[50] However, due to frustrations in satisfying its needs and internal psychological pressures (primarily a growing desire for autonomy), it begins to realize that its mother is a separate person, an other. This discovery is accompanied simultaneously by panic and exhilaration, for while the child knows it is still dependent on the mother, it also begins to realize that autonomy requires separation. Accompanying separation is an increased sense of both power and vulnerability. One reaction and defense to the discovery of separateness is narcissism, in which the outside world is seen purely as a creation of and an object for the self.

Through "good enough"[51] social relations this stage is transformed into a genuine reciprocity in which separateness and mutuality (interdependence) exist simultaneously. However, denial of separateness, of the individual integrity of the object (mother) will lead to the adoption of narcissism as a permanent character structure,[52] precisely the type of solipsistic isolated self with delusions of omnipotence which Descartes' cogito displays.

Furthermore, underlying the narcissistic position, the fear and wish for regression to the helpless infantile state remains. The longings for symbiosis with the mother are not resolved. Therefore, one's own wishes, body, women and anything like them (nature) must be partially objectified, depersonalized and rigidly separated from the core self in order to be controlled.[53] Once this position is established, the relationship between the self (subject) and object (other persons, nature, the body) becomes extremely problematic, perhaps unresolvable. This frozen posture is one of the social roots of the subject-object dichotomy and its persistence within modern philosophy.[54] It is an abstract expression of a deeply felt dilemma in psychological development under patriarchy and thus cannot be resolved by philosophy alone.

DO WOMEN AND CHILDREN EXIST IN THE STATE OF NATURE?

Modern political philosophy also conceals a denial of early infantile experience, although in a different way. Especially important is the denial of the primary relatedness to and dependence upon the caretaker present in infancy and the consequences of this denial for conceptions of human nature. It is noteworthy that both Hobbes and Rousseau, despite their many differences, assume that "man" is a solitary creature by nature, and that dependence, indeed any social interaction, inevitably leads to power struggles which ultimately result in either domination or submission. These assumptions, which are not unique to them, shape each theorist's conception of the original state of human beings (the state of nature). An analysis of Hobbes' and Rousseau's concepts of human nature will show both some of the effects of patriarchal forms of psychological development on political life and the dilemmas these effects introduce into political theory. In Hobbes, although not in Rousseau, the state of nature is marked by the prevalence of anxiety and insecurity. Significantly, the anxiety is centered on the fear of wounds to the body and the deprivation of needed and desired objects — paralleling the paranoid aspect of the separation process.[55] "Natural man" attributes this fear to an external "bad object" — to fear of aggression from other persons who will not respect his autonomy. Aggression and separateness are viewed as innate in humans rather than as problems with social roots.

1. Hobbes: The State of Nature

It is only possible to view people in this way if an earlier period of nurturance and dependence is unsatisfactory and/or denied and repressed. The state of nature seems to be primarily populated by adult, single males whose behavior is taken as constitutive of human nature and experience as a whole. Hobbes is clearly puzzled about how to fit the family into his state of nature. There are only a few contradictory comments on it in the *Leviathan*.

In the state of nature, Hobbes argues, men and women have equal claim to possession and control of children, but this is impossible, "for no man can obey two Masters":[56]

And whereas some have attributed the Dominion to the man onely, as being of the more excellent Sex; they misreckon in it. Fore there is not always that difference of strength or prudence between the man and the woman, as that the right can be determined without War.[57]

Dominion over children can be settled by contract. If there is conflict or no contract exists (and no contract can be guaranteed in the state of nature), children should obey the mother, since parentage can only be ascertained with certainty for her. On the other hand, Hobbes says, the allegiance children owe to their parents is derived *not* from generation, "but from the Child's Consent, either expresse, or by other sufficient arguments declared."[58] He does not seem to notice the contradiction between this argument and the previous one.

In a startling paragraph which reveals the fear infantile dependence can induce (of the power of the mother to virtually annihilate the child), Hobbes introduces a different argument. The child in the state of nature ultimately owes its allegiance to the mother because of her power over it:

Again, seeing the Infant is first in the power of the Mother, so as she may either nourish or expose it, if she nourish it, it oweth its life to the Mother; and is therefore obliged to obey her, rather than any other; and by consequence the Dominion over it is hers. But if she expose it, and another nourish it, the Dominion is in him that nourish it. For it ought to obey him by whom it is preserved; because preservation of life being the end, for which one man becomes subject to another, every man is supposed to promise obedience to him, in whose power it is to save or destroy him.[59]

Despite (or because of) what we might call this original natural obligation to women as (at least the potential) preservers of life, in civil society dominion usually belongs to the father, because, "for the most part Common-wealths have been erected by the Fathers, not by the Mothers of families."[60] Later in the argument, this acknowledgment of original dependence on women (mothers) seems to be completely forgotten or repressed. Hobbes states flatly that in civil society, the family is a lawful private body, in which:

the Father, or Master ordereth the whole Family ... For the Father, and Master being before the Institution of Commonwealth, absolute Sovereigns in their own Families, they lose afterward to more of their Authority, than the Law of the Commonwealth taketh from them.[61]

Hobbes' mechanistic model of human nature does not include the female, that is, it excludes the traits culturally attributed to females — sociability, nurturance, concern for dependent and helpless persons. Humans are said to be motivated only by passion, especially fear and the wish to have no impediments to the gratification of desire, which is insatiable and asocial. Human beings are basically greedy infants driven by:

a perpetuall and restlesse desire of Power after power that ceaseth onely in Death. And the cause of this, is not alwayes that a man hopes for a more intensive delight, than he has already attained to; or that he cannot be content with a moderate power; but because he cannot assure the power and the means to live well, which he hath present, without the acquisition of more.[62]

In other words, without infantile omnipotence one cannot be certain that one will continue to be nourished at all.

The problem, of course (as for the infant), is that the gratification being sought (complete symbiotic security) can never be regained, "for there is no such thing as perpetuall Tranquility of mind while we live here; because Life itselfe is but Motion, and can never be without Desire, nor without Feare, no more than without sense."[63] Furthermore, despite the fact that felicity consists in "obtaining those things which a man from time to time desireth;"[64] "Passions unguided, are for the most part meere Madness."[65]

The only ways out of this dilemma are to either resolve the wishes by acknowledging the autonomy of the other person and entering into reciprocal relationships or to retain the wishes and control them defensively. Hobbes adopts the second alternative. Given his premises about the passions and human motivation (and the assumption of scarcity), the state of war inevitably follows. This state can only be abolished (or perhaps contained) by the creation of the Leviathan, a sort of externalized supergo, "One Will" who unites all these unruly passions and (literally) incorporates their power in his Person:

This is more than consent, or Concord; it is a reall Unitie of them all, in one and in the same person, made by Covenant of every man with every man, in such a manner, as if every man should say to every man, *I Authorize and give up my Right of Governing my selfe, to this Man, or to this Assembly of men, on this condition, that thou give up thy right to him, and Authorize all his Actions in like manner.*[66]

Since the character of the passions makes it impossible for any man to govern himself or to cooperate with others, an "artificial person" must be created. This artificial person is a good patriarchal father. By controlling and channeling the passions of his sons, he creates the possibility of civil society, morality and culture, none of which can exist in the state of nature.[67]

Thus we pass from the dominion of the Mothers, in which nourishment could be refused and life alternates between intense fear, desire and gratification, to the dominion of the Father, in which the preservation of life, "peace and security" are guaranteed by obedience to his will (Law) and renunciation of the absolute right to gratify any passion. This artificial person is superior to any real one. Like Descartes, Hobbes believed that the use of right reason (modeled on mathematics) could conquer death, or at least the death of states:

> So, long time after men have begun to constitute Commonwealths, imperfect, and apt to relapse into disorder, there may be found out, by industrious meditation, to make their constitution (excepting by externall violence) everlasting. And such are those which I have in this discourse set forth.[68]

Mortal fathers create their immortal Father, and their sovereignty over the once powerful Mother and all that she represents is now assured.

2. Rousseau: Escape from Desire

Rousseau's version of the state of nature in *A Discourse on the Origin of Inequality*[69] appears to be quite different than Hobbes'. Despite Rousseau's criticisms of Hobbes, he also shares many of Hobbes' assumptions concerning "natural man."[70] While acknowledging their profound differences, I wish to investigate their similarities, since it is especially remarkable that certain fundamental assumptions are shared by such otherwise differing theorists.

Rousseau's natural man is "alone, idle, and always near danger"[71] (from the attacks of animals, not other persons, at least in the initial, uncorrupted state of nature). Human beings are "ingenious machines" who are perfectly adapted to their environment, like other animals. However, the faculty that distinguishes them from animals is also "the source of all man's misfortunes."[72] This is "the faculty of self-perfection . . . which, with the aid of circumstances, successfully develops all the others."[73]

Like Hobbes, Rousseau assumes that people are solitary by nature and that culture and social institutions are not natural. Both assume that natural man is driven by passion and desire. According to Rousseau, the only goods natural man "knows in the universe are nourishment, a female and repose; the only evils he fears are pain and hunger."[74] In this "primitive state" men had brief, purely sexual encounters with women and returned to their solitary ways; "males and females united fortuitously, depending on encounter, occasion or desire ... they left each other with the same ease."[75] Even childbearing gave rise only to brief, utilitarian relations between mother and child. A mother would nurse her children, first:

for her own need; then, habit having endeared them to her, she nourished them afterward for their need. As soon as they had the strength to seek their food, they did not delay in leaving the mother herself; and ... they were soon at the point of not even recognizing each other.[76]

Rousseau's natural man (like Hobbes') has an intense dislike and fear of dependence. Dependence on another person inevitably leads to servitude:

The bonds of servitude are formed only from the mutual dependence of men and the reciprocal needs that unite them, it is impossible to enslave a man without first putting him in the position of being unable to do without another; a situation which, as it did not exist in the state of nature, leaves each man there free of the yoke. ... [77]

What could compel man to leave this natural state? Unlike Hobbes, Rousseau does not posit that the state of nature degenerated *immediately* into the state of war, for each person is motivated both by the desire for self-preservation (which leads him to avoid others) and the feeling of compassion (which leads him to avoid harming others).[78]

The answer seems to be a combination of the social and technological innovations made possible by man's faculty for self-improvement and the evil manipulation of women. As population increases, people encounter each other more frequently. Gradually they developed crude tools to acquire subsistence and defend against animals. The "repeated utilization of various beings to himself and of some beings in relation to others, must naturally have engendered in man's mind perceptions of certain relations."[79] This realization developed a new intelligence which in turn increased human superiority over animals, by making them sensible of it. This was the first source of pride and also of the ability to distinguish similarities between the self and other humans (who were also not animals). Once these similarities were noticed, it occurred to people that they might cooperate together, since each person was motivated by the same interest. This created the possibility of a mutual interest and also of conflicting interests.

"These first advances finally put man in a position to make more rapid ones. The more the mind was enlightened, the more industry was perfected."[80] People learned how to build huts. These advances prepared the way for the "first revolution, which produced the establishment and differentiation of families, and which introduced a sort of property — from which perhaps many quarrels and fights already rose."[81] Domestic life in turn led to an expansion of the human heart to include "the sweetest sentiments known to man, conjugal love and paternal love."[82]

These developments, while progressive, had a negative aspect as well. Although this period of human history "must have been the happiest and most durable epoch,"[83] it contained the roots of its own degeneration. Life became softer and more sedentary. People created conveniences which became necessities, they "were unhappy to lose them without being happy to possess them."[84] People began to compete for love and the esteem of others:

A tender and gentle sentiment is gradually introduced into the soul and at the least obstacle becomes an impetuous fury. Jealousy awakens with love; discord triumphs, and the gentlest of all passions receives sacrifices of human blood.[85]

The mutual dependence of one person on another and the desire for one person to have enough for two led to the disappearance of equality; "property was introduced, labor became necessary . . . slavery and misery were soon seen to germinate and grow with the crops."[86] Civil society and law were introduced by wealthy persons to protect their property,[87] and "all ran to meet their chains, thinking they secured their freedom."[88]

Three aspects of Rousseau's account are especially significant. (1) While he sees human development as a process involving the interaction of nature and human capacities,[89] social interaction is not considered essential, in fact it occurs quite late in the process. Ultimately, it has a destructive effect.

(2) Domestic life in the family is seen as the source of "sweet sentiment", as one of the primary bases of civilization and of human unhappiness. Women seem to benefit far more than men from this new way of life, despite the fact that it is really the root of their subordination to men. Rousseau says "it was then that the first difference was established in the way of life of the two sexes, which until this time had had but one. Women became more sedentary and grew accustomed to tend the hut and the children, while the man went to seek their common subsistence."[90] As Okin argues,[91] by Rousseau's own logic this change means that women have become dependent on men for their very subsistence and thus are no longer their equal, despite Rousseau's claim that "reciprocal affection and freedom"[92] were the family's only bonds. Thus, even within the ideal period of the state of nature, there has already emerged a fundamental form of inequality, although Rousseau

does not acknowledge it as such. Paradoxically, Rousseau argues that mono-gamous love enables women to control men: "it is easy to see that the moral element of love is an artifical sentiment, born of the usage of society and extolled with much skill and care by women in order to establish their ascendency and make dominant the sex which ought to obey." [93]

(3) Social interaction is assumed to lead to dependence, which in turn leads to slavery. Just like a small child, Rousseau's natural man seems to have only two choices: isolation or total engulfment. The fear of dependence is so strong that any acknowledgment of it must be totally denied — like the child in the state of nature who escapes from its mother as soon as possible and is unable to recognize her from then on. The passions aroused by love are also like those of an infant — desire to possess the person totally, to be esteemed above all others and an "impetuous fury" if those desires are denied.

The only solution Rousseau proposes for this dilemma is to seek mastery. Patriarchal authority in the family and in the state is the necessary counter to the power women exercise over and through the passions. Reciprocity, in the sense of mutual interdependence and independence is not possible. [94] How could it be, given the assumptions he makes about human nature, women and the character of the passions? It is odd that compassion dis-appears as natural men degenerates, while self-preservation retains its power. Perhaps this is because Rousseau's implicit assumptions about human nature and development are more similar to Hobbes' than Rousseau consciously realizes. In both cases, desire, stimulated by and rooted in part in interactions with women, is a main cause of human misery.

Once natural man is corrupted, he cannot return to his happy state, [95] although women retain their natural status, that is, they are to remain under the authority of men. The only cure for the misery of civil society is the creation of a Republic in which men are citizen-soldiers whose primary bond is the deliberately impersonal one of the social contract. [96] Women are to be excluded from sovereignty and thus from the general will. [97] As in the *Leviathan*, creation of political society represents the triumph of law over desire: [98]

only then when the voice of duty replaced physical impulse and right replaced appetite, does man, who until that time only considered himself, find himself forced to act up on other principles and to consult his reason before heeding his inclinations . . . what man loses by the social contract is his natural freedom and an unlimited right to every-thing that tempts him and that he can get; what he gains is civil freedom and proprietor-ship of everything he possesses . . . to the foregoing acquisitions of the civil state could be added moral freedom, which alone makes him truly master of himself. For the impulse of appetite alone is slavery, and obedience to the law one has prescribed for oneself is freedom.[99]

Thus not only does Rousseau deny any sort of primary relatedness, but he establishes the Republic out of an impersonal, depersonalized interdependency (the social contract). The citizens are free precisely because they are not dependent on any person, while simultaneously they are a part of a whole which transcends all particular wills and passions and makes their sublimation possible. The danger and corruption brought about by personal dependence will be abolished through the tutelage of the Legislator, "a superior intelligence who saw all of men's passions yet experienced none of them."[100] Once his authority is internalized, the general will emerges as a moral force. External authority and internal authority are merged into a collective super-ego and both the public and the individual are "free" (from desire).

In conclusion, then, philosophy reflects the fundamental division of the world according to gender and a fear and devaluation of women characteristic of patriarchal attitudes. The concrete form these pervasive attitudes take varies among the philosophers. In Plato and Descartes it manifests itself in the radical disjunction between mind and body and an identification of the passions with chaos and error. In both, the body is to be placed firmly under the control of a desexualized reason. In Plato this form of reason is to be developed and reinforced by the state, which in turn is to be ruled only by those capable of so reasoning. In Hobbes, the work that only women do (childrearing) and the qualities it demands — relatedness, sociability, nurturance, concern for others — are not seen as part of human nature or the human condition. While Rousseau includes compassion among the natural impulses, he shares many of Hobbes' other assumptions about human nature. In both philosophers, childhood experience is repressed on a social and individual level. Only thus is it possible to deny the most fundamental proof of the necessity of human bonding and its effects, which extend far beyond mere utility, and reverberate throughout adult life.

This denial is an essential element of patriarchy, since, as we have seen, male identity is created out of a rejection of the mother, including the female parts of the male self. The female represents all that is either not civilized and/or not rational or moral. In turn, the denial becomes a justification for relegating women to the private sphere and devaluing what women are allowed to do and be.[101] Not only is individual psychological development distorted, but these distortions are elevated into abstract theories of human nature, the character of politics, of the self and of knowledge which reflect, it is then claimed, unchangeable and inevitable aspects of human existence (and/or the structure of the external world).

TOWARDS A FEMINIST EPISTEMOLOGY

The task of feminist epistemology is to uncover how patriarchy has permeated both our concept of knowledge and the concrete content of bodies of knowledge, even that claiming to be emancipatory. Without adequate knowledge of the world and our history within it (and this includes knowing how to know), we cannot develop a more adequate social practice. A feminist epistemology is thus both an aspect of feminist theory and a preparation for and a central element of a more adequate theory of human nature and politics.

The prevailing forms of rationality and consciousness and our present accumulated knowledge reflect all aspects of human history including the existence of a sex/gender system in which biological characteristics are transformed into different and unequal social statuses and women are devalued. Reason is seen as a triumph over the senses, of the male over the female. In Hobbes, Freud and Rousseau, for example, reason can only emerge as a secondary process, under the authority and pressure of the patriarchal father. In Plato and Descartes, reason emerges only when nature (the female)[102] is posited as the other with an "inevitable" moment of domination. Yet all these theorists fail to locate this process within its social and historical context. This context is specific — because historically women have been the caretakers as well as the bearers of children, they represent both the body and our first encounter with the sometimes terrifying, sometimes gratifying viscitudes of social relations. They become the embodiments of the unconscious, just as men become the embodiments of reason and law (the ego and the superego).

The following theses are offered as the beginning of a feminist theory of knowledge: (1) All human knowledge serves (among others) a defensive function. Analysis reveals an arrested stage of human development, or as Hegel calls it, "the unhappy consciousness' behind most forms of knowledge and reason.[103] Separation-individuation cannot be completed and true reciprocity emerge if the "other" must be dominated and/or repressed rather than incorporated into the self while simultaneously acknowledging difference. An unhealthy self projects its own dilemmas on the world and posits them as the "human condition."

(2) The apparently irresolvable dualisms of subject-object, mind-body,

inner-outer, reason-sense, reflect this dilemma. Philosophies which locate such dualisms in the domination of the commodity form,[104] the dialectic of enlightenment,[105] the opposition of instinct and culture,[106] or the history of the monad[107] are incomplete and abstract (that is, not adequately grounded in human experience). What is lacking is an account of the earliest period of individual history in which the self emerges within the context of a relation with a woman (or women) which is itself overdetermined by patriarchy and class relations. Only certain forms of the self and of philosophy can emerge under these conditions.

(3) Feminists should analyze the epistemology of all bodies of knowledge which claim to be emancipatory including psychoanalysis and Marxism. There is a danger that the "female" dimensions of experience will be lost in philosophies developed under patriarchy. The relation between content and method is often not accidental. For example, the relationship between the positivistic aspects of Marxist theory and the disappearance of women and the "relations of reproduction" within it could be investigated.[108] It is necessary to develop an autonomous feminist viewpoint(s).

(4) Women's experience, which has been excluded from the realm of the known, of the rational, is not in itself an adequate ground for theory. As the other pole of the dualities it must be incorporated and transcended. Women, in part because of their own history as daughters, have problems with differentiation and the development of a true self and reciprocal relations.[109] Feminist theory and practice must thus include a therapeutic aspect, with consciousness raising as a model and an emphasis on process as political.

(5) Feminism is a revolutionary theory and practice. It requires simultaneously an incorporation, negation and transformation of all human history, including existing philosophies. Nothing less than a new stage of human development is required in which reciprocity can emerge for the first time as the basis of social relations. The destruction of class systems and the "critique of domination" alone cannot bring this about, nor will they be possible unless the analysis of gender-based power relations is pushed far beyond its present forms and a new feminist practice emerges.

(6) All forms of social relations and knowledge which arise out of them including the concept of liberation must be rethought and reformulated. This will only be possible when the development of theory is seen in relation to practice and knowledge itself is demystified, traced back to the life histories and purposes (conscious or not) of those who produce it.

(7) All concepts must be relational and contextual. Ways of thinking and thinking about thinking must be developed which do justice to the

multiplicity of experience, the many layers of any instant in time and space.

(8) Dialectics is a way of beginning to think in terms of process, history and interrelationships. In Hegel, for example, knowing is treated (although somewhat abstractly) as an activity. This activity constitutes being in and through social relations which themselves have a history, just as individuals do.

(9) Knowledge and method must be self-reflective and self-critical. We do not just *experience* (at least not most of the time) but need and create concepts to filter and shape experience. For conceptualization to avoid rigidity, we must be members of a self-reflective society in which social relations (and relations with nature) are not organized on a principle of domination (of race, class, gender, and/or "expertise in light of institutional necessity").[110]

(10) Not all ways of thinking do justice to our experience or can be adequately connected to, informed by and inform practice. Claims of objectivity or neutrality are not more privileged than any others as we re-evaluate knowledge and experience.

It seems ironic and paradoxical that feminism, a political expression of women's desire for liberation, must take on these philosophical tasks. Women have represented being, as the bearers 'and nurturers of life. Yet precisely because knowing and being cannot be separated, we must know how to be. To do so requires a transformation of knowledge adequate to our being and which points us beyond its present distorted forms.

NOTES

* The ideas presented in this paper are a product of an ongoing discourse among many persons including Kirstin Dahl, Sandra Harding, Nancy Hartsock, Jill Lewis and Phyllis Palmer. I would especially like to thank Nannerl Keohane and Roger Masters for their detailed comments on an earlier draft. Discourse, of course, does not imply agreement.
** Reprinted by permission of Faber and Faber Ltd. and Random House Inc. from *The English Auden: Poems, Essays and Dramatic Writings 1927–1939* by W. H. Auden.
1 On this point, see Michelle Rosaldo, 'Theoretical Overview' in Michelle Zimbalist Rosaldo and Louise Lamphere (eds.), *Women, Culture and Society* (Stanford: Stanford University Press, 1974).
2 Nancy Chodorow, *The Reproduction of Mothering: Psychoanalysis and the Sociology of Gender* (Berkeley and Los Angeles: University of California Press, 1978), pp. 3–39; pp. 173–209.
3 I read Freud's account of the Oedipal situation as an analysis of the mechanisms and psychic costs of the intitiation of the male into patriarchal society. For Freud's own account, see Sigmund Freud, *An Outline of Psycho-Analysis* (New York: Norton, 1949),

pp. 44–51; *The Ego and the Id* (New York: Norton, 1960), pp. 18–29; and *Civilization and Its Discontents* (New York: Norton, 1960), pp. 46–53. On the use of gender as a means of social organization, see Gayle Rubin, 'The Traffic in Women: Notes on the Political Economy of Sex,' in Rayna Reiter (ed.), *Towards an Anthropology of Women* (New York: Monthly Review Press, 1975).

4 For a further development of this argument see Dorothy Dinnerstein, *The Mermaid and the Minotaur: Sexual Arrangements and the Human Malaise* (New York: Harper and Row, 1976), especially Chapter Ten.

5 On object splitting and the internal world, see Harry Guntrip, *Personality Structure and Human Interaction* (New York: International Universities Press, 1961), pp. 356–444.

6 After carefully examining an extensive body of historical and anthropological evidence, Rosaldo concludes (Rosaldo and Lamphere, p. 3): "Whereas some anthropologists argue that there are, or have been, truly egalitarian societies . . . and all agree that there are societies in which women have achieved considerable social recognition and power, none has observed a society in which women have publicly recognized power and authority surpassing that of men. . . . Everywhere we find that women are excluded from certain crucial economic or political activities, that their roles as wives and mothers are associated with fewer powers and prerogatives than are the roles of men. It seems fair to say then, that all contemporary societies are to some extent male-dominated, and although the degree and expression of female subordination vary greatly, sexual asymmetry is presently a universal fact of human life . . . the evidence of contemporary anthropology gives scant support for matriarchy . . . there is little reason to believe that early sexual orders were substantially different from those observed around the world today." By patriarchy I mean any system in which men as a group oppress women as a group, even though there may be hierarchies among men. Typically in patriarchal societies, men have more access to and control over the most highly valued and esteemed resources and social activities, e.g., in a religious society, men will be priests and women excluded from the most important religious functions (if not considered polluting to them). Patriarchy has a material base in men's control of women's labor power and reproductive power and a psychodynamic base as a defense against the infantile mother and men's fear of women. It has assumed many different historical forms, but it still remains a dynamic force today. On the economic base of patriarchy, see Heidi Hartmann, 'Capitalism, Patriarchy, and Job Segregation by Sex' in Zillah R. Eisenstein, ed., *Capitalist Patriarchy and the Case for Socialist Feminism* (New York: Monthly Review Press, 1979). On reproduction, see Linda Gordon, *Women's Body, Women's Right* (New York: Grossman, 1976), especially pp. 3–25; pp. 403–418. On psychodynamics, see Gregory Zilboory, 'Masculine and Feminine: Some Biological and Cultural Aspects,' in Jean Baker Miller, ed., *Psychoanalysis and Women* (Baltimore: Penguin 1973); Karen Horney, 'The Flight from Womanhood' in her *Feminine Psychology* (New York: Norton, 1967); Dinnerstein and Chodorow.

7 Frederick M. Watkins, 'Political Theory as a Datum of Political Science,' in Roland Young, ed., *Approaches to the Study of Politics* (Evanston: Northwestern University Press, 1958), p. 154. Cited in Nannerl O. Keohane, 'Female Citizenship', a paper presented at the Meeting of the Conference for the Study of Political Thought, April, 1979. Keohane argues that "the history of ideas is the psychiatry of social belief" (p. 34), although of course this is not its only purpose.

[8] Immanuel Kant, *Critique of Pure Reason*, trans. F. Max Muller (Garden City: Doubleday Anchor, 1966). See for example, p. 18: "The most important consideration in the arrangement of such a science is that no concepts should be admitted which contain anything empirical, and that a priori knowledge shall be perfectly pure. Therefore, although the highest principles of morality and their fundamental concepts are a priori knowledge, they do not belong to transcendental philosophy, because the concepts of pleasure and pain, desire, inclination, free-will, etc., which are all of empirical origin, must here be presupposed. Transcentental philosophy is the wisdom of pure speculative reason. Everything practical, so far as it contains motives, has reference to sentiments, and these belong to empirical sources of knowledge." For an interesting psychoanalytic interpretation of Kant and other philosophers, see Ben-Ami Scharfstein and Mortimer Ostow, 'The Need to Philosophize' in Charles Hanly and Morris Lazerowitz, eds., *Psychoanalysis and Philosophy* (New York: International Universities Press, 1970).

[9] A clear and poignant example of this is the work of Max Weber. See especially his essays 'Science as a Vocation' and 'Politics as a Vocation' in *From Max Weber*, eds. H. H. Gerth and C. W. Mills (New York: Oxford University Press, 1958). Weber worked within a neo-Kantian philosophical framework.

[10] This is of course not true of all philosophers. Much of Habermas' work has focused on precisely this possibility. See, for example, Jurgen Habermas, *Knowledge and Human Interests* (Boston: Beacon, 1968), especially the Appendix.

[11] For a similar view of political theory, see Norman Jacobson, *Pride and Solace: The Functions and Limits of Political Theory* (Berkeley: University of California Press, 1978).

[12] Of course, the social relations of childrearing are not the only determinant of human experience. They interact with and are partially determined by class relations and culture in the broadest sense: art, politics, religion, ideology, language, etc. For a more detailed working out of these relations, see Jane Flax, 'A Materialist Theory of Women's Status' in the *Psychology of Women Quarterly* (forthcoming, 1981) and Zillah Eisenstein, 'Developing a Theory of Capitalist Patriarchy and Socialist Feminism' in Eisenstein.

[13] On object relations theory see, D. W. Winnicott, *The Maturational Processes and the Facilitating Environment* (New York: International Universities Press, 1965); Margaret Mahler, Fred Pine and Anni Bergman, *The Psychological Birth of the Human Infant* (New York: Basic Books, 1975; and Guntrip, *Personality Structure* and also his *Psychoanalytic Theory, Therapy and the Self* (New York: Basic Books, 1971). There is much controversy about psychoanalysis and its status as a science, therapy and its relevance to feminism as well as the implications of psychoanalysis for the empiricist account of science. On psychoanalysis as a science (and what this might mean), see Guntrip, *Personality*, pp. 15–21; the essays by Salmon, Glymour, Alexander, Mischel, and Wisdom in Richard Wollheim, ed., *Freud: A Collection of Critical Essays* (Garden City: Anchor/Doubleday, 1974); and Jurgen Habermas, *Knowledge and Human Interests*, especially pp. 214–245. Object relations theory is based on close observation of healthy and disturbed children, and a reconstruction of childhood psychodynamics through clinical work with adults, including psychotics who have regressed to childhood. Its insights are confirmed by the success of therapy based on the theory and by the investigations of other researchers such as Piaget who do not work from a psychoanalytic perspective. See Jean Piaget, *The Construction of Reality in the Child* (New York: Basic Books, 1954), especially Chapter One. See also the clinical material in Jane Flax, 'The Conflict

Between Nurturance and Autonomy in Mother-Daughter Relationships and within Feminism,' in *Feminist Studies*, v. 4, no. 2, June 1978; and my 'Mother-Daughter Relationships: Psychodynamics, Politics and Philosophy' in Hester Eisenstein (ed.) *The Future of Difference* (Boston: G. K. Hall, 1980). Robert J. Stoller in *Splitting: A Case of Female Masculinity* (New York: Delta, 1974), reports a detailed case history which provides information on both psychoanalytic technique and psychodynamics. My account of object relations theory may be more political and feminist than some of the theorists would like, although both Guntrip and Winnicott discuss political as well as philosophical issues related to their theories.

14 Behaviorism denies the existence of the unconscious. I find this unacceptable on both empirical and moral grounds. Empirically, behaviorism cannot explain dreams, psychosomatic illness, the resistance of persons to strong stimuli even when it might be in their interest to respond to them (e.g., under torture). It assumes a direct link between body and behavior with no complex mediation. In sociology, for example, biological sexuality is conflated with a sex/gender system, that is, "the set of arrangements by which a society transforms biological sexuality into products of human activity, and in which these transformed sexual needs are satisfied" (Rubin, p. 159). Morally, behaviorism allows one to ignore the unique individuality of each person and to see people as subject to endless manipulation. It denies the possibility of innate characteristics and/or capacities (such as the ability to speak or reason), thus denying any non-environmental grounds for resisting authority.

15 On psychoanalytic method, see Sigmund Freud, *Therapy and Technique*, ed. Philip Rieff (New York: Collier, 1963); Otto Fenichel, *The Psychoanalytic Theory of Neurosis* (New York: W. W. Norton, 1945), especially pp. 3–32; pp. 547–588; Frieda Fromm-Reichmann, *Principles of Intensive Psychotherapy* (Chicago: University of Chicago Press, 1950); Harry Guntrip; and Ralph R. Greenson, *The Technique and Practice of Psychoanalysis* (New York: International Universities Press, 1967).

16 This is Mahler's phrase.

17 Psychoanalysts tend to call other persons objects. This terminology is meant to do justice to the ways in which we do objectify persons – through projection and introjection, for example – and to point to the process through which the cluster of feelings, experience and fantasies we have with and about other persons become *our* object, that is, part of our internal mental life and structure. In turn, these now internal processes can become an object for consciousness as we attempt to uncover their social roots, in analysis for example. In this sense, subject and object are aspects of one continuous process.

18 Some of this material first appeared in a different form in Flax, 'Nurturance.' I am using the word mother rather than parent in this account because *women* do this work and I want to emphasize this fact. However, this does not imply that men could not do it (see n. 28).

19 See John Money and Anke A. Ehrhardt, *Man and Woman, Boy and Girl* (Baltimore: The Johns Hopkins University Press, 1972), especially pp. 176–194.

20 Mahler, et al., p. 44.

21 Mahler, et al., p. 48.

22 Flax, 'Nurturance,' pp. 172–184; Chodorow, pp. 114–140.

23 Mahler, et al., p. 102.

24 Sigmund Freud discusses this in 'Some Psychical Consequences of the Anatomical

Distinction between the Sexes' reprinted in Jean Strouse, ed., *Women and Analysis* (New York: Dell, 1974), especially p. 10. See also Chodorow, Chapter 11.

25 Dinnerstein, Chapter Four.

26 Little work has been done which does justice to the complexity of the relationship between individual development and social relations. But see an exemplary study which focuses on class, Lillian Breslow Rubin, *Worlds of Pain: Life in the Working Class Family* (New York: Basic Books, 1976).

27 On the universality of this aspect of the sexual division of labor, Margaret Mead, 'On Freud's View of Female Psychology,' in Strouse, especially pp. 121–122.

28 The desire and capacity to mother should be separated from the capacity to give birth. Despite Rossi's claims and social ideology, there is no evidence that the capacity to care for young children is a female, genetically linked trait. See Alice S. Rossi, 'A Biosocial Perspective on Parenting' in Alice S. Rossi, Jerome Kagan and Tamara K. Hareven, *The Family* (New York: W. W. Norton, 1977). As critics of her article have pointed out, Rossi cites evidence misleadingly and draws conclusions from clinical material in direct opposition to the results of the work. See Wini Breines, Margaret Cerullo and Judith Stacey, 'Social Biology, Family Studies and Anti-Feminist Backlash,' in *Feminist Studies* 4 (1978), 45–51; and Chodorow, pp. 18–21. Burton White somewhat reluctantly concludes, "When you look closely at what it means to be a child-rearer in the child's first three years, you find that most of the factors involved do not seem to be sex linked . . . I see nothing that a mother does (except breast feeding) that a father could not do," in *The First Three Years of Life* (New York: Avon, 1975), p. 256. It is indicative of the power of patriarchal ideas that psychologists conclude from the study of child development not that children need reliable, consistent and loving relationships (which they obviously do) but that only their *mother* can provide such a relationship. A recent example of this confusion is Selma Fraiberg, *Every Child's Birthright: In Defense of Mothering* (New York: Bantam, 1977).

29 On the relationship between attitudes toward women and toward nature, see Susan Griffin, *Women and Nature* (New York: Harper and Row, 1978); Adrienne Rich, *Of Woman Born: Motherhood as Experience and Institution* (New York: W. W. Norton, 1976), chapters 3–6; and Max Horkheimer and Theodor Adorno, *Dialectic of Enlightenment* (New York: Herder and Herder, 1972), pp. 70–80.

30 Plato, *The Republic*, trans. Francis MacDonald Cornford (New York: Oxford University Press, 1947), especially pp. 120–143. There is much dispute over exactly how "feminist" Plato's philosophy is. Some argue that the evidence in the *Republic* is ambiguous. See, for example, Sarah B. Pomeroy, 'Feminism in Book V of Plato's Republic' in *Apeiron* VIII (1974), pp. 33–35. Susan Moller Okin in *Women in Western Political Thought* (Princeton: Princeton University Press, 1979), pp. 15–70, argues that once Plato abolishes the family for the guardians, he must question traditional Greek ideas about women. Arlene Saxonhouse in 'Comedy in Callipolis: Animal Imagery in the Republic,' *American Political Science Review* 72 (1978), 888–901, considers the fifth book of the *Republic* as a sort of "detour" whose main purpose is to suggest Plato's deep conviction that justice is not possible in the city. My own position is that the latent content of Plato's theory, even in the *Republic*, undercuts his apparent inclusion of women in the guardian class. The crucial step in the *Republic* is not the abolition of the family as Okin argues, but the repression and sublimation of sexuality. However, I agree with Okin's critique, especially in her 'Appendix to Chapter Two,' of the reconstruction

of Plato's philosophy by Leo Strauss and Alan Bloom. Saxonhouse's arguments are dependent upon this reconstruction; thus although they are elegantly developed, I find them unconvincing.

31 Jacobson, p. 3.

32 Plato, p. 115. Of course, part of the better system of nurture for the philosopher kings is the separation of parent and child. This would reduce bonding and the problems and possibilities of psychological attachment which ensue.

33 Plato, p. 115.

34 Plato, p. 183.

35 Plato, p. 128.

36 Plato, p. 213.

37 Plato, p. 183.

38 Plato, pp. 198–204.

39 Plato, pp. 227–235. See Philip E. Slater, *The Glory of Hera* (Boston: Beacon Press, 1968), especially pp. 3–122, on the relationship between persistent themes in Greek culture and Greek family structure.

40 Plato, the *Symposium* in Edith Hamilton and Huntington Cairns, eds., *Plato: The Collected Dialogues* (Princeton: Princeton University Press, 1961), especially pp. 553–563. See also the *Phaedrus* in Hamilton and Cairns, pp. 494–502. In the *Phaedo*, in Hamilton and Cairns, p. 48, Socrates says, "In despising the body and avoiding it, and endeavoring to become independent, the philosopher's soul is ahead of all the rest."

41 See for example, the distinctions between earthly, younger Aphrodite and elder, heavenly Aphrodite in the *Symposium*, pp. 534–535.

42 Plato, the *Symposium*, p. 563.

43 Aristophanes' account of the original hermaphrodite sex is especially significant in this regard. After the man-woman is split apart by Zeus, the two halves are left with a deep and incomprehensible longing. What they really want is to be welded together "to live two lives in one . . . to be merged . . . into an utter oneness with the beloved," the *Symposium*, p. 545. The gender of the two lovers is much less important than the character of the longing itself. E. R. Dodd's comment, in *The Greeks and the Irrational* (Berkeley: University of California Press, 1951), pp. 218–219, is also relevant here: (eros) "spans the whole compass of human personality, and makes the one empirical bridge between man as he is and man as he might be. Plato, in fact comes very close here to the Freudian concept of *libido* and sublimation. But he never, as it seems to me, fully integrated this line of thought with the rest of his philosophy; had he done so, the notion of the intellect as a self-sufficient entity independent of the body might have been imperiled, and Plato was not going to risk that." However, I doubt Dodds would agree with my explanation of why eros is not integrated into the rest of Plato's thought. Socrates' hatred of the body and the corruption it brings is so great that he suggests (in the *Phaedo*, p. 49) that true knowledge is possible only after death, when the soul is finally free of the body.

44 Just as Plato says on p. 55 of the *Republic*.

45 Rene Descartes, *Discourse on Method* (Baltimore: Penguin, 1968).

46 Descartes, p. 54.

47 Descartes, p. 31.

48 Descartes, p. 78.

49 Descartes, p. 79.

50 Mahler, et al., p. 41–120.

51 See Winnicott, pp. 56–63.

52 On narcissism and the need to deny the separateness of the object, see Otto Kernberg, *Borderline Conditions and Pathological Narcissism* (New York: Jason Aronson, 1975), especially pp. 3–47; pp. 213–243.

53 For a further development of this argument, see Evelyn Fox Keller, 'Gender and Science' in *Psychoanalysis and Contemporary Thought* 1, reprinted in this volume.

54 Other social roots in Western capitalist societies include the domination of the commodity form, see Georg Lukacs, 'Reification and the Consciousness of the Proletariat' in *History and Class Consciousness* (Cambridge: The MIT Press, 1971); relations of domination, see G. W. F. Hegel, 'Independence and Dependence of Self-Consciousness: Lordship and Bondage,' in *The Phenomenology of Mind*, trans. J. B. Baillie (New York: Harper, 1967); and alienation from nature, see Horkheimer and Adorno, *Dialectic of Enlightenment*, especially pp. 81–120.

55 See Melanie Klein, *Envy, Gratitude and Other Works 1946–1963* (New York: Delta, 1977), especially papers 1–3. This psychoanalytic analysis supports Wolin's argument that "liberalism was a philosophy of sobriety, born in fear, nourished by disenchantment, and prone to believe that the human condition was and was likely to remain one of pain and anxiety." Sheldon Wolin, *Politics and Vision: Continuity and Innovation in Western Political Thought* (Boston: Little, Brown and Company, 1960), pp. 293–294. However, I think this insight applies to Hobbes as well as Locke; "political society as a system of rules" (Wolin's description of Hobbes) can be seen as, in part, a defense against the same anxieties he attributed to Locke and other liberals.

56 Thomas Hobbes, *Leviathan*, ed. C. B. Macpherson (Baltimore: Penguin, 1968), p. 253. Gordon J. Schochet, in *Patriarchalism in Political Thought* identifies some of the overtly patriarchal aspects of Hobbes' thought (see pp. 225–243). However his definition of patriarchalism is more literal than mine (rule of *the* father). He does not seem to notice that although the liberal sons could overthrow the authority of the father, this privilege was not extended to the daughters or wives. From a feminist perspective, rule of the *fathers* is not all that different from rule of a single father. For feminism, it is not true that "genetic justification and the identification of familial and political power were becoming dead issues" (Schochet, p. 276). See, for example, Mary Wollstonecraft, *A Vindication of the Rights of Woman*, ed., Charles W. Hagelman (New York: W. W. Norton, 1967; originally published 1791), especially chapter 5.

57 Hobbes, p. 253.

58 Hobbes, p. 253.

59 Hobbes, p. 254.

60 Hobbes, p. 253.

61 Hobbes, p. 285. This argument also seems ironic in that all other rights of governing are given over absolutely to the Leviathan. On the character of the Leviathan's authority, see Hanna Pitkin, *The Concept of Representation* (Berkeley: University of California Press, 1967), pp. 14–37.

62 Hobbes, p. 161. The parallels between Hobbes' and Freud's assumptions concerning the character of basic instincts and the political consequences following from them are quite striking. See Sigmund Freud, *Civilization and Its Discontents* (New York: W. W. Norton, 1962), especially pp. 33–81. It is also noteworthy that this essay begins with a discussion of Freud's inability to grasp an oceanic felling – "of something limitless,

unbounded" — which Romain Rolland had described to him. After Freud decides that the feeling of "oneness with the universe" (which is similar to symbiotic unity) cannot be investigated by or included in psychoanalysis, he then develops his theory of culture and its conflict with instinct.

63 Hobbes, p. 130. I believe C. B. Macpherson is right, in *The Political Theory of Possessive Individualism* (New York: Oxford University Press, 1962), p. 79, when he argues that "Hobbes' materialism was neither an afterthought nor a window-dressing but an essential part of his political theory." Indeed his theory of human nature is a "necessary condition of his theory of political obligation" (p. 79). However the additional necessary postulate, that "the motion of every individual is necessarily opposed to the motion of every other" (p. 79) is derived not only from the "market assumption," but from the infantile level at which Hobbes was able to conceptualize human needs. This is one of the unconscious sources of Hobbes' "refusal to impose more differences on men's wants, his acceptance of the equal need for continued motion" (p. 78) since for a very young child all needs are intensely felt, lack of gratification is experienced as pain and needs are experienced as insatiable. Thus possessive individualism is, in part, a defense against disappointment and deprivation.

64 Hobbes, p. 129.

65 Hobbes, p. 142.

66 Hobbes, p. 227, italics in the original. The Leviathan, like the super ego, retains many of the irrational aspects of the process by which it is formed. For further development of this argument, see Jacobson, pp. 53–92. Although Jacobson does not explicitly root his interpretation in psychoanalysis, he does argue that, "Anxiety, despair and dread of annihilation are the most prominent features of Hobbes' state of nature. And since these are always with us, his state of nature cannot be merely historical or analytic. It resides within us and is perpetual" (p. 61). However, unlike Jacobson, I am not convinced that these feelings (or those of Camus) must always be with us, at least in such overwhelming and determinant intensity. Dinnerstein's arguments provide an interesting challenge to Jacobson's last chapter.

67 Hobbes, p. 186; p. 188. Compare to Freud, 'Civilization,' p. 44.

68 Hobbes, p. 378.

69 Jean-Jacques Rousseau, *Discourse on the Origin and Foundations of Inequality* in Jean-Jacques Rousseau, *The First and Second Discourses*, ed. Roger D. Masters, trans., Roger D. and Judith R. Masters (New York: St. Martins Press, 1964).

70 The important differences between Hobbes and Rousseau include: (1) Rousseau argues that there are two fundamental human instincts — self-preservation *and* compassion. (2) Many faults in human behavior Hobbes attributes to man's fundamental nature, Rousseau attributes to the consequences of living in civil society which are read back into the state of nature. (3) In Hobbes, the Leviathan is not rational while Rousseau argues authority can be legitimate only if it is rational. The Republic is established to escape arbitrariness, not to be governed by it.

71 Rousseau, p. 112.

72 Rousseau, p. 115.

73 Rousseau, p. 114.

74 Rousseau, p. 116.

75 Rousseau, pp. 120–121.

76 Rousseau, p. 121. In note (1), especially pp. 218–220, Rousseau argues, contrary

to Locke, that pregnancy would not give rise to women's dependency on men or to the family although this assertion is undercut by his positing a sexual division of labor in the state of nature.

77 Rousseau, p. 140.

78 Rousseau, pp. 132–141.

79 Rousseau, p. 143. This bears an interesting relation to Winnicott's notion of mother as "mirror," see D. W. Winnicott, *Playing and Reality* (New York: Basic Books, 1971), pp. 111–118.

80 Rousseau, p. 146.

81 Rousseau, p. 146.

82 Rousseau, pp. 146–147.

83 Rousseau, p. 150.

84 Rousseau, p. 147.

85 Rousseau, p. 148–149.

86 Rousseau, pp. 151–152.

87 Rousseau, pp. 158–159.

88 Rousseau, p. 159.

89 In fact his account is similar to cultural anthropologists such as S. L. Washburn, see his 'Behavior and Human Evolution,' in *Classification and Human Evolution* (Chicago: Aldine Publishing Company, 1963). The feminist critique of "man the hunter" theories could be fruitfully applied to Rousseau as well. See Nancy Tanner and Adrienne Zihlman, 'Women in Evolution. Part I: Innovation and Selection in Human Origins' in *Signs* 1, no. 3, part 1 (Spring 1976), 585–608; and Adrienne L. Zihlman, 'Women in Evolution, Part II: Subsistence and Social Organization among Early Hominids' in *Signs* 4, no. 1 (Autumn 1978), 4–20. There are also interesting parallels between Marx's and Rousseau's accounts of human development. See Lucio Colletti, 'Rousseau as Critic of "Civil Society"' in *From Rousseau to Lenin: Studies in Ideology and Society* (New York: Monthly Review Press, 1972).

90 Rousseau, p. 147.

91 Okin, pp. 108–115. Okin (pp. 115–194) also shows how this notion of women as "naturally" subordinate to men permeates Rousseau's subsequent writings.

92 Rousseau, p. 147.

93 Rousseau, p. 135.

94 On this point, see also Elizabeth Rapaport, 'On the Future of Love: Rousseau and the Radical Feminists' in Carol C. Gould and Marx W. Wartofsky, *Women and Philosophy* (New York: G. P. Putnam, 1976) and Jacobson, pp. 104–105.

95 On the radical disjunction of natural and civil man, see Nannerl O. Keohane, 'The Masterpiece of Policy in Our Century: Rousseau on the Morality of the Enlightenment,' in *Political Theory* 6, no. 4, November 1978.

96 Rousseau, *On the Social Contract*, ed. Roger D. Masters, trans. Judith R. Masters (New York: St. Martins Press, 1978), pp. 52–58; pp. 67–70.

97 Or at least so one infers from this "praise" of the daughters of Geneva in the 'Dedication to the Republic of Geneva' of his *Discourse on the Origin of Inequality*, pp. 78–90. The "chaste power" exercised by the "virtuous daughters" of Geneva is contrasted to the pernicious effects of the "debauched women" of other countries. It is hard to understand what this influence really means, since although Rousseau says it will always be the lot of women to govern men this is only true in the family and the influence is to

be exercised solely "for the glory of the State and the public happiness" (p. 189). But if liberty consists in obeying laws which the individual has made and women are excluded from the general will, how are they to know what the happiness of the public would be? No private interest is to enter into the public will. In fact, Rousseau is consigning women to silence. Wollstonecraft aptly criticizes the contradictions in Rousseau's view of women. See especially, chapter one of 'Vindication.'

[98] Although, of course Hobbes' citizen orients his life far more around the pursuit of private interest than does Rousseau's, and law in Rousseau is meant to be an expression of the general will, not just the Leviathan's will, which must be *accepted* as the general will (except under limited conditions, i.e., when the Leviathan is unable to preserve his life or the citizens').

[99] Rousseau, *Social Contract*, pp. 55–56.

[100] Rousseau, *Social Contract*, p. 67. Compare to Sigmund Freud's account of a group unified through identification, in *Group Psychology and the Analysis of the Ego* (New York: W. W. Norton, 1959), pp. 52–55: "If one cannot be the favourite oneself, at all events nobody else shall be the favourite . . . originally rivals, they have succeeded in identifying themselves with one another by means of a similar love for the same object . . . Social justice means that we deny ourselves many things so that other may have to do without them as well, or, what is the same thing, may not be able to ask for them. This demand for equality is the root of social conscience and the sense of duty . . . thus social feeling is based upon the reversal of what was first a hostile feeling into a positively-toned tie in the nature of an identification."

[101] See also, Aristotle, the *Politics*, trans. Ernest Barker (New York: Oxford University Press, 1962), Book I.

[102] Donna Haraway discusses the patriarchal reconstruction of nature and biology in 'Animal Sociology.' Nature is frequently identified with the female both in terms of the gender given to the noun (Die Natur, la nature) and imagery (mother earth). Griffin shows that attitudes towards women and towards nature are inextricably linked and that the liberation of women requires a radical rethinking of our relationship with nature.

[103] Hegel, *The Phenomenology*, pp. 242–267. This chapter of the Spirit's life history is a brilliant philosophical account of a self frozen in ambivalence because it is unable to recognize the other in itself and the self in the other: "But for its self, action and its *own* concrete action remain something miserable and insignificant, its enjoyment pain and the sublimation of these, positively considered, remains a mere 'beyond' " (p. 267).

[104] Lukacs, 'Reification,' for example.

[105] For example, Horkheimer and Adorno.

[106] For example, Freud.

[107] For example, Sartre's early work, *Being and Nothingness* (New York: Pocket Books, 1966), in which the self is in a perpetual state of anxiety and can never be certain that an "other" ever exists (and of course precisely for this reason can never overcome its anxiety). Or the tragic struggle in Edmund Husserl, *The Crisis of European Sciences and Transcendental Phenomenology*, trans. David Carr (Evanston: Northwestern University Press, 1970), in which the transcendental ego discovers its grounding in the "life world," but is trapped precisely in Hegel's unhappy consciousness, because it is unable to see and feel itself there. Merleau-Ponty also realized the lack of an adequate account of the "life world" within phenomenology and the challenge psychoanalysis posed to it. See his preface to A. Hesnard, *L'Oeuvre de Freud et Son Importance pour la Monde Moderne* (Paris: Payot, 1960).

[108] On this point, see Nancy Hartsock, 'Response to "What Causes Gender Privilege and Class Privilege?" by Sandra Harding,' a paper presented at the 1978 meetings of the American Philosophical Association.

[109] Flax, 'Nurturance and Autonomy.'

[110] This seems to be the clearest and more sensible aspect of Habermas' recent work on a "universal pragmatics." See Jurgen Habermas, *Communication and the Evolution of Society* (Boston: Beacon, 1976), essays one and four. Habermas' always problematic synthesis of Hegelian and Kantian rationalism, and its consequences, seem most evident in his recent work.

NANCY C. M. HARTSOCK

THE FEMINIST STANDPOINT: DEVELOPING THE GROUND FOR A SPECIFICALLY FEMINIST HISTORICAL MATERIALISM*

The power of the Marxian critique of class domination stands as an implicit suggestion that feminists should consider the advantages of adopting a historical materialist approach to understanding phallocratic domination. A specifically feminist historical materialism might enable us to lay bare the laws of tendency which constitute the structure of patriarchy over time and to follow its development in and through the Western class societies on which Marx's interest centered. A feminist materialism might in addition enable us to expand the Marxian account to include all human activity rather than focussing on activity more characteristic of males in capitalism. The development of such a historical and materialist account is a very large task, one which requires the political and theoretical contributions of many feminists. Here I will address only the question of the epistemological underpinnings such a materialism would require. Most specifically, I will attempt to develop, on the methodological base provided by Marxian theory, an important epistemological tool for understanding and opposing all forms of domination – a feminist standpoint.

Despite the difficulties feminists have correctly pointed to in Marxian theory, there are several reasons to take over much of Marx's approach. First, I have argued elsewhere that Marx's method and the method developed by the contemporary women's movement recapitulate each other in important ways.[1] This makes it possible for feminists to take over a number of aspects of Marx's method. Here, I will adopt his distinction between appearance and essence, circulation and production, abstract and concrete, and use these distinctions between dual levels of reality to work out the theoretical forms appropriate to each level when viewed not from the standpoint of the proletariat but from a specifically feminist standpoint. In this process I will explore and expand the Marxian argument that socially mediated interaction with nature in the process of production shapes both human beings and theories of knowledge. The Marxian category of labor, including as it does both interaction with other humans and with the natural world can help to cut through the dichotomy of nature and culture, and, for feminists, can help to avoid the false choice of characterizing the situation of women as either "purely natural" or "purely social". As embodied humans we are of course

283

Sandra Harding and Merrill B. Hintikka (eds.), Discovering Reality, 283–310.
Copyright © 1983 by D. Reidel Publishing Company.

inextricably both natural and social, though feminist theory to date has, for important strategic reasons, concentrated attention on the social aspect.

I set off from Marx's proposal that a correct vision of class society is available from only one of the two major class positions in capitalist society. On the basis of this meta-theoretical claim, he was able to develop a powerful critique of class domination. The power of Marx's critique depended on the epistemology and ontology supporting this meta-theoretical claim. Feminist Marxists and materialist feminists more generally have argued that the position of women is structurally different from that of men, and that the lived realities of women's lives are profoundly different from those of men.[2] They have not yet, however, given sustained attention to the epistemological consequences of such a claim. Faced with the depth of Marx's critique of capitalism, feminist analysis, as Iris Young has correctly pointed out, often

accepts the traditional Marxian theory of production relations, historical change, and analysis of the structure of capitalism in basically unchanged form. It rightly criticizes that theory for being essentially gender-blind, and hence seeks to supplement Marxist theory of capitalism with feminist theory of a system of male domination. Taking this route, however, tacitly endorses the traditional Marxian position that 'the woman question' is auxiliary to the central questions of a Marxian theory of society.[3]

By setting off from the Marxian meta-theory I am implicitly suggesting that this, rather than his critique of capitalism, can be most helpful to feminists. I will explore some of the epistemological consequences of claiming that women's lives differ structurally from those of men. In particular, I will suggest that like the lives of proletarians according to Marxian theory, women's lives make available a particular and privileged vantage point on male supremacy, a vantage point which can ground a powerful critique of the phallocratic institutions and ideology which constitute the capitalist form of patriarchy. After a summary of the nature of a standpoint as an epistemological device, I will address the question of whether one can discover a feminist standpoint on which to ground a specifically feminist historical materialism. I will suggest that the sexual division of labor forms the basis for such a standpoint and will argue that on the basis of the structures which define women's activity as contributors to subsistence and as mothers one could begin, though not complete, the construction of such an epistemological tool. I hope to show how just as Marx's understanding of the world from the standpoint of the proletariat enabled him to go beneath bourgeois ideology, so a feminist standpoint can allow us to understand patriarchal institutions and ideologies as perverse inversions of more humane social relations.

THE NATURE OF A STANDPOINT

A standpoint is not simply an interested position (interpreted as bias) but is interested in the sense of being engaged. It is true that a desire to conceal real social relations can contribute to an obscurantist account, and it is also true that the ruling gender and class have material interests in deception. A standpoint, however, carries with it the contention that there are some perspectives on society from which, however well-intentioned one may be, the real relations of humans with each other and with the natural world are not visible. This contention should be sorted into a number of distinct epistemological and political claims: (1) Material life (class position in Marxist theory) not only structures but sets limits on the understanding of social relations. (2) If material life is structured in fundamentally opposing ways for two different groups, one can expect that the vision of each will represent an inversion of the other, and in systems of domination the vision available to the rulers will be both partial and perverse. (3) The vision of the ruling class (or gender) structures the material relations in which all parties are forced to participate, and therefore cannot be dismissed as simply false. (4) In consequence, the vision available to the oppressed group must be struggled for and represents an achievement which requires both science to see beneath the surface of the social relations in which all are forced to participate, and the education which can only grow from struggle to change those relations. (5) As an engaged vision, the understanding of the oppressed, the adoption of a standpoint exposes the real relations among human bengs as inhuman, points beyond the present, and carries a historically liberatory role.

The concept of a standpoint structures epistemology in a particular way. Rather than a simple dualism, it posits a duality of levels of reality, of which the deeper level or essence both includes and explains the "surface" or appearance, and indicates the logic by means of which the appearance inverts and distorts the deeper reality. In addition, the concept of a standpoint depends on the assumption that epistemology grows in a complex and contradictory way from material life. Any effort to develop a standpoint must take seriously Marx's injunction that "all mysteries which lead theory to mysticism find their rational solution in human practice and in the comprehension of this practice."[4] Marx held that the source both for the proletarian standpoint and the critique of capitalism it makes possible is to be found in practical activity itself. The epistemological (and even ontological) significance of human activity is made clear in Marx's argument not only that persons are active but that reality itself consists of "sensuous human activity, practice."[5]

Thus Marx can speak of products as crystallized or congealed human activity or work, of products as conscious human activity in another form. He can state that even plants, animals, light, etc. constitute theoretically a part of human consciousness, and a part of human life and activity.[6] As Marx and Engels summarize their position.

As individuals express their life, so they are. What they are, therefore, coincides with their production, both with *what* they produce and with *how* they produce. The nature of individuals thus depends on the material conditions determining their production.[7]

This starting point has definite consequences for Marx's theory of knowledge. If humans are not what they eat but what they do, especially what they do in the course of production of subsistence, each means of producing subsistence should be expected to carry with it *both* social relations *and* relations to the world of nature which express the social understanding contained in that mode of production. And in any society with systematically divergent practical activities, one should expect the growth of logically divergent world views. That is, each division of labor, whether by gender or class, can be expected to have consequences for knowledge. Class society, according to Marx, does produce this dual vision in the form of the ruling class vision and the understanding available to the ruled.

On the basis of Marx's description of the activity of commodity exchange in capitalism, the ways in which the dominant categories of thought simply express the mystery of the commodity form have been pointed out. These include a dependence on quantity, duality and opposition of nature to culture, a rigid separation of mind and body, intention and behavior.[8] From the perspective of exchange, where commodities differ from each other only quantitatively, it seems absurd to suggest that labor power differs from all other commodities. The sale and purchase of labor power from the perspective of capital is simply a contract between free agents, in which "the agreement [the parties] come to is but the form in which they give legal expression of their common will." It is a relation of equality,

because each enters into relation with the other, as with a simple owner of commodities, and they exchange equivalent for equivalent. . . . The only force that brings them together and puts them in relation with each other, is the selfishness, the gain and the private interests of each. Each looks to himself only, and no one troubles himself about the rest, and just because they do so, do they all, in accordance with the pre-established harmony of things, or under the auspices of an all shrewd providence, work together to their mutual advantage, for the common weal and in the interest of all.

This is the only description available within the sphere of circulation or exchange of commodities, or as Marx might put it, at the level of appearance. But at the level of production, the world looks far different. As Marx puts it,

On leaving this sphere of simple circulation or of exchange of commodities . . . we can perceive a change in the physiognomy of our *dramatis personae*. He who before was the money-owner, now strides in front as capitalist; the possessor of labor-power follows as his laborer. The one with an air of importance, smirking, intent on business; the other timid and holding back, like one who is bringing his own hide to market and has nothing to expect but – a hiding.

This is a vastly different account of the social relations of the buyer and seller of labor power.[9] Only by following the two into the realm of production and adopting the point of view available to the worker could Marx uncover what is really involved in the purchase and sale of labor power, i.e. – uncover the process by which surplus value is produced and appropriated by the capitalist, and the means by which the worker is systematically disadvantaged.[10]

If one examines Marx's account of the production and extraction of surplus value, one can see in it the elaboration of each of the claims contained in the concept of a standpoint. First, the contention that material life structures understanding points to the importance of the epistemological consequences of the opposed models of exchange and production. It is apparent that the former results in a dualism based on both the separation of exchange from use, and on the positing of exchange as the only important side of the dichotomy. The epistemological result if one follows through the implications of exchange is a series of opposed and hierarchical dualities – mind/body, ideal/material, social/natural, self/other – even a kind of solipsism – replicating the devaluation of use relative to exchange. The proletarian and Marxian valuation of use over exchange on the basis of involvement in production, in labor, results in a dialectical rather than dualist epistemology: the dialectical and interactive unity (distinction within a unity) of human and natural worlds, mind and body, ideal and material, and the cooperation of self and other (community).

As to the second claim of a standpoint, a Marxian account of exchange vs. production indicates that the epistemology growing from exchange not only inverts that present in the process of production but in addition is both partial and fundamentally perverse. The real point of the production of goods and services is, after all, the continuation of the species, a possibility dependent on their use. The epistemology embodied in exchange then, along with the social relations it expresses, not only occupies only one side of the dualities it constructs, but also reverses the proper ordering of any hierarchy in the dualisms: use is primary, not exchange.

The third claim for a standpoint indicates a recognition of the power realities operative in a community, and points to the ways the ruling group's vision may be *both* perverse *and* made real by means of that group's power to define the terms for the community as a whole. In the Marxian analysis, this power is exercised in both control of ideological production, and in the real participation of the worker in exchange. The dichotomous epistemology which grows from exchange cannot be dismissed either as simply false or as an epistemology relevant to only a few: the worker as well as the capitalist engages in the purchase and sale of commodities, and if material life structures consciousness, this cannot fail to have an effect. This leads into the fourth claim for a standpoint – that it is achieved rather than obvious, a mediated rather than immediate understanding. Because the ruling group controls the means of mental as well as physical production, the production of ideals as well as goods, the standpoint of the oppressed represents an achievement both of science (analysis) and of political struggle on the basis of which this analysis can be conducted.

Finally, because it provides the basis for revealing the perversion of both life and thought, the inhumanity of human relations, a standpoint can be the basis for moving beyond these relations. In the historical context of Marx's theory, the engaged vision available to the producers, by drawing out the potentiality available in the actuality, that is, by following up the possibility of abundance capitalism creates, leads towards transcendence. Thus, the proletariat is the only class which has the possibility of creating a classless society. It can do this simply (!) by generalizing its own condition, that is, by making society itself a propertyless producer.[11]

These are the general characteristics of the standpoint of the proletariat. What guidance can feminists take from this discussion? I hold that the powerful vision of both the perverseness and reality of class domination made possible by Marx's adoption of the standpoint of the proletariat suggests that a specifically feminist standpoint could allow for a much more profound critique of phallocratic ideologies and institutions than has yet been achieved. The effectiveness of Marx's critique grew from its uncompromising focus on material life activity, and I propose here to set out from the Marxian contention that not only are persons active, but that reality itself consists of "sensuous human activity, practice". But rather than beginning with men's labor, I will focus on women's life activity and on the institutions which structure that activity in order to raise the question of whether this activity can form the ground for a distinctive standpoint, that is, to determine whether it meets the requirements for a feminist standpoint. (I use the term, "feminist" rather than "female" here to indicate both the achieved character of a standpoint and that a standpoint by definition carries a liberatory potential.)

Women's work in every society differs systematically from men's. I intend to pursue the suggestion that this division of labor is the first and in some societies the only division of labor, and moreover, that it is central to the organization of social labor more generally. On the basis of an account of the sexual division of labor, one should be able to begin to explore the oppositions and differences between women's and men's activity and their consequences for epistemology. While I cannot attempt a complete account, I will put forward a schematic and simplified account of the sexual division of labor and its consequences for epistemology. I will sketch out a kind of ideal type of the social relations and world view characteristic of male and female activity in order to explore the epistemology contained in the institutionalized sexual division of labor. In so doing, I do not mean to attribute this vision to individual women or men any more than Marx (or Lukacs) meant their theory of class consciousness to apply to any particular worker or group of workers. My focus is instead on institutionalized social practices and on the specific epistemology and ontology manifested by the institutionalized sexual division of labor. Individuals, as individuals, may change their activity in ways which move them outside the outlook embodied in these institutions, but such a move can be significant only when it occurs at the level of society as a whole.

I will discuss the "sexual division of labor" rather than the "gender division of labor" to stress, first my belief that the division of labor between women and men cannot be reduced to purely social dimensions. One must distinguish between what Sara Ruddick has termed "invariant and *nearly* unchangeable" features of human life, and those which despite being "*nearly* universal" are "certainly changeable."[12] Thus, the fact that women and not men *bear* children is not (yet) a social choice, but that women and not men rear children in a society structured by compulsory heterosexuality and male dominance is clearly a societal choice. A second reason to use the term "sexual division of labor" is to keep hold of the bodily aspect of existence — perhaps to grasp it over-firmly in an effort to keep it from evaporating altogether. There is some biological, bodily component to human existence. But its size and substantive content will remain unknown until at least the certainly changeable aspects of the sexual division of labor are altered.

On a strict reading of Marx, of course, my enterprise here is illegitimate. While on the one hand, Marx remarked that the very first division of labor occurred in sexual intercourse, he argues that the division of labor only becomes "truly such" when the division of mental and manual labor appears. Thus, he dismisses the sexual division of labor as of no analytic importance. At the same time, a reading of other remarks — such as his claim that the mental/manual division of labor is based on the "natural" division of labor in the family — would seem to support the legitimacy of my attention to the sexual division of labor and even add weight to the radical feminist argument that capitalism is an outgrowth of male dominance, rather than vice versa.

On the basis of a schematic account of the sexual division of labor, I will begin to fill in the specific content of the feminist standpoint and begin to specify how women's lives structure an understanding of social relations, that is, begin to follow out the epistemological consequences of the sexual division of labor. In addressing the institutionalized sexual division of labor, I propose to lay aside the important differences among women across race and class boundaries and instead search for central commonalities. I take some justification from the fruitfulness of Marx's similar strategy in constructing a simplified, two class, two man model in which everything was exchanged at its value. Marx's schematic account in Volume I of *Capital* left out of account such factors as imperialism, the differential wages, work, and working conditions of the Irish, the differences between women, men, and children, and so on. While all of these factors are important to the analysis of contemporary capitalism, none changes either Marx's theories of surplus value or alienation, two of the most fundamental features of the Marxian analysis of capitalism. My effort here takes a similar form in an attempt to move toward a theory of the extraction and appropriation of women's activity and women themselves. Still, I adopt this strategy with some reluctance, since it contains the danger of making invisible the experience of lesbians or women of color.[13] At the same time, I recognize that the effort to uncover a feminist standpoint assumes that there are some things common to all women's lives in Western class societies.

The feminist standpoint which emerges through an examination of women's activities is related to the proletarian standpoint, but deeper going. Women and workers inhabit a world in which the emphasis is on change rather than stasis, a world characterized by interaction with natural substances rather than separation from nature, a world in which quality is more important than quantity, a world in which the unification of mind and body is inherent in the activities performed. Yet, there are some important differences, differences marked by the fact that the proletarian (if male) is immersed in this world only during the time his labor power is being used by the capitalist. If, to paraphrase Marx, we follow the worker home from the factory, we can once again perceive a change in the *dramatis personae*. He who before followed behind as the worker, timid and holding back, with nothing to expect but a hiding, now strides in front while a third person, not specifically present in Marx's account of the transaction between capitalist and worker (both of whom are male) follows timidly behind, carrying groceries, baby and diapers.

THE SEXUAL DIVISION OF LABOR

Women's activity as institutionalized has a double aspect − their contribution to subsistence, and their contribution to childrearing. Whether or not all of us do both, women as a sex are institutionally responsible for producing both goods and human beings and all women are forced to become the kinds of people who can do both. Although the nature of women's contribution to subsistence varies immensely over time and space, my primary focus here is on capitalism, with a secondary focus on the Western class societies which preceded it.[14] In capitalism, women contribute both production for wages and production of goods in the home, that is, they like men sell their labor power and produce both commodities and surplus value, and produce use-values in the home. Unlike men, however, women's lives are institutionally defined by their production of use-values in the home.[15] And here we begin to encounter the narrowness of the Marxian concept of production. Women's production of use-values in the home has not been well understood by socialists. It is no surprise to feminists that Engels, for example, simply asks how women can continue to do the work in the home and also work in production outside the home. Marx too takes for granted women's responsibility for household labor. He repeats, as if it were his own, the question of a Belgian factory inspector: If a mother works for wages, "how will [the household's] internal economy be cared for; who will look after the young children; who will get ready the meals, do the washing and mending?"[16]

Let us trace both the outlines and the consequences of woman's dual contribution to subsistence in capitalism. Women's labor, like that of the male worker, is contact with material necessity. Their contribution to subsistence, like that of the male worker, involves them in a world in which the relation to nature and to concrete human requirements is central, both in the form of interaction with natural substances whose quality, rather than quantity is important to the production of meals, clothing, etc., and in the form of close attention to the natural changes in these substances. Women's labor both for wages and even more in household production involves a unification of mind and body for the purpose of transforming natural substances into socially defined goods. This too is true of the labor of the male worker.

There are, however, important differences. First, women as a group work more than men. We are all familiar with the phenomenon of the "double day," and with indications that women work many more hours per week than men.[17] Second, a larger proportion of women's labor time is devoted to the production of use-values than men's. Only some of the goods women produce are commodities (however much they live in a society structured by commodity production and exchange). Third, women's production is structured by repetition in a different way than men's. While repetition for both the woman and the male worker may take the form of production of the same object, over and over — whether apple pies or brake linings — women's work in housekeeping involves a repetitious cleaning.[18]

Thus, the male worker in the process of production, is involved in contact with necessity, and interchange with nature as well as with other human beings but the process of production or work does not consume his whole life. The activity of a woman in the home as well as the work she does for wages keeps her continually in contact with a world of qualities and change. Her immersion in the world of use — in concrete, many-qualitied, changing material processes — is more complete than his. And if life itself consists of sensuous activity, the vantage point available to women on the basis of their contribution to subsistence represents an intensification and deepening of the materialist world view and consciousness available to the producers of commodities in capitalism, an intensification of class consciousness. The availability of this outlook to even non-working-class women has been strikingly formulated by Marilyn French in *The Women's Room.*

Washing the toilet used by three males, and the floor and walls around it, is, Mira thought, coming face to face with necessity. And that is why women were saner than men, did not come up with the mad, absurd schemes men developed; they were in touch with necessity, they had to wash the toilet bowl and floor.[19]

The focus on women's subsistence activity rather than men's leads to a model in which the capitalist (male) lives a life structured completely by commodity exchange and not at all by production, and at the furthest distance from contact with concrete material life. The male worker marks a way station on the path to the other extreme of the constant contact with material necessity in women's contribution to subsistence. There are of course important differences along the lines of race and class. For example, working class men seem to do more domestic labor than men higher up in the class structure — car repairs, carpentry, etc. And until very recently, the wage work done by most women of color replicated the housework required by their own households. Still, there are commonalities present in the institutionalized sexual division of labor which make women responsible for both housework and wage work.

The female contribution to subsistence, however, represents only a part of women's labor. Women also produce/reproduce men (and other women) on both a daily and a long-term basis. This aspect of women's "production" exposes the deep inadequacies of the concept of production as a description of women's activity. One does not (cannot) produce another human being in anything like the way one produces an object such as a chair. Much more is involved, activity which cannot easily be dichotomized into play or work. Helping another to develop, the gradual relinquishing of control, the experience of the human limits of one's action — all these are important features of women's activity as mothers. Women as mothers even more than as workers, are institutionally involved in processes of change and growth, and more than workers, must understand the importance of avoiding excessive control in order to help others grow.[20] The activity involved is far more complex than the instrumental working with others to transform objects. (Interestingly, much of women's wage work — nursing, social work, and some secretarial jobs in particular — requires and depends on the relational and interpersonal skills women learned by being mothered by someone of the same sex.)

This aspect of women's activity too is not without consequences. Indeed, it is in the production of men by women and the appropriation of this labor and women themselves by men that the opposition between feminist and masculinist experience and outlook is rooted, and it is here that features of the proletarian vision are enhanced and modified for the woman and diluted for the man. The female experience in reproduction represents a unity with nature which goes beyond the proletarian experience of interchange with nature. As another theorist has put it," reproductive labor might be said to combine the functions of the arthitect and the bee: like the architect, parturitive woman knows what she is doing; like the bee, she cannot help what she is doing." And just as the worker's acting on the external world changes both the world and the worker's nature, so too "a new life changes the world and the consciousness of the woman."[21] In addition, in the process of producing human beings, relations with others may take a variety of forms with deeper significance than simple cooperation with others for common goals — forms which range from a deep unity with another through the many-leveled and changing connections mothers experience with growing children. Finally, the female experience in bearing and rearing children involves a unity of mind and body more profound than is possible in the worker's instrumental activity.

Motherhood in the large sense, i.e., motherhood as an institution rather than experience, including pregnancy and the preparation for motherhood almost all female children receive as socialization, results in the construction of female existence as centered with a complex relational nexus.[22] One aspect of this relational existence is centered on the experience of living in a female rather than male body. There are a series of boundary challenges inherent in the female physiology — challenges which make it impossible to maintain rigid separation from the object world. Menstruation, coitus, pregnancy, childbirth, lactation — all represent challenges to bodily boundaries.[23] Adrienne Rich has described the experience of pregnancy as one in which the embryo was both inside and

daily more separate, on its way to becoming separate from me and of-itself. In early pregnancy the stirring of the fetus felt like ghostly tremors of my own body, later like the movements of a being imprisoned in me; but both sensations were *my* sensations, contributing to my own sense of physical and psychic space.[24]

In turn, the fact that women but not men are primarily responsible for young children means that the infant first experiences itself as not fully differentiated from the mother, and then as an I in relation to an It that it later comes to know as female.[25]

Jane Flax and Nancy Chodorow have argued that the object relations school of psychoanalytic theory puts forward a materialist psychology, one which I propose to treat as a kind of empirical hypothesis. If the account of human development provided by object relations is correct, one ought to expect to find consequences — both psychic, and social. According to object relations theory, the process of differentiation from a woman by both male and female children reinforces boundary confusion in female egos and boundary strengthening in males. Individuation is far more conflictual for male than for female children, in part because both mother and son experience the other as a definite "other." The experience of oneness on the part of both mother and infant seems to last longer with girls.[26]

The complex relational world inhabited by women has its start in the experience and resolution of the oedipal crisis, cleanly resolved for the boy, whereas the girl is much more likely to retain both parents as love objects. The nature of the crisis itself differs by sex: the boy's love for the mother is an extension of mother-infant unity and thus essentially threatening to his ego and independence. Male ego-formation necessarily requires repressing this first relation and negating the mother.[27] In contrast, the girls' love for the father is less threatening both because it occurs outside this unity and because it occurs at a later stage of development. For boys, the central issue to be resolved concerns gender identification; for girls the issue is psycho-sexual development.[28] Chodorow concludes that girls' gradual emergence from the oedipal period takes place in such a way that empathy is built into their primary definition of self, and they have a variety of capacities for experiencing another's needs or feelings as their own. Put another way girls, because of female parenting, are less differentiated from others than boys, more continuous with and related to the external object world. They are differently oriented to their inner object world as well.[29]

The more complex female relational world is reinforced by the process of socialization. Girls learn roles from watching their mothers; boys must learn roles from rules which structure the life of an absent male figure. Girls can identify with a concrete example present in daily life; boys must identify with an abstract set of maxims only occasionally concretely present in the form of the father. Thus, not only do girls learn roles with more interpersonal and relational skills, but the process of role learning itself is embodied in the concrete relation with the mother. The male, in contrast, must identify with an abstract, cultural stereotype and learn abstract behaviors not attached to a well-known person. Masculinity is idealized by boys whereas femininity is concrete for girls.[30]

Women and men, then, grow up with personalities affected by different boundary experiences, differently constructed and experienced inner and outer worlds, and preoccupations with different relational issues. This early experience forms an important ground for the female sense of self as connected to the world and the male sense of self as separate, distinct, and even disconnected. By retaining the preoedipal attachment to the mother, girls come to define and experience themselves as continuous with others. In sum, girls enter adulthood with a more complex layering of affective ties and a rich, ongoing inner set of object relations. Boys, with a simpler oedipal situation and a clear and early resolution, have repressed ties to another. As a result, women define and experience themselves relationally and men do not.[31]

ABSTRACT MASCULINITY AND THE FEMINIST STANDPOINT

This excursion into psychoanalytic theory has served to point to the differences in the male and female experience of self due to the sexual division of
labor in childrearing. These different (psychic) experiences both structure and
are reinforced by the differing patterns of male and female activity required
by the sexual division of labor, and are thereby replicated as epistemology
and ontology. The differential male and female life activity in class society
leads on the one hand toward a feminist standpoint and on the other toward
an abstract masculinity.

Because the problem for the boy is to distinguish himself from the mother
and to protect himself against the real threat she poses for his identity, his
conflictual and oppositional efforts lead to the formation of rigid ego boundaries. The way Freud takes for granted the rigid distinction between the "me
and not-me" makes the point well: "Normally, there is nothing of which we
are more certain than the feeling of ourself, of our own ego. This ego appears
to us as something autonomous and unitary, marked off distinctly from
everything else." At least toward the outside, "the ego seems to maintain
clear and sharp lines of demarcation."[32] Thus, the boy's construction of
self in opposition to unity with the mother, his construction of identity as
differentiation from the other, sets a hostile and combative dualism at the
heart of both the community men construct and the masculinist world view
by means of which they understand their lives.

I do not mean to suggest that the totality of human relations can be
explained by psychoanalysis. Rather I want to point to the ways male rather
than female experience and activity replicates itself in both the hierarchical
and dualist institutions of class society and in the frameworks of thought
generated by this experience. It is interesting to read Hegel's account of the
relation of self and other as a statement of male experience: the relation of
the two consciousness takes the form of a trial by death. As Hegel describes
it, "each seeks the death of the other."

Thus, the relation of the two self-conscious individuals is such that they provide themselves and each other through a life-and-death struggle. They must engage in this struggle,
for they must raise their certainty *for themselves* to truth, both in the case of the other
and in their own case.[33]

The construction of the self in opposition to another who threatens one's
very being reverberates throughout the construction of both class society
and the masculinist world view and results in a deepgoing and hierarchical

dualism. First, the male experience is characterized by the duality of concrete versus abstract. Material reality as experienced by the boy in the family provides no model, and is unimportant in the attainment of masculinity. Nothing of value to the boy occurs with the family, and masculinity becomes an abstract ideal to be achieved over the opposition of daily life.[34] Masculinity must be attained by means of opposition to the concrete world of daily life, by escaping from contact with the female world of the household into the masculine world of public life. This experience of two worlds, one valuable, if abstract and deeply unattainable, the other useless and demeaning, if concrete and necessary, lies at the heart of a series of dualisms — abstract/ concrete, mind/body, culture/nature, ideal/real, stasis/change. And these dualisms are overlaid by gender: only the first of each pair is associated with the male.

Dualism, along with the dominance of one side of the dichotomy over the other, marks phallocentric society and social theory. These dualisms appear in a variety of forms — in philosophy, technology, political theory, and the organization of class society itself. One can, for example, see them very clearly worked out in Plato, although they appear in many other forms.[35] There, the concrete/abstract duality takes the form of an opposition of material to ideal, and a denial of the relevance of the material world to the attainment of what is of fundamental importance: love of knowledge, or philosophy (masculinity). The duality between nature and culture takes the form of a devaluation of work or necessity, and the primacy instead of purely social interaction for the attainment of undying fame. Philosophy itself is separate from nature, and indeed, exists only on the basis of the domination of (at least some) of the philosopher's own nature.[36] Abstract masculinity, then, can be seen to have structured Western social relations and the modes of thought to which these relations give rise at least since the founding of the *polis*.

The oedipal roots of these hierarchical dualisms are memorialized in the overlay of female and male connotations: it is not accidental that women are associated with quasi-human and non-human nature, that the female is associated with the body and material life, that the lives of women are systematically used as examples to characterize the lives of those ruled by their bodies rather than their minds.[37]

Both the fragility and fundamental falseness of the masculinist ideology and the deeply problematic nature of the social relations from which it grows are apparent in its reliance on a series of counterfactual assumptions and contentions. Consider how the following contentions are contrary to lived experience: the body is both irrelevant and in opposition to the (real) self, an impediment to be overcome by the mind; the female mind either does not exist (Do women have souls?) or works in such incomprehensible ways as to be unintelligible (the "enigma of woman"); what is real and primary is imperceptible to the senses and impervious to nature and natural change. What is remarkable is not only that these contentions have absorbed a great deal of philosophical energy, but, along with a series of other counterfactuals, have structured social relations for centuries.

Interestingly enough the epistemology and society constructed by men suffering from the effects of abstract masculinity have a great deal in common with that imposed by commodity exchange. The separation and opposition of social and natural worlds, of abstract and concrete, of permanence and change, the effort to define only the former of each pair as important, the reliance on a series of counter factual assumptions — all this is shared with the exchange abstraction. Abstract masculinity shares still another of its aspects with the exchange abstraction: it forms the basis for an even more problematic social synthesis. Hegel's analysis makes clear the problematic social relations available to the self which maintains itself by opposition: each of the two subjects struggling for recognition risks its own death in the struggle to kill the other, but if the other is killed the subject is once again alone.[38] In sum, then, the male experience when replicated as epistemology leads to a world conceived as, and (in fact) inhabited by, a number of fundamentally hostile others whom one comes to know by means of opposition (even death struggle) and yet with whom one must construct a social relation in order to survive.

The female construction of self in relation to others leads in an opposite direction — toward opposition to dualisms of any sort, valuation of concrete, everyday life, sense of a variety of connectednesses and continuities both with other persons and with the natural world. If material life structures consciousness, women's relationally defined existence, bodily experience of boundary challenges, and activity of transforming both physical objects and human beings must be expected to result in a world view to which dichotomies are foreign. Women experience others and themselves along a continuum whose dimensions are evidenced in Adrienne Rich's argument that the child carried for nine months can be defined "*neither* as me or as not-me," and she argues that inner and outer are not polar opposites but a continuum.[39] What the sexual division of labor defines as women's work turns on issues of change rather than stasis, the changes involved in producing both use-values and commodities, but more profoundly in the activity of rearing human beings who change in both more subtle and more autonomous ways than any inanimate object. Not only the qualities of things but also the qualities of people are important in women's work: quantity becomes peripheral. In addition, far more than the instrumental cooperation of the workplace is required; the mother-child relation and the maintenance of the family, while it has instrumental aspects, is not defined by them. Finally, the unity of mental and manual labor, and the directly sensuous nature of much of women's work leads to a more profound unity of mental and manual labor, social and natural worlds, than is experienced by the male worker in capitalism. The unity grows from the fact that women's bodies, unlike men's, can be themselves instruments of production: in pregnancy, giving birth or lactation, arguments about a division of mental from manual labor are fundamentally foreign.

That this is indeed women's experience is documented in both the theory and practice of the contemporary women's movement and needs no further development here.[40] The more important question here is whether female experience and the world view constructed by female activity can meet the criteria for a standpoint. If we return to the five claims carried by the concept of a standpoint, it seems clear that women's material life activity has important epistemological and ontological consequences for both the understanding and construction of social relations. Women's activity, then, does satisfy the first requirement of a standpoint.

I can now take up the second claim made by a standpoint: that the female experience not only inverts that of the male, but forms a basis on which to expose abstract masculinity as both partial and fundamentally perverse, as not only occupying only one side of the dualities it has constructed, but reversing the proper valuation of human activity. The partiality of the masculinist vision and of the societies which support this understanding is evidenced by its confinement of activity proper to the male to only one side of the dualisms. Its perverseness, however, lies elsewhere. Perhaps the most dramatic (though not the only) reversal of the proper order of things characteristic of the male experience is the substitution of death for life.

The substitution of death for life results at least in part from the sexual division of labor in childrearing. The self-surrounded by rigid ego-boundaries, certain of what is inner and what is outer, the self experienced as walled city, is discontinuous with others. Georges Bataille has made brilliantly clear the ways in which death emerges as the only possible solution to this discontinuity and has followed the logic through to argue that reproduction itself must be understood not as the creation of life, but as death. The core experience to be understood is that of discontinuity and its consequences. As a consequence of this experience of discontinuity and aloneness, penetration of ego-boundaries, or fusion with another is experienced as violent. Thus, the desire for fusion with another can take the form of domination of the other. In this form, it leads to the only possible fusion with a threatening other: when the other ceases to exist as a separate, and for that reason, threatening being. Insisting that another submit to one's will is simply a milder form of the destruction of discontinuity in the death of the other since in this case one is no longer confronting a discontinuous and opposed will, despite its discontinuous embodiment. This is perhaps one source of the links between sexual activity, domination, and death.

Bataille suggests that killing and sexual activity share both prohibitions and religious significance. Their unity is demonstrated by religious sacrifice since the latter:

is intentional like the act of the man who lays bare, desires and wants to penetrate his victim. The lover strips the beloved of her identity no less than the bloodstained priest his human or animal victim. The woman in the hands of her assailant is despoiled of her being . . . loses the firm barrier that once separated her from others . . . is brusquely laid open to the violence of the sexual urges set loose in the organs of reproduction; she is laid open to the impersonal violence that overwhelms her from without.[41]

Note the use of the term "lover" and "assailant" as synonyms and the presence of the female as victim.

The importance of Bataille's analysis lies in the fact that it can help to make clear the links between violence, death, and sexual fusion with another, links which are not simply theoretical but actualized in rape and pornography. Images of women in chains, being beaten, or threatened with attack carry clear social messages, among them that "the normal male is sexually aggressive in a brutal and demeaning way."[42] Bataille's analysis can help to understand why "men advertise, even brag, that their movie is the 'bloodiest thing that ever happened in front of a camera'."[43] The analysis is supported by the psychoanalyst who suggested that although one of the important dynamics of pornography is hostility, "one can raise the possibly controversial question whether in humans (especially males) powerful sexual excitement can ever exist without brutality also being present."[44]

Bataille's analysis can help to explain what is erotic about "snuff" films, which not only depict the torture and dismemberment of a woman, but claim that the actress is *in fact* killed. His analysis suggests that perhaps she is a sacrificial victim whose discontinuous existence has been succeeded in her death by "the organic continuity of life drawn into the common life of the beholders."[45] Thus, the pair "lover-assailant" is not accidental. Nor is the connection of reproduction and death.

"Reproduction," Bataille argues, "implies the existence of *discontinuous* beings." This is so because, "Beings which reproduce themselves are distinct from one another, and those reproduced are likewise distinct from each other, just as they are distinct from their parents. Each being is distinct from all others. His birth, his death, the events of his life may have an interest for others, but he alone is directly concerned in them. He is born alone. He dies alone. Between one being and another, there is a *gulf*, a discontinuity."[46] (Clearly it is not just a gulf, but is better understood as a chasm.) In reproduction sperm and ovum unite to form a new entity, but they do so from the death and disappearance of two separate beings. Thus, the new entity bears within itself "the transition to continuity, the fusion, fatal to both, of two separate beings."[47] Thus, death and reproduction are intimately linked, yet Bataille stresses that "it is only death which is to be identified with continuity." Thus, despite the unity of birth and death in this analysis, Bataille gives greater weight to a "tormenting fact: the urge towards love, pushed to its limit, is an urge toward death."[48] Bataille holds to this position despite his recognition that reproduction is a form of growth. The growth, however, he dismisses as not being "ours," as being only "impersonal."[49] This is not the female experience, in which reproduction is hardly impersonal, nor experienced as death. It is, of course, in a literal sense, the sperm which is cut off from its source, and lost. No wonder, then, at the masculinist occupation with death, and the feeling that growth is "impersonal," not of fundamental concern to oneself. But this complete dismissal of the experience of another bespeaks a profound lack of empathy and refusal to recognize the very being of another. It is a manifestation of the chasm which separates each man from every other being and from the natural world, the chasm which both marks and defines the problem of community.

The preoccupation with death instead of life appears as well in the argument that it is the ability to kill (and for centuries, the practice) which sets humans above animals. Even Simone de Beauvoir has accepted that "it is not in giving life but in risking life that man is raised above the animal: that is why superiority has been accorded in humanity not to the sex that brings forth but to that which kills."[50] That superiority has been accorded to the sex which kills is beyond doubt. But what kind of experience and vision can take reproduction, the creation of new life, and the force of life in sexuality, and turn it into death — not just in theory but in the practice of rape, pornography, and sexual murder? Any why give pride of place to killing? This is not only an inversion of the proper order of things, but also a refusal to recognize the real activities in which men as well as women are engaged. The producing of goods and the reproducing of human beings are certainly life-sustaining activities. And even the deaths of the ancient heroes in search of undying fame were pursuits of life, and represented the attempt to avoid death by attaining immortality. The search for life, then, represents the deeper reality which lies beneath the glorification of death and destruction.

Yet one cannot dismiss the substitution of death for life as simply false. Men's power to structure social relations in their own image means that women too must participate in social relations which manifest and express abstract masculinity. The most important life activities have consistently been held by the powers that be to be unworthy of those who are fully human most centrally because of their close connections with necessity and life: motherwork (the rearing of children), housework, and until the rise of capitalism in the West, any work necessary to subsistence. In addition, these activities in contemporary capitalism are all constructed in ways which systematically degrade and destroy the minds and bodies of those who perform them.[51] The organization of motherhood as an institution in which a woman is alone with her children, the isolation of women from each other in domestic labor, the female pathology of loss of self in service to others — all mark the transformation of life into death, the distortion of what could have been creative and communal activity into oppressive toil, and the destruction of the possibility of community present in women's relational self-definition. The ruling gender's and class's interest in maintaining social relations such as these is evidenced by the fact that when women set up other structures in which the mother is not alone with her children, isolated from others — as is frequently the case in working class communities or communities of people of color — these arrangements are categorized as pathological deviations.

The real destructiveness of the social relations characteristic of abstract masculinity, however, is now concealed beneath layers of ideology. Marxian theory needed to go beneath the surface to discover the different levels of determination which defined the relation of capitalist and (male) worker. These levels of determination and laws of motion or tendency of phallocratic society must be worked out on the basis of female experience. This brings me to the fourth claim for a standpoint — its character as an achievement of both analysis and political struggle occurring in a particular historical space. The fact that class divisions should have proven so resistant to analysis and required such a prolonged political struggle before Marx was able to formulate the theory of surplus value indicates the difficulty of this accomplishment. And the rational control of production has certainly not been achieved.

Feminists have only begun the process of revaluing female experience, searching for common threads which connect the diverse experiences of women, and searching for the structural determinants of the experiences. The difficulty of the problem faced by feminist theory can be illustrated by the fact that it required a struggle even to define household labor, if not done for wages, as work, to argue that what are held to be acts of love instead must be recognized as work whether or not wages are paid.[52] Both the valuation of women's experience, and the use of this experience as a ground for critique are required. A feminist standpoint may be present on the basis of the common threads of female experience, but it is neither self-evident nor obvious.

Finally, because it provides a way to reveal the perverseness and inhumanity of human relations, a standpoint forms the basis for moving beyond these relations. Just as the proletarian standpoint emerges out of the contradiction between appearance and essence in capitalism, understood as essentially historical and constituted by the relation of capitalist and worker, the feminist standpoint emerges both out of the contradiction between the systematically differing structure of male and female life activity in Western cultures. It expresses female experience at a particular time and place, located within a particular set of social relations. Capitalism, Marx noted, could not develop fully until the notion of human equality achieved the status of universal truth.[53] Despite women's exploitation both as unpaid reproducers of the labor force and as a sex-segregated labor force available for low wages, then, capitalism poses problems for the continued oppression of women. Just as capitalism enables the proletariat to raise the possibility of a society free from class domination, so too, it provides space to raise the possibility of a society free from all forms of domination. The articulation of a feminist standpoint based on women's relational self-definition and activity exposes the world men have constructed and the self-understanding which manifests these relations as partial and perverse. More importantly, by drawing out the potentiality available in the actuality and thereby exposing the inhumanity of human relations, it embodies a distress which requires a solution. The experience of continuity and relation — with others, with the natural world, of mind with body — provides an ontological base for developing a non-problematic social synthesis, a social synthesis which need not operate through the denial of the body, the attack on nature, or the death struggle between the self and other, a social synthesis which does not depend on any of the forms taken by abstract masculinity.

What is necessary is the generalization of the potentiality made available by the activity of women – the defining of society as a whole as propertyless producer both of use-values and of human beings. To understand what such a transformation would require we should consider what is involved in the partial transformation represented by making the whole of society into propertyless producers of use-values – i.e. socialist revolution. The abolition of the division between mental and manual labor cannot take place simply by means of adopting worker-self-management techniques, but instead requires the abolition of provate property, the seizure of state power, and lengthy post-revolutionary class struggle. Thus, I am not suggesting that shared parenting arrangements can abolish the sexual division of labor. Doing away with this division of labor would of course require institutionalizing the participation of both women and men in childrearing; but just as the rational and conscious control of the production of goods and services requires a vast and far-reaching social transformation, so the rational and conscious organization of reproduction would entail the transformation both of *every* human relation, and of human relations to the natural world. The magnitude of the task is apparent if one asks what a society without institutionalized gender differences might look like.

CONCLUSION

An analysis which begins from the sexual division of labor – understood not as taboo, but as the real, material activity of concrete human beings – could form the basis for an analysis of the real structures of women's oppression, an analysis which would not require that one sever biology from society, nature from culture, an analysis which would expose the ways women both participate in and oppose their own subordination. The elaboration of such an analysis cannot but be difficult. Women's lives, like men's, are structured by social relations which manifest the experience of the dominant gender and class. The ability to go beneath the surface of appearances to reveal the real but concealed social relations requires both theoretical and political activity. Feminist theorists must demand that feminist theorizing be grounded in women's material activity and must as well be a part of the political struggle necessary to develop areas of social life modeled on this activity. The outcome could be the development of a political economy which included women's activity as well as men's, and could as well be a step toward the redefining and restructuring of society as a whole on the basis of women's activity.

Generalizing the activity of women to the social system as a whole would raise, for the first time in human history, the possibility of a fully human community, a community structured by connection rather than separation and opposition. One can conclude then that women's life activity does form the basis of a specifically feminist materialism, a materialism which can provide a point from which both to critique and to work against phallocratic ideology and institutions.

My argument here opens a number of avenues for future work. Clearly, a systematic critique of Marx on the basis of a more fully developed understanding of the sexual division of labor is in order. And this is indeed being undertaken by a number of feminists. A second avenue for further investigation is the relation between exchange and abstract masculinity. An exploration of Mauss's *The Gift* would play an important part in this project, since he presents the solipsism of exchange as an overlay on and substitution for a deeper going hostility, the exchange of gifts as an alternative to war. We have seen that the necessity for recognizing and receiving recognition from another to take the form of a death struggle memorializes the male rather than female experience of emerging as a person in opposition to a woman in the context of a deeply phallocratic world. If the community of exchangers (capitalists) rests on the more overtly and directly hostile death struggle of self and other, one might be able to argue that what underlies the exchange abstraction is abstract masculinity. One might then turn to the question of whether capitalism rests on and is a consequence of patriarchy. Perhaps then feminists can produce the analysis which could amend Marx to read: "Though class society appears to be the source, the cause of the oppression of women, it is rather its consequence." Thus, it is "only at the last culmination of the development of class society [that] this, its secret, appear[s] again, namely, that on the one hand it is the *product* of the oppression of women, and that on the other it is the *means* by which women participate in and create their own oppression".[55]

The Johns Hopkins University

NOTES

* I take my title from Iris Young's call for the development of a specifically feminist historical materialism. See 'Socialist Feminism and the Limits of Dual Systems Theory,' in *Socialist Review* 10, 2/3 (March-June, 1980). My work on this paper is deeply indebted to a number of women whose ideas are incorporated here, although not always

used in the ways they might wish. My discussions with Donna Haraway and Sandra Harding have been intense and ongoing over a period of years. I have also had a number of important and useful conversations with Jane Flax, and my project here has benefitted both from these contacts, and from the opportunity to read her paper, 'Political Philosophy and the Patriarchal Unconscious: A Psychoanalytic Perspective on Epistemology and Metaphysics'. In addition I have been helped immensely by collective discussions with Annette Bickel, Sarah Begus, and Alexa Freeman. All of these people (along with Iris Young and Irene Diamond) have read and commented on drafts of this paper. I would also like to thank Alison Jaggar for continuing to question me about the basis on which one could claim the superiority of a feminist standpoint and for giving me the opportunity to deliver the paper at the University of Cincinnati Philosophy Department Colloquium; and Stephen Rose for taking the time to read and comment on a rough draft of the paper at a critical point in its development.

¹ See my 'Feminist Theory and the Development of Revolutionary Strategy,' in Zillah Eisenstein, ed., *Capitalist Patriarchy and the Case for Socialist Feminism* (New York: Monthly Review, 1978).

² The recent literature on mothering is perhaps the most detailed on this point. See Dorothy Dinnerstein, *The Mermaid and the Minotaur* (New York: Harper and Row, 1976); Nancy Chodorow, The *Reproduction of Mothering* (Berkeley: University of California Press, 1978).

³ Iris Young, 'Socialist Feminism and the Limits of Dual Systems Theory,' in *Socialist Review* 10, 2/3 (March-June, 1980), p. 180.

⁴ Eighth Thesis on Feuerbach, in Karl Marx, 'Theses on Feuerbach,' in *The German Ideology*, C. J. Arthur, ed. (New York: International Publishers, 1970), p. 121.

⁵ *Ibid*. Conscious human practice, then, is at once both an epistemological category and the basis for Marx's conception of the nature of humanity itself. To put the case even more strongly, Marx argues that human activity has both an ontological and epistemological status, that human feelings are not "merely anthropological phenomena," but are "truly ontological affirmations of being." See Karl Marx, *Economic and Philosophic Manuscripts of 1844*, Dirk Struik, ed. (New York: International Publishers, 1964), pp. 113, 165, 188.

⁶ Marx, *1844*, p. 112. Nature itself, for Marx, appears as a form of human work, since he argues that humans duplicate themselves actively and come to contemplate themselves in a world of their own making. (*Ibid.*, p. 114). On the more general issue of the relation of natural to human worlds see the very interesting account by Alfred Schmidt, *The Concept of Nature in Marx*, tr. Ben Foukes (London: New Left Books, 1971).

⁷ Marx and Engels, *The German Ideology*, pp. 42.

⁸ See Alfred Sohn-Rethel, *Intellectual and Manual Labor: A Critique of Epistemology* (London: MacMillan, 1978). I should note that my analysis both depends on and is in tension with Sohn-Rethel's. Sohn-Rethel argues that commodity exchange is a characteristic of all class societies – one which comes to a head in capitalism or takes its most advanced form in capitalism. His project, which is not mine, is to argue that (a) commodity exchange, a characteristic of all class societies, is an original source of abstraction, (b) that this abstraction contains the formal element essential for the cognitive faculty of conceptual thinking and (c) that the abstraction operating in exchange, an abstraction in practice, is the source of the ideal abstraction basic to Greek philosophy and to modern science. (See *Ibid*., p. 28).In addition to a different purpose,

I should indicate several major differences with Sohn-Rethel. First, he treats the productive forces as separate from the productive relations of society and ascribes far too much autonomy to them. (See, for example, his discussions on pp. 84–86, 95.) I take the position that the distinction between the two is simply a device used for purposes of analysis rather than a feature of the real world. Second, Sohn-Rethel characterizes the period preceding generalized commodity production as primitive communism. (See p. 98.) This is however an inadequate characterization of tribal societies.

[9] Karl Marx, *Capital*, I (New York: International Publishers, 1967), p. 176.

[10] I have done this elsewhere in a systematic way. For the analysis, see my discussion of the exchange abstraction in *Money, Sex, and Power: An Essay on Domination and Community* (New York: Longman, Inc., 1983).

[11] This is Iris Young's point. I am indebted to her persuasive arguments for taking what she terms the "gender differentiation of labor" as a central category of analysis (Young, 'Dual Systems Theory,' p. 185). My use of this category, however, differs to some extent from hers. Young's analysis of women in capitalism does not seem to include marriage as a part of the division of labor. She is more concerned with the division of labor in the productive sector.

[12] See Sara Ruddick, 'Maternal Thinking,' *Feminist Studies* 6, 2 (Summer, 1980), p. 364.

[13] See, for discussions of this danger, Adrienne Rich, 'Disloyal to Civilization: Feminism, Racism, Gynephobia,' in *On Lies, Secrets, and Silence* (New York: W. W. Norton & Co., 1979), pp. 275–310; Elly Bulkin, 'Racism and Writing: Some Implications for White Lesbian Critics,' in *Sinister Wisdom*, No. 6 (Spring, 1980).

[14] Some cross-cultural evidence indicates that the status of women varies with the work they do. To the extent that women and men contribute equally to subsistence, women's status is higher than it would be if their subsistence-work differed profoundly from that of men; that is, if they do none or almost all of the work of subsistence, their status remains low. See Peggy Sanday, 'Female Status in the Public Domain,' in Michelle Rosaldo and Louise Lamphere, eds., *Women, Culture, and Society* (Stanford: Stanford University Press, 1974), p. 199. See also Iris Young's account of the sexual division of labor in capitalism, mentioned above.

[15] It is irrelevant to my argument here that women's wage labor takes place under different circumstances than men's – that is, their lower wages, their confinement to only a few occupational categories, etc. I am concentrating instead on the formal, structural features of women's work. There has been much effort to argue that women's domestic labor is a source of surplus value, that is, to inclue it within the scope of Marx's value theory as productive labor, or to argue that since it does not produce surplus value it belongs to an entirely different mode of production, variously characterized as domestic or patriarchal. My strategy here is quite different from this. See, for the British debate, Mariarosa Dalla Costa and Selma James, *The Power of Women and the Subversion of the Community* (Falling Wall Press, Bristol, 1975); Wally Secombe, 'The Housewife and Her Labor Under Capitalism,' *New Left Review* 83 (January-February, 1974); Jean Gardiner, 'Women's Domestic Labour,' *New Left Review* 89 (March, 1975); and Paul Smith, 'Domestic Labour and Marx's Theory of Value,' in Annette Kuhn and Ann Marie Wolpe, eds., *Feminism and Materialism* (Boston: Routledge and Kegal Paul, 1978). A portion of the American debate can be found in Ira Gerstein, 'Domestic Work and Capitalism,' and Lisa Vogel, 'The Earthly Family,' *Radical America* 7, 4/5

(July-October, 1973); Ann Ferguson, 'Women as a New Revolutionary Class,' in Pat Walker, ed., *Between Labor and Capital* (Boston: South End Press, 1979).

16 Frederick Engels, *Origins of the Family, Private Property and the State* (New York: International Publishers, 1942); Karl Marx, *Capital*, Vol. I, p. 671. Marx and Engels have also described the sexual division of labor as natural or spontaneous. See Mary O'Brien, 'Reproducing Marxist Man,' in Lorenne Clark and Lynda Lange, eds., *The Sexism of Social and Political Theory: Women and Reproduction from Plato to Nietzsche* (Toronto: University of Toronto Press, 1979).

17 For a discussion of women's work, see Elise Boulding, 'Familial Constraints on Women's Work Roles,' in Martha Blaxall and B. Reagan, eds., *Women and the Workplace* (Chicago, University of Chicago Press, 1976), esp. the charts on pp. 111, 113.

An interesting historical note is provided by the fact that even Nausicaa, the daughter of a Homeric king, did the household laundry. (See M. I. Finley, *The World of Odysseus* (Middlesex, England: Penguin, 1979), p. 73.) While aristocratic women were less involved in actual labor, the difference was one of degree. And as Aristotle remarked in *The Politics*, supervising slaves is not a particularly uplifting activity. The life of leisure and philosophy, so much the goal for aristocratic Athenian men, ten, was almost unthinkable for any woman.

18 Simone de Beauvoir holds that repetition has a deeper significance and that women's biological destiny itself is repetition. (See *The Second Sex*, tr. H. M. Parshley (New York: Knopf, 1953), p. 59.) But see also her discussion of housework in *Ibid.*, pp. 434ff. There her treatment of housework is strikingly negative. For de Beauvoir, transcendence is provided in the hstorical struggle of self with other and with the natural world. The oppositions she sees are not really stasis vs. change, but rather transcendence, escape from the muddy concreteness of daily life, from the static, biological, concrete repetition of "placid femininity."

19 Marilyn French, *The Women's Room* (New York: Jove, 1978), p. 214.

20 Sara Ruddick, 'Maternal Thinking,' presents an interesting discussion of these and other aspects of the thought which emerges from the activity of mothering. Although I find it difficult to speak the language of interests and demands she uses, she brings out several valuable points. Her distinction between maternal and scientific thought is very intriguing and potentially useful (see esp. pp. 350–353).

21 O'Brien, 'Reproducing Marxist Man,' p. 115, n. 11.

22 It should be understood that I am concentrating here on the experience of women in Western culture. There are a number of cross-cultural differences which can be expected to have some effect. See, for example, the differences which emerge from a comparison of childrearing in ancient Greek society with that of the contemporary Mbuti in central Africa. See Phillip Slater, *The Glory of Hera* (Boston: Beacon, 1968) and Colin Turnbull, 'The Politics of Non-Aggression,' in Ashley Montagu, ed., *Learning Non-Aggression* (New York: Oxford University Press, 1978).

23 See Nancy Chodorow, 'Family Structure and Feminine Personality,' in Michelle Rosaldo and Louise Lamphere, *Woman, Culture, and Society* (Stanford: Stanford University Press, 1974), p. 59.

24 *Of Woman Born* (New York: Norton, 1976), p. 63.

25 See Chodorow, *The Reproduction of Mothering*, and Flax, 'The Conflict Between Nurturance and Autonomy in Mother-Daughter Relations and in Feminism,' *Feminist Studies* 4, 2 (June, 1978). I rely on the analyses of Dinnerstein and Chodorow but there are difficulties in that they are attempting to explain why humans, both male and

female, fear and hate the female. My purpose here is to invert their arguments and to attempt to put forward a positive account of the epistemological consequences of this situation. What follows is a summary of Chodorow, *The Reproduction of Mothering*.

[26] Chodorow, *Reproduction*, pp. 105–109.

[27] This is Jane Flax's point.

[28] Chodorow, *Reproduction*, pp. 127–131, 163.

[29] *Ibid.*, p. 166.

[30] *Ibid.*, pp. 174–178. Chodorow suggest a correlation between father absence and fear of women (p. 213), and one should, treating this as an empirical hypotheses, expect a series of cultural differences based on the degree of father absence. Here the ancient Greeks and the Mbuti provide a fascinating contrast. (See above, note 22.)

[31] *Ibid.*, p. 198. The flexible and diffuse female ego boundaries can of course result in the pathology of loss of self in responsibility for and dependence on others. (The obverse of the male pathology of experiencing the self as walled city.)

[32] Sigmund Freud, *Civilization and Its Discontents* (New York: Norton, 1961), pp. 12–13.

[33] Hegel, *Phenomenology of Spirit* (New York: Oxford University Press, 1979), trans. A. V. Miller, p. 114. See also Jessica Benjamin's very interesting use of this discussion in 'The Bonds of Love: Rational Violence and Erotic Domination,' *Feminist Studies* 6, 1 (June, 1980).

[34] Alvin Gouldner has made a similar argument in his contention that the Platonic stress on hierarchy and order resulted from a similarly learned opposition to daily life which was rooted in the young aristocrat's experience of being taught proper behavior by slaves who could not themselves engage in this behavior. See *Enter Plato* (New York: Basic Books, 1965), pp. 351–355.

[35] One can argue, as Chodorow's analysis suggests, that their extreme form in his philosophy represents an extreme father-absent (father-deprived?) situation. A more general critique of phallocentric dualism occurs in Susan Griffin, *Woman and Nature* (New York: Harper & Row, 1978).

[36] More recently, of course, the opposition to the natural world has taken the form of destructive technology. See Evelyn Fox Keller, 'Gender and Science,' *Psychoanalysis and Contemporary Thought* 1, 3 (1978), reprinted in this volume.

[37] See Elizabeth Spelman, 'Metaphysics and Misogyny: The Soul and Body in Plato's Dialogues,' mimeo. One analyst has argued that its basis lies in the fact that "the early mother, monolithic representative of nature, is a source, like nature, of ultimate distress as well as ultimate joy. Like nature, she is both nourishing and disappointing, both alluring and threatening . . . The infant loves her . . . and it hates her because, like nature, she does not perfectly protect and provide for it . . . The mother, then – like nature, which sends blizzards and locusts as well as sunshine and strawberries – is perceived as capricious, sometimes actively malevolent." Dinnerstein, p. 95.

[38] See Benjamin, p. 152. The rest of her analysis goes in a different direction than mine, though her account of *The Story of O* can be read as making clear the problems for any social synthesis based on the Hegelian model.

[39] *Of Woman Born*, p. 64, p. 167. For a similar descriptive account, but a dissimilar analysis, see David Bakan, *The Duality of Human Existence* (Boston: Beacon, 1966).

[40] My arguments are supported with remarkable force by both the theory and practice of the contemporary women's movement. In theory, this appears in different forms in the work of Dorothy Riddle, 'New Visions of Spiritual Power,' *Quest: a Feminist*

Quarterly 1, 3 (Spring, 1975); Susan Griffin, *Woman and Nature*, esp. Book IV: 'The Separate Rejoined'; Adrienne Rich, *Of Woman Born*, esp. pp. 62–68; Linda Thurston, 'On Male and Female Principle,' *The Second Wave* 1, 2 (Summer, 1971). In feminist political organizing, this vision has been expressed as an opposition of leadership and hierarchy, as an effort to prevent the development of organizations divided into leaders and followers. It has also taken the form of an insistence on the unity of the personal and the political, a stress on the concrete rather than on abstract principles (an opposition to theory), and a stress on the politics of everyday life. For a fascinating and early example, see Pat Mainardi, 'The Politics of Housework,' in Leslie Tanner, ed., *Voices of Women's Liberation* (New York: New American Library, 1970).

[41] George Bataille, *Death and Sensuality* (New York: Arno Press, 1977), p. 90.

[42] Women Against Violence Against Women Newsletter, June, 1976, p. 1.

[43] *Aegis: A Magazine on Ending Violence Against Women*, November/December, 1978, p. 3.

[44] Robert Stoller, *Perversion: The Erotic Form of Hatred* (New York: Pantheon, 1975), p. 88.

[45] Bataille, p. 91. See pp. 91ff for a more complete account of the commonalities of sexual activity and ritual sacrifice.

[46] *Death and Sensuality*, p. 12 (italics mine). See also de Beauvoir's discussion in *The Second Sex*, pp. 135, 151.

[47] Bataille, p. 14.

[48] *Ibid.*, p. 42. While Adrienne Rich acknowledges the violent feelings between mothers and children, she quite clearly does not put these at the heart of the relation (*Of Woman Born*).

[49] Bataille, pp. 95–96.

[50] *The Second Sex*, p. 58. It should be noted that killing and risking life are ways of indicating one's contempt for one's body, and as such are of a piece with the Platonic search for disembodiment.

[51] Consider, for example, Rich's discussion of pregnancy and childbirth, Ch. VI and VII, *Of Woman Born*. And see also Charlotte Perkins Gilman's discussion of domestic labor in *The Home* (Urbana, Ill.: The University of Illinois Press, 1972).

[52] The Marxist-feminist efforts to determine whether housework produces surplus value and the feminist political strategy of demanding wages for housework represent two (mistaken) efforts to recognize women's non-wage activity at work. Perhaps domestic labor's non-status as work is one of the reasons why its wages – disproportionately paid to women of color – are so low, and working conditions so poor.

[53] *Capital*, Vol. I, p. 60.

[54] The phrase is O'Brien's, p. 113.

[55] See Marx, *1844*, p. 117.

SANDRA HARDING

WHY HAS THE SEX/GENDER SYSTEM BECOME VISIBLE ONLY NOW?

During the last decade of feminist inquiry, a new "object" for scientific scrutiny has emerged into visibility: the sex/gender system.* Sex/gender is a system of male-dominance made possible by men's control of women's productive and reproductive labor, where "reproduction" is broadly construed to include sexuality, family life, and kinship formations, as well as the birthing which biologically reproduces the species. However, the "discovery" of the sex/gender system has implications beyond the need for revisions in our scientific understandings. While many feminists have argued that this discovery calls for new morals and new politics, I intend to show why its discovery at this particular moment in history also calls for a revolution in epistemology. The new epistemology must be one which is not fettered by the self-imposed limitations of empiricist, functionalist/relativist, or marxist epistemologies. We shall see, within the all too brief limits of so short a paper, what the main limitations of these existing epistemologies are, and distinguish the pre-conditions for an adequate theory of belief production from the epistemological goals of feminist inquirers which lean too heavily on these inadequate epistemological programs. The feminist discovery of the sex/ gender system certainly is more than the expression of socially unobstructed "natural talents and abilities," of functionally adequate beliefs, and of changes in the division of labor by class. But an insufficiently critical stance toward the existing epistemologies has obscured for us just what this "more" is. We need to investigate more fully why it is that only now can we understand "patriarchy," "misogyny," "sex-roles," "discrimination against women," and "the first division of labor — by sex" as mere appearances of the underlying reality of the sex/gender system. Let us begin by first looking at the newly visible size and shape of the sex/gender system, and then examining the self-imposed limitations of empiricist, functionalist/relativist, and marxist epistemologies.

What is the sex/gender system? If one looks over the vast array of studies during the last decade which have been animated by feminist concerns, one can virtually see the emergence into visibility of a widely existing object in nature/history. In retrospect, it appears clear that every study animated by feminist concerns has been trying to clarify and deepen our understanding of

311

Sandra Harding and Merrill B. Hintikka (eds.), Discovering Reality, 311–324.
Copyright © 1983 *by D. Reidel Publishing Company.*

how this or that aspect of the sex/gender system structures social life and social thought in this or that society today or in the past.[1] The authors of this research often say that what they are describing is a facet of "patriarchy," "misogyny," "sex-roles," "discrimination against women," or "the first division of labor — by sex." But in retrospect we can understand the explicit objects of these studies as "appearances" of the underlying "reality" of the sex/gender system.

The size and shape of this newly visible object are becoming clearer. The sex/gender system appears to be a fundamental variable organizing social life throughout most recorded history and in every culture today. Like racism and classism, it is an *organic* social variable — it is not merely an "effect" of other, more primary, causes. Of course, the sex/gender system is expressed in differing intensities and forms in different cultures and classes. Men's and women's "natures" and relative abilities to determine their own social, economic, and political lives appear very different if one looks from matrilineal to patrilineal societies,[2] from pre-capitalist to capitalist formations,[3] from aristocratic to democratic cultures,[4] and, of course, from wealthy to poverty-level and white to black lives in America today. However, beneath this considerable variation in the intensities and forms the sex/gender system takes, its underlying dynamic is detectable. Like racism and classism, the sex/gender system appears to limit and create opportunities within which are constructed the social practices of daily life, the characteristics of social institutions, and all of our patterns of thought. Not only are the "macro" social institutions the way they are in the vast majority of societies because the sex/gender system is interacting with other organic social variables to structure them that way, but also the very existence and design of characteristics of daily life to which sex and/or gender seemed irrelevant now appear suffused with sex/gender. Now we can detect sex/gender in the details of domestic and public architecture,[5] in what *the* problems of philosophy are supposed to be,[6] in the forms of technology a culture chooses,[7] in the intensity and forms of the very distinction between nature and culture,[8] and even in the forms of the state.[9] The genes of the sex/gender system now can be detected in most of the social interactions which have ever occurred between humans of any sex, age, class, race, culture. Each historic detail now can be read as a clue to the particular cultural forms the sex/gender system takes in that particular society. It may be that there has only rarely, anywhere, been a *human* act performed or a *human* thought produced, for acts and thoughts have had to occur within the differential opportunities and limits set by the sex/gender system. Thus, cutting across and shaping the variations in social action and social thought produced by racist and classist variables are the variations produced by the sex/gender system.[10] Thus this newly visible object has this kind of immense social dimension.

Furthermore, we can begin to see the gross morphological outlines of this system. As indicated above, it is a system of male-dominance made possible by men's control of women's productive and reproductive labor, where "reproduction" is broadly construed to include sexuality, family life and kinship formations as well as the birthing which biologically reproduces the species. Even anthropologists who are critical of feminist tendencies to universalize what are really only culturally-specific features of the sex/gender system argue that male-dominance, in the form of men's direct control of women's productive and reproductive labor through control of a broad array of social institutions, appears to be an *organic* feature of most recorded social life. For instance, in the context of just such a criticism, M. Z. Rosaldo nevertheless cites five kinds of evidence for the existence of male dominance as an organic social variable:

Male dominance is evidenced, I believe, when [1] we observe that women almost everywhere have daily responsibilities to feed and care for children, spouse, and kin, while men's economic obligations tend to be less regular and more bound up with extrafamilial sorts of ties; [2] certainly men's work within the home is not likely to be sanctioned by a spouse's use of force. Even in those groups in which the use of physical violence is avoided, a man can say, "She is a good wife, I don't have to beat her," wheres no woman evokes violent threats when speaking of her husband's work. [3] Women will, in many societies, discover lovers and enforce their will to marry as they choose; but, again, we find in almost every case that the formal initiation and arrangement of permanent heterosexual bonds is something organized by men. [4] Women may have ritual powers of considerable significance to themselves as well as men, but women never dominate in rites requiring the participation of the community as a whole. And even though men everywhere are apt to listen to and be influenced by their wives, I know of no case where men are required to serve as an obligatory audience to female ritual or political performance. [5] Finally, women often form organizations of real and recognized political and economic strength; at times they rule as queens, acquire followings of men, beat husbands who prefer strange women to their wives, or perhaps enjoy a sacred status in their role as mothers. But, again, I know of no political system in which women individually or as a group are expected to hold more offices or have more political clout than their male counterparts.[11]

Even this critic of feminist over-universalizing concludes: "Male dominance, in short, does not inhere in any isolated or measurable set of omnipresent facts. Rather it seems to be an aspect of the *organization of collective life.* . . ."[12]

Thus it is these kinds of considerations which suggest that what feminist research has been producing during the last decade is the "discovery" of the sex/gender system, where the latter is understood as an organic social variable which has been functioning in varying intensities and forms throughout most recorded history. However, if it is fair to describe the feminist research of the last decade in this way, then there is another question which we can ask in order to appreciate why this scientific discovery requires not merely revisions in many existing scientific theories, in morals and in politics, but, more fundamentally, a revolution in epistemology: why has the sex/gender system become visible only now? What "causes" its "discovery" at this moment in history rather than in 1776, 1848, or 1919? Lest I disappoint the reader, let me make clear here that I do not intend to try to answer this question. Rather I want to show why it is a reasonable and neglected question with important epistemological implications, and why existing epistemologies can not ask it. Of course, as recent re-evaluations of the history of science have revealed, scientific discoveries are not the "Eureka!" accomplishments of individual geniuses which traditional history and philosophy of science would lead us to believe.[13] Rather, they seem to be slow and collective processes of "paradigm shift" marked by dawning recognitions of the following sorts. The known problems for available theories remain unsolvable within those theories. Observations which cannot be accounted for in a systematic way by the existing theories probably should be regarded as significant indicators that the existing theories' concepts and methodologies are too impoverished to enable us to grasp important regularities in nature or social life. Alternative theories can be developed which will account systematically for the previously observed regularities as well as for the new and "recalcitrant" observations, and which will open new and fruitful research issues. Paradigm shifts have frequently occurred in the context of broad social movements aimed at redistributing political power. Of course, the "discovery" of the sex/gender system has occurred in the context of the Second Women's Movement.

Intuitively it appears reasonable to ask why the sex/gender system has become visible only now. However, asking this question requires three assumptions, and at least one of these appears unintelligible from the perspective of the three dominant existing epistemologies.[14] Let us turn to examine the self-imposed limitations of empiricist, functionalist/relativist, and marxist epistemologies to see why this is so.

Empiricist epistemologies. The empiricist-derived epistemology which has directed most social and natural scientific inquiry for the last three centuries explicitly holds that historical social relations can only distort our "natural," trans-historical abilities to arrive at reliable beliefs. Historically specific social relations cannot "improve" these abilities to provide progressively more complete and undistorted belief ("true belief about reality" as epistemologists say, or "truth-like claims about nature's regularities and underlying causal determinants," as philosophers of science say). Thus this epistemology assumes that attempts to provide causal, scientific accounts of the social conditions tending to improve the production of "true belief" are unintelligible. We shall see that from this perspective can emerge only bizarre accounts of why the discovery of the sex/gender system has occurred at this moment in history.

David Bloor has pointed out how this assumption is justified by appeal to one or more of four faulty arguments.[15] The first two arguments fail to exhibit analytic impartiality. They mistakenly assume that there is no need for causal, scientific analyses of the changing social relations which have contributed to making visible a heliocentric planetary system, the evolution of species, the class system, or the sex/gender system. All that requires analysis is the social causes which produced the errors in thought and the obstacles to the unimpeded exercise of human "natural talents" which are to be found in social life prior to these discoveries. The last two arguments trivialize and abort thought about the role of social relations in the production of desirable belief by drawing absurd but unfounded conclusions from the hypothetically held premise that social relations could play a causal role in the production of "true belief."

The first two arguments appeal to a particular theory of human nature and claim that the socially unobstructed functioning of trans-historical "natural talents and abilities" is the only possible "cause" of the production of "true belief." The "autonomy of knowledge argument" claims "that truth, rationality and validity are man's natural goal and the direction of certain natural tendencies with which he is endowed," and hence that it is only errors and obstacles to this natural tendency which need explanation.[16] The "argument from empiricism" claims that while all belief is indeed caused, it is "only social influences which produce distortions in our beliefs, whilst the uninhibited use of our faculties of perception and our sensory-motor apparatus produce true beliefs."[17] We need to explain scientifically only the distortions produced by social influences which inhibit our faculties of perception and our sensory-motor apparatuses. Given "free play," the latter will combine to produce true belief.

The next two arguments, the "argument from self-refutation" and the "argument from future knowledge," assume the same empiricist-derived theory of human nature and try to convince us that it is either self-defeating or absurd to hold that there can be social causes of more complete and less distorted beliefs. The "argument from self-refutation" begins by saying, let us assume hypothetically that all of our beliefs are totally caused. Then "if there is necessarily within these causes a component provided by society, then . . . these beliefs are bound to be false."[18] Therefore, the epistemologist's or sociologist of knowledge's beliefs are also bound to be false. They, like the beliefs he explains, must also partially be socially caused and therefore false. Thus, arguing that all beliefs have social causes is self-defeating. Finally, the "argument from future knowledge" claims that since any truly scientific account enables accurate prediction at least in principle, the possibility of a truly scientific, causal account of the production of true belief would result in the logical absurdity that the epistemologist or sociologist of knowledge could in principle "discover" now all the true beliefs which science will produce in the future.[19]

The flaws in these arguments can be grasped quickly by noting the intuitively bizarre accounts defenders of each argument would have to give to the discovery of the sex/gender system. The autonomy of knowledge argument would imply that only the incomplete and distorted sexist accounts require scientific explanation. The more adequate feminist accounts are simply the result of unspecifiable "natural human tendencies" which, evidently, feminists can actualize more effectively than can non-feminists. (Which tendencies? Why can feminists actualize them now but not in the past?) The argument from empiricism would lead us to the conclusion that it is the *lack* of "social influences" on feminists which have allowed free play for our "faculties of perception" and "sensory-motor apparatuses" to produce these accounts. (So much for the role of the Women's Movement in guiding feminist inquiry.) The argument from self-refutation would claim that our question commits us to holding that there are no criteria which can be used in evaluating the relative adequacy of feminist accounts and sexist accounts: since both are in part the product of social causes, both are equally false or equally only relatively true. (We must hold that it is both equally false and equally true that the sex/gender system is an organic social variable?) And the argument from future knowledge commits us to holding that if we could indeed identify social pre-conditions for the discovery of the sex/gender system, we could replace substantive scientific inquiry with epistemology and the sociology of knowledge. (Accounts of the social pre-conditions for the emergence into visibility of the sex/gender system would logically oblige us to find biology, history, anthropology, etc. unnecessary?)

These conceptual problems damage the general efficacy of empiricist-derived epistemology, as well as the usefulness of this epistemology for providing explanations of why the sex/gender system has emerged into visibility only at this moment in history. This epistemology must argue either that feminists' "natural talents and abilities" are superior to those of past inquirers; or that the sudden flourishing of feminist talents needs no causal explanation and that all that requires explanation is the unfortunate social influences blocking earlier inquirers' exercise of their natural talents and abilities. Philosophers and historians have often provided just these kinds of analyses of why modern science emerged in Europe when it did. These accounts imply that Copernicus, Galileo and Newton were just smarter than their predecessors. Or they argue that, unlike medieval scientists, the eyes of the Great Men were not blinded nor their intellects clouded by feudal social values.

Sometimes feminist researchers and theoreticians also imply or state this kind of analysis. They suggest that women's talents and abilities are superior to men's,[20] or that feminism wipes the sexist blinders from our eyes so that our natural talents may flourish unimpeded by social myths.[21] These claims do have obvious political appeal, and there are elements of truth probably in the first claim and certainly in the second. However, baldly stated, they do not increase our understanding of just what has made feminism and the new research it inspires possible. They do not help us to understand just *which* social relations contribute to the formation of more adequate belief. They stop thought at unanalyzed "natural talents" and unanalyzed historical facts. They do not permit scientific analyses of the changing social relations which have made more adequate belief about sex/gender possible now but not in 1776, 1848, or 1919.

Functionalist/Relativist epistemologies. Epistemologies which restrict scientific causal accounts of the production of belief to functionalist accounts do not permit us to sort beliefs, in at least a hypothetical way, into more and less complete and undistorted beliefs. Many of the recent antipositivist, social and historical studies of science do this implicitly.[22] At least one of these is explicit about this restriction: the "strong programme" in the sociology of knowledge recently developed by David Bloor, Barry Barnes, and others.[23] Let us look at Bloor's claims.

As a remedy for the flaws we saw Bloor point out in the empiricist-derived epistemologies, Bloor's own program for understanding the production of belief requires that causal scientific accounts be provided of the production of beliefs regarded as true and rational as well as of beliefs regarded as false or irrational. Thus a value-neutral sociology of knowledge "would be impartial with respect to truth and falsity, rationality or irrationality, success or failure. Both sides of these dichotomies will require explanation."[24] However, Bloor argues that other scientific disciplines, such as physics and chemistry, *value* value-neutrality; hence the only fully *scientific* accounts are ones which are value-neutral.[25] A strong program in the sociology of knowledge must be value-neutral if it is to "embody the same values which are taken for granted in other scientific disciplines".[26] The problem here is that Bloor takes his desirable goal of analytic impartiality also to require epistemic impartiality. His position results in epistemological relativism. Thus his program forbids our hypothesizing which particular beliefs in the history of science have been the more complete and less distorted beliefs. Bloor himself notes that this commitment to value-neutrality makes his program relativist and unable to go beyond a sociological account of how appeals to objectivity function as resources in science,[27] but he does not regard this as a serious problem with his program.

There are three kinds of weaknesses in functionalist defenses of epistemological relativism. First, Bloor's kind of program cannot explain why Newton's physics, Darwin's biology, or Marx's class theory all increased understanding of key aspects of the regularities of nature and/or social life and their underlying causes. Since the reasons for these theories' greater explanatory powers cannot be provided, there is no reason for us to take these intuitively reasonable sets of beliefs as scientifically preferable to their predecessors'. We want to understand the social causes of the emergence of these beliefs, but there is more to be understood about these beliefs than the fact that they have social causes. Why is it unintelligible to go on and hypothesize on epistemic grounds that these theories increase our understanding of nature/ social life, and therefore that both these beliefs and perhaps some of their social causes are epistemically preferable? Second, any program which simultaneously insists on scientific and functionalist accounts restricts what should count as scientific more rigidly than any field of scientific inquiry does or can do. What should count as an adequate kind of scientific theory is itself a matter for empirical and theoretical inquiry, and these assessments will and should change over time.[28] Obviously, functionalist accounts do not exhaust the category of valuable types of social theories. Finally, any program which simultaneously insists on scientific and functionalist accounts is unable to justify its own program. We are led to conclude that empiricist-derived sociologies of knowledge are functionally adequate for those who hold them; an alternative such as Bloor's program is functionally adequate for him. But functional adequacy is not identical with epistemological or scientific adequacy. On what grounds other than idle curiosity should anyone listen to arguments for an epistemology? Why should one be convinced by Bloor's appeals to criteria of scientific adequacy? These considerations reveal that epistemological relativism, like ethical relativism, is not so much false as it is incoherent. Its substantive relativist stance prevents its defenders from articulating any epistemic grounds upon which its "absolutist" support of relativism can be defended.

From the perspective of this kind of relativist epistemology, belief in the existence of the sex/gender system is simply functionally adequate for feminists; belief in the natural inferiority of women and in the moral correctness of maintaining women in only inferior social statuses is simply functionally adequate for Aristotle and for contemporary misogynists. Scientific inquiry can reveal the social reasons why each belief is functionally adequate for its holders, but cannot reveal the greater explanatory power provided by hypothetically holding that the sex/gender system exists.

Occasionally, echoes of this functionalist relativism appear in feminist writings. For instance, Francine Blau, an economist, argues that since values play an "important and unavoidable role" in all social science research, feminists should disregard the explicit methodological prescriptions to eliminate values from social inquiry. They, too, should draw upon their "values" in selecting the problems to be studied, in making research decisions, and in interpreting findings.[29] While this is undoubtedly true, it leaves unexamined the reasons that research directed by some "social values" is scientifically more valuable than research directed by other "social values." Is research directed by feminist values (and which "feminist values"?), anti-racist, and anti-classist values only equally scientifically fruitful to research directed by sexist, racist, and classist "values"? Obviously not, but why is this so? Relativist epistemologies provide us with no tools to understand why some but not all beliefs which are made possible by changes in historical social relations do indeed mark advances in our understanding of nature and social life.

Marxist epistemologies. In contrast to empiricist epistemologies, marxist epistemology assumes that historical changes in social relations create the possibilities for more complete and less distorted belief as well as for less complete and more distorted belief. In contrast to relativist epistemologies, it holds that there are scientific grounds upon which we can sort belief into these two categories. But it denies the existence of the sex/gender system which changing social relations and feminist inquiry make visible.

The marxist rules of inquiry prescribe that the regularities of nature/ social life and their underlying causal tendencies can be understood only if one tries to describe and explain these from the politically-activated perspective of those dominated by the division of labor by class. It is (only) the domination of modern thought by the interested and limited perspective of the bourgeoisie which distorts our understandings of the regularities and underlying causal tendencies in nature and social life. From this perspective, and in contrast to empiricist and relativist epistemologies, marxism offers explanations for how historical changes in social relations can create new possibilities for "improvements" in our "natural talents and abilities" to provide scientifically progressive beliefs. For Engels, the class system was undetectable to earlier thinkers not because they were lacking in intellectual brilliance or just because they were under the sway of false social myths, but because the class system had not yet appeared in forms which could reveal its existence to anyone. Thus "the great thinkers of the Eighteenth Century could, no more than their predecessors, go beyond the limits imposed upon them by their epoch."[30] Engels argued that only with the emergence in Nineteenth Century industrializing societies of a "conflict between productive forces and modes of production" — a conflict which "exists, in fact, objectively, outside us, independently of the will and actions even of the men that have brought it on"[31] — could the organic class structure of earlier societies be detected for the first time. "Modern socialism is nothing but the reflex, in thought, of this conflict in fact; its ideal reflection in the minds, first, of the class directly suffering under it, the working class."[32]

However, for marxists, women as a "sex-class" do not exist and sex/ gender cannot be an organic social variable. This is because the social conditions creating the oppression of working-class women and the social conditions creating privileges for bourgeoise women are claimed to be different, and both are claimed to be entirely the consequences of the shifting dynamics of the class system. Hence there can be no "object" which changing social relations and feminist inquiry makes visible. From this perspective, male bias in the history of thought and social life clearly exists, but it is simply an ideological consequence — an epiphenomenon — of attempts to maintain the division of labor by class. And since it is claimed that there are no explanatorily-significant social relations shared by women cross-class, there cannot be distinctive "women's experience", analogous to working-class experience, upon which to ground a distinctive feminist scientific and epistemological analysis.

Feminists point out that this epistemology is too closely tied to the inadequate marxist theory of political economy.[33] One problem with this theory is that male domination exists in cultures which are clearly pre-class societies. Since male-domination thus appears earlier than class society, it is hard to understand how the former can be a consequence of the latter. Another problem is that the sex/gender system has persisted through every recent economic/political/social change, including the recent socialist revolutions. Thus socialist transformations do not in themselves end the oppression of women. Furthermore, systematically absent from marxist theory are categories and concepts capable of even detecting the sources of male domination which lie in the sphere of reproduction. For instance, "class" so defined is not a category which can explain why it is men who control all the institutions of modern social life and women who have the primary responsibility for infant care in every society. The distinctions between owners and workers, bourgeoisie and proletariat, middle-class and working-class all obscure the profound respective commonalities of women's labor and of men's labor across these distinctions.[34] Thus, taking only the "standpoint of the proletariat" necessarily results in gaps and distortions in our understanding of the regularities of both men's and women's lives and their underlying causes cross-culturally and historically. As several theoreticians have pointed out, a feminist standpoint must begin from the politically-activated perspective of women in the division of labor by sex/gender. It must be sensitive to the differences as well as to the commonalities of women's labor across class, culture, and race divisions. Such a standpoint can provide more complete and less distorted understandings than can a science restricted to the standpoint of the proletariat.[35]

However, there are still some problems to be resolved in the feminist perspective. This kind of defense of a distinctive feminist standpoint has still left unexamined why it is that the discovery of the sex/gender system occurs only at this time in history. Of what conflict "in fact, objectively, outside us" is feminism "the reflex in thought"? Of what conflict is feminism "the ideal reflection in the minds, first, of the class directly suffering under it"? Further historical analysis is required in order to develop a feminist epistemology which corrects the errors of the three epistemological programs we examined above. That is to say, the feminist standpoint must be analytically impartial and epistemically non-relativist. It must also be able to understand sex/gender as an organic social variable which has become visible to us only because of changes in historical social relations. These changes are beginning to be investigated in a number of recent writings.[36] But the implications of new conflicts in social relations for the emergence of a feminist science and epistemology remain largely unexplored.

University of Delaware

NOTES

* Research for this essay was supported by summer fellowships from the National Endowment for the Humanities in 1979 and the University of Delaware in 1980. I have benefitted from discussions with many people on the issues of this paper over the last few years. I am indebted to Margaret Andersen, Pamela Armstrong, and especially to Nancy Hartsock and Sarah Begus for improving my thinking on these issues. The inadequacies of the arguments are all my own.

1 The best source to use to begin to grasp the main focuses of the last decade's feminist research is *Signs: Journal of Women in Culture and Society* (Chicago: University Press, 1975 et seq.), especially the review essays for each discipline regularly published there.
2 Cf., E. G. Ruby Rohrlich Leavitt, 'Women in Other Cultures', in V. Gornick and B. Moran (eds.), *Woman in Sexist Society: Studies in Power and Powerlessness* (New York: Basic Books, 1971).
3 Cf. e.g., Eleanor Leacock, 'Women's Status in Egalitarian Society: Implications for Social Evolution,' *Current Anthropology* 19, 2 (1978).
4 Cf., e.g., Joan Kelly-Gadol, 'Did Women Have a Renaissance,' and Marilyn Arthur, ' "Liberated" Women: The Classical Era,' both in R. Bridenthal and C. Koonz (eds.), *Becoming Visible: Women in European History* (Boston: Houghton Mifflin, 1977).
5 Cf. *Heresies* 11 (1981), special issue on 'Making Room: Women and Architecture.'
6 Cf. Jane Flax, 'Political Philosophy and the Patriarchal Unconscious: A Psychoanalytic Perspective on Epistemology and Metaphysics' in this volume; Nancy Hartsock, 'Social Life and Social Science: The Signficance of the Naturalist/Intentionalist Dispute,' and Sandra Harding, 'The Norms of Social Inquiry and Masculine Experience,' both in *PSA 1980*, Vol. II, P. D. Asquith and R. N. Giere (eds.) (East Lansing: Philosophy of Science Association, 1980).
7 Cf. Barbara Ehrenreich and Deirdre English, *For Her Own Good: 150 Years of the Experts' Advice to Women* (New York: Doubleday, 1978).
8 Cf. Carolyn Merchant, *The Death of Nature: Women, Ecology and the Scientific Revolution* (New York: Harper and Row, 1980); Susan Griffin, *Woman and Nature* (New York: Harper and Row, 1978); Sherry Ortner, 'Is Female to Male as Nature is to Culture?', in M. Z. Rosaldo and L. Lamphere (eds.), *Woman, Culture and Society* (Stanford: University Press, 1974).
9 Cf. e.g., Zillah Eisenstein, *The Radical Future of Liberal Feminism* (New York: Longmans, 1980); Isaac Balbus, *Marxism and Domination* (Princeton: University Press,

1982); June Nash and Ruby Rohrlich, 'Patriarchal Puzzle: State Formation in Meso-potamia and Mesoamerica,' *Heresies* 13 (1981).

[10] However, it seems unwise methodologically as well as theoretically to go so far as to assume that anything like the sex/gender system we know is a *universal* trait of human social life. Such an assumption obscures and mystifies the culturally-differing features of men's and women's lives. It may well be that the appearance of the familiar dynamics of the sex/gender system in some societies distant from our own is a product only of male-dominated inquiry, as Eleanor Leacock argues in *Myths of Male Dominance* (New York: Monthly Review Press, 1981). Refusing to adopt the universalist assumption leaves open important empirical questions such as to what extent changes in reproduc-tion should be understood as part of, or partially a product of, changes in production; and how biology is shaped by culture. Nevertheless, stopping short of claims that the sex/gender system is universal, it is still now hypothesized that the vast majority of cultures to which we will ever have historical access have been ones in which the sex/gender system organically structures social interactions to some degree or other.

[11] M. Z. Rosaldo, 'The Use and Abuse of Anthropology: Reflections on Feminism and Cross-Cultural Understanding,' *Signs* 5 (1980), 394–395. The enumeration in the quotation is mine.

[12] *Ibid.*, p. 394 (my emphasis).

[13] See Thomas S. Kuhn, *The Structure of Scientific Revolutions* (Chicago: University Press, 1970).

[14] *At least one* appears unintelligible. In the following three sections I focus only on the assumption which most broadly challenges each epistemology from a feminist perspective. Individual authors often adopt positions which make more than one of these assumptions unintelligible.

[15] David Bloor, *Knowledge and Social Imagery* (London: Routledge and Kegan Paul, 1977).

[16] *Ibid.*, p. 8. Bloor notes where this argument can be found in the writings of Gilbert Ryle, D. W. Hamlyn, R. S. Peters, I. Lakatos, and even Karl Mannhein on pp. 5–10.

[17] *Ibid.*, p. 10. Bloor notes that this argument has been made by empiricists from Bacon through the present day. Bloor specifies this argument alone as the "argument from empiricism," however I am suggesting what I cannot here argue: that all four of these arguments have roots in empiricism.

[18] *Ibid.*, p. 13. This argument is made by Mannheim's critic Grunwald, by A. Lovejoy, and by T. Bottomore.

[19] *Ibid.*, p. 15. Bloor points out where Karl Popper, for one, makes this argument.

[20] Radical Feminists sometimes make this claim.

[21] Feminists writing within a Liberal discourse often imply this. Cf., e.g., Marcia Millman and Rosabeth Moss Kanter's otherwise illuminating 'Introduction' to *Another Voice: Feminist Perspectives on Social Life and Social Science* (New York: Anchor/Doubleday, 1975).

[22] Hilary Rose *op. cit.*, discusses this problem in 'Hyper-Reflexivity – A New Danger for the Counter-Movements,' in H. Nowotny and H. Rose (eds.), *Counter-Movements in the Sciences*, Sociology of the Sciences Volume III (Dordrecht: D. Reidel, 1979).

[23] Bloor, *op. cit.*; Barry Barnes, *Interests and the Growth of Knowledge* (London: Routledge & Kegan Paul) 1977. Extensions and criticisms of the 'Strong Programme' have created a new industry. Among the key recent writings are Bruno Latour and

Steve Woolgar, *Laboratory Life: The Social Construction of Scientific Facts* (Beverly Hills: Sage, 1979); Mary Hesse, Chapter Two of *Revolutions and Reconstructions in the Philosophy of Science* (Bloomington: Indiana University Press, 1980); Larry Laudan, 'The Pseudo-Science of Science?' and Bloor's response, 'The Strengths of the Strong Programme,' in *Philosophy of the Social Sciences* 11 (1981); Hugo Meynell, 'On the Limits of the Sociology of Knowledge,' *Social Studies of Science* 7 (1977); Erik Millstone, 'A Framework for the Sociology of Knowledge,' *Social Studies of Science* 8 (1978). Hilary Rose also briefly discusses Bloor's work in an illuminating way, *op. cit.*

[24] Bloor, *op. cit.*, p. 4.

[25] *Ibid.*, p. 141.

[26] *Ibid.*, p. 4.

[27] *Ibid.*, p. 141ff.

[28] Kuhn's type of analysis (*op. cit.*) makes this clear.

[29] Francine Blau, 'On the Role of Values in Feminist Scholarship,' *Signs* 6 (1981), 538–540.

[30] F. Engels, 'Socialism: Utopian and Scientific,' in R. Tucker (ed.), *The Marx and Engels Reader* (New York: Norton, 1972), p. 606.

[31] *Ibid.*, p. 624.

[32] *Ibid.*

[33] Cf. Jane Flax, *op. cit.*; Nancy Hartsock, 'The Feminist Standpoint: Developing the Ground for a Specifically Feminist Historical Materialism,' in this volume; and Dorothy Smith, 'Women's Perspective as a Radical Critique of Sociology,' *Sociological Inquiry* 44 (1974).

[34] These criticisms can be found in Hilda Scott, *Does Socialism Liberate Women?* (Boston: Beacon Press, 1974); Batya Weinbaum, *The Curious Courtship of Women's Liberation and Socialism* (Boston: South End Press, 1978); Jane Flax, 'Do Feminists Need Marxism,' *Quest* 3, 1 (1976); Gayle Rubin, 'The Traffic in Women: Notes on the "Political Economy" of Sex,' in *Toward an Anthropology of Women*, ed. R. Reiter (New York: Monthly Review Press, 1975); Heidi Hartmann, 'The Unhappy Marriage of Marxism and Feminism: Towards a More Progressive Union,' and many of the other papers in the volume in which it appears: *Women and Revolution*, ed. Lydia Sargent (Boston: South End Press, 1981); many of Nancy Hartsock's and Jane Flax's papers, including those in this volume.

[35] Hartsock in her paper in this volume has developed the concept of a "feminist standpoint"; the point can also be found in Dorothy Smith, *op. cit.*, and in Jane Flax's paper in this volume.

[36] Carol Brown, 'Mothers, Fathers, and Children: From Private to Public Patriarchy,' in *Women and Revolution*, ed. Lydia Sargent (Boston: South End Press, 1981); Isaac Balbus, *op. cit.*; Zillah Eisenstein, 'Antifeminism in the Politics and Election of 1980' and Rosalind Pollack Petchesky, 'Antiabortion, Antifeminism, and the Rise of the New Right,' both in *Feminist Studies* 7, 2 (1981); E. Currie, R. Dunn, and D. Fogarty, 'The New Immiseration: Staglation, Inequality, and the Working Class,' *Socialist Review* 10, 6 (1980), and the response to this paper by the Washington Area Marxist-Feminist Theory Study Group, 'None Dare Call It Patriarchy: A Critique of "The New Immisera-tion"', *Socialist Review* 12:1 (1982).

INDEX OF NAMES*

* Our thanks to Margaret Pyle Hassert for her heroic labor in providing the Index of Names. – S. H., M.B.H.

325